Palgrave Studies in Medieval and Early Modern Medicine

Series Editors
Jonathan Barry
Department of History
University of Exeter
Exeter, UK

Fabrizio Bigotti
Institute for the History of Medicine
Julius Maximilian University
Würzburg, Germany

CENTRE FOR THE STUDY OF MEDICINE AND THE BODY IN THE RENAISSANCE

The series focuses on the intellectual tradition of western medicine as related to the philosophies, institutions, practices, and technologies that developed throughout the medieval and early modern period (500-1800). Partnered with the Centre for the Study of Medicine and the Body in the Renaissance (CSMBR), it seeks to explore the range of interactions between various conceptualisations of the body, including their import for the arts (e.g. literature, painting, music, dance, and architecture) and the way different medical traditions overlapped and borrowed from each other. The series particularly welcomes contributions from young authors. The editors will consider proposals for single monographs, as well as edited collections and translations/editions of texts, either at a standard length (70-120,000 words) or as Palgrave Pivots (up to 50,000 words).

More information about this series at
http://www.palgrave.com/gp/series/16206

Ruben E. Verwaal

Bodily Fluids, Chemistry and Medicine in the Eighteenth-Century Boerhaave School

palgrave
macmillan

Ruben E. Verwaal
Institute for Medical Humanities
Durham University
Durham, UK

ISSN 2524-7387 ISSN 2524-7395 (electronic)
Palgrave Studies in Medieval and Early Modern Medicine
ISBN 978-3-030-51540-9 ISBN 978-3-030-51541-6 (eBook)
https://doi.org/10.1007/978-3-030-51541-6

Cover illustration: Design by eStudio Calamar

This Palgrave Macmillan imprint is published by the registered company Springer Nature Switzerland AG.
The registered company address is: Gewerbestrasse 11, 6330 Cham, Switzerland

For David

In loving memory,
Ernst Verwaal (1934–2018)

Acknowledgements

This book originates in a PhD dissertation I began in 2013 at the University of Groningen, as part of the research project *Vital Matters: Boerhaave's Chemico-Medical Legacy and Dutch Enlightenment Culture* (2012–2017), funded by the Dutch Research Council (NWO). I would therefore like to express my gratitude to Rina Knoeff for her encouragement and expert comments. I could not have wished for a better supervisor. Heartfelt thanks also to Marieke Hendriksen for her helpful feedback over the years—and I am delighted that we continue working together on a most fascinating and fun project, Boerhaave's little furnace.

Many people have been instrumental in making this book possible, and I am grateful for their feedback on my papers and draft chapters. I would particularly like to thank Raingard Esser for her sharp eye and the many discussions we had; Catrien Santing for her broad knowledge on all matters sanguine; and Hasok Chang for hosting me at the Department of History and Philosophy of Science, Cambridge, and sharpening my thinking about this project. My work has also benefited from conversations with Rebekah Ahrendt, Anita Guerrini, Ludmilla Jordanova, Natalie Köhle, Inger Leemans, Elizabeth Williams, Klaas van Berkel, Fabrizio Bigotti, Andrew Cunningham, Frank Huisman, James Kennaway, Lionel Laborie, Lissa Roberts, Michael Stolberg and Charles Wolfe. Other colleagues generously shared documents, unpublished work, and their palaeography and translation skills: Lieke van Deinsen, Raf Praet, Eric Jorink and Wouter Klein.

Societies and institutes have been crucial to conducting research for this book. I thank friends and colleagues at *Gewina*, the *History of Science*

Society, *Werkgroep Achttiende Eeuw*, the *Society History of Alchemy and Chemistry* and the Huizinga Institute for their stimulating discussions. I would also like to thank the librarians and curators at museums and libraries across Europe, in particular the Wellcome Library, the Special Collections of Leiden and Groningen, Rijksmuseum Boerhaave, and the University Museums. Special thanks also to the members of the Descartes Centre and the *Isis* journal editorial office in Utrecht.

Research for this book was funded by a number of grants, for which I am deeply grateful: the History of Science Society early career travel grant; the Van de Sande Fellowship offered by the Scaliger Institute at Leiden University for a research trip in 2016; the Herzog-Ernst-Scholarship Programme of the Fritz Thyssen Foundation for consulting sources at the Research Library Gotha and having inspiring conversations with fellows at the Gotha Research Centre of Erfurt University in 2017; and the Santorio Fellowship for Medical Humanities and Science of the Fondazione Comel for joining the *Humours, Mixtures, Corpuscles* conference at the Centre for the Study of Medicine and the Body in the Renaissance in Pisa in 2017.

I am grateful for the permission to use the figures shown in this book and to reuse material found in earlier versions of some chapters. Chapter 3 previously appeared as 'The Nature of Blood: Debating Haematology and Blood Chemistry in the Eighteenth-Century Dutch Republic', *Early Science and Medicine*, 22 (2017), 271–300. Parts of Chap. 5 were published in 'Increasing and Reducing: Breastmilk Flows and Female Health', in *Lifestyle and Medicine in the Enlightenment*, edited by James Kennaway and Rina Knoeff (London, 2020), 223–39.

I am very appreciative to all the work put in by Molly Beck and Joseph Johnson, editors at Palgrave Macmillan, by the anonymous reviewers, and of course by Fabrizio Bigotti and Jonathan Barry, the series editors for Palgrave Studies in Medieval and Early Modern Medicine.

To my old colleagues and students at University of Groningen, I owe much gratitude for my wonderful years there. I thank my colleagues at the Erasmus University Medical Center in Rotterdam for their support and the opportunity to continue my line of work in medical heritage and medical humanities. More recently, my new colleagues at the dynamic Institute for Medical Humanities and the History Department, Durham University, have greatly motivated me to finish this book. In particular, I thank Jane Macnaughton for her support and encouragement.

Finally, I am very grateful to my family and friends for their encouragement and support in this project. Very special thanks, of course, go to

David, without whom this would not have been possible. You joined me on this journey from the very beginning, and I am excited to see where it will take us in the future!

I dedicate this book to my grandfather Ernst.

Nijmegen, 11 June 2020

Ruben E. Verwaal

Contents

List of Figures

CHAPTER 1

Introduction

In a chemistry class on the nature of blood, Herman Boerhaave (1668–1738), professor of medicine, botany, and chemistry at Leiden University, shared a hazardous incident in the laboratory with his students. 'I had long entertain'd an opinion that phosphorus might be procured from human blood', he explained. In an attempt to confirm his idea, Boerhaave had taken a quantity of blood and let it separate into the red and watery parts through coagulation. He had then started distilling the serum in a retort, that is, a glass container with long neck, so that it would yield salt on the sides of the receiver, and yellow oily liquor at the bottom of the vessel, called 'the spirit of human blood'. Boerhaave was called away during the experiment. At night, when he returned with a candle to see the effect, he saw that the neck of the retort was completely blocked with a thick matter. He immediately backed away and left the room. And not a moment later, the force of the ascending oil burst the vessel into a million pieces and set the whole laboratory on fire. 'Had I been by, I had doubtless perish'd'.[1]

[1] Herman Boerhaave, *Institutiones et experimenta chemiae*, 2 vols ('Paris', 1724), 49; idem, *A New Method of Chemistry: Including the Theory and Practice of that Art: Laid down on Mechanical Principles, and Accommodated to the Uses of Life*, trans. Peter Shaw and Ephraim Chambers, 2 vols (London, 1727), vol. 1, 37; vol. 2, 215. On Boerhaave and the Dutch Republic, see Harold J. Cook, *Matters of Exchange: Commerce, Medicine, and Science in the Dutch Golden Age* (New Haven, 2007), 378–409.

R. E. Verwaal, *Bodily Fluids, Chemistry and Medicine in the Eighteenth-Century Boerhaave School*, Palgrave Studies in Medieval and Early Modern Medicine,
https://doi.org/10.1007/978-3-030-51541-6_1

The fact that Boerhaave performed chemical experiments was nothing out of the ordinary. Leiden had a chemical laboratory in its botanical garden since 1669, and at the turn of the eighteenth century, all universities in the Low Countries cultivated chemistry by establishing laboratories and chairs in the medical faculty. This correlated with the growing interest in new remedies and chemical preparations. Still this anecdote raises numerous questions: why did Boerhaave subject bodily fluids like blood to a chemical examination? If medical students needed to know the normal functioning of the body, it would make sense for them to visit an anatomical theatre to learn about the body's internal structures. But why would they also be sent into smoke-filled chemical laboratories and be covered in ash while maintaining fires, distilling fluids, and synthesising substances? What results did they gain from their practice, and how did they use this knowledge to better understand the functioning of the body?

It is important to note that this anecdote was not communicated by Boerhaave himself, but by his students.[2] Boerhaave's students diligently wrote down their professor's lectures, and even published their lecture notes as textbooks, thus playing a key role in the dissemination of knowledge. Furthermore, a plethora of medical dissertations and treatises preserve the thoughts and actions of these physicians' work with fluids in the laboratory.[3] From this rich corpus of sources it is not only clear that bodily fluids played an essential role in medical education, but also that early modern physicians in the Dutch Republic were seriously reconsidering the nature and function of blood and other bodily fluids. And rightly so. For these natural flows and fluids performed crucial tasks in the healthy functioning of the human body. Saliva lubricated chewing and swallowing food; blood continuously nourished all parts of the body; the bladder collected and stored urine until it was released through the urethra; mothers expressed their milk to feed their new-borns; perspiration exuded through the innumerable pores of the skin; and at moments of sexual climax, male bodies emitted the reproductive fluid of semen. That the discharges and interflows of bodily fluids were directly related to the body's vital

[2] On the multiple editions of Boerhaave's chemistry textbook, see J.R.R. Christie, 'Historiography of Chemistry in the Eighteenth Century: Hermann Boerhaave and William Cullen', *Ambix*, 41 (1994), 4–19.

[3] On the Dutch universities as centres of scholarly publication and the important and lucrative role of students in the book market, see Andrew Pettegree and Arthur der Weduwen, *The Bookshop of the World: Making and Trading Books in the Dutch Golden Age* (New Haven, 2019), 172–94.

functions had been known for centuries. But it was at the turn of the eighteenth century that physicians like Boerhaave took a new interest in the fluids and investigated them with a new set of instruments and ideas.

This book argues that at the turn of the eighteenth century, new research methods and instruments crucially changed the perception of bodily fluids and that this contributed to a new system of medicine. For centuries physicians had based their knowledge on the ancient humoral pathology, which had its origins in the writings of Hippocrates (c. 460–377 BC) and Galen (129–c. 210 AD), and was further developed in the medieval period by scholars such as Ibn Sina (known as Avicenna, 980–1037). The humoral theory proposed that there were four bodily humours—blood, phlegm, black and yellow bile—which needed to be balanced in relation to each other to sustain health. Blood regularly expelled via nosebleeds, haemorrhoids, menstruation, or surgically by phlebotomy, to sustain the humoral balance. In the Renaissance, however, physicians started questioning the humoral theory. While some progressed learned medicine by emulating the ancients, others protested against the establishment of Galenic medicine.[4] Most notably Theophrastus von Hohenheim (1493–1541), better known as Paracelsus, presented a rival model based on three primary substances of salt, sulphur, and mercury, envisioned disease as having external causes and presented a new set of more drastic remedies that included mercury, antimony, and arsenic.[5] In the seventeenth century, Galenic medicine received another blow when influential figures such as René Descartes (1596–1650), Giovanni Alfonso Borelli (1608–1678), Marcello Malpighi, and Archibald Pitcairne (1652–1713), reduced bodily processes to motion and mechanical actions, anatomical structures, and mathematical calculation. In their hydraulic analyses of bodily streams, organs like hearts and kidneys were referred to as pumps

[4] On Galenism and its protracted decline, see R.J. Hankinson, ed., *The Cambridge Companion to Galen* (Cambridge, 2008); Petros Bouras-Vallianatos and Barbara Zipser, eds., *Brill's Companion to the Reception of Galen* (Leiden, 2019); and Matteo Favaretti Camposampiero, 'Galenism in Early Modern Philosophy and Medicine', in *Encyclopedia of Early Modern Philosophy and the Sciences*, ed. Dana Jalobeanu and Charles T. Wolfe (Cham, 2020).

[5] On the medicine, philosophy and impact of Paracelsus, see Walter Pagel, *Paracelsus: An Introduction to Philosophical Medicine in the Era of the Renaissance*, 2nd ed. (Basel, 1982); Charles Webster, *Paracelsus: Medicine, Magic, and Mission at the End of Time* (New Haven, 2008).

and sieves.[6] In the early modern period, in other words, new anatomical observations, mathematics, and chemical theories of the body transformed the ways in which philosophers and physicians thought about the functioning of the healthy, living body.[7] But while Paracelsian and Cartesian theories had completely dismissed the humoral theory, Boerhaave and his students developed a new appreciation for the nature of the fluids. Indeed, the application of new instruments and chemical methods to bodily fluids led to an extraordinary set of changes, which reinvented ancient humoral theory and established a new and irenic physiology of the fluids.

Histories of Physiology and Chemistry

When in the 1770s it came to defining physiology in a new edition of Noël Chomel's (1632–1712) famous household dictionary, the French-Frisian editor Jacques de Chalmot (*c.* 1730–1801) came straight to the point: 'The best books on *physiology* in Latin are those of Boerhaave'.[8] The word physiology derived from the Greek *physis* ('nature') and *logia* ('knowledge'), and therefore initially referred to the philosophical inquiry into the nature of things. In the early modern period, however, physiology came to

[6] On mechanical philosophy in medicine, or iatromechanism, see Theodore M. Brown, 'Physiology and the Mechanical Philosophy in Mid-seventeenth Century England', *Bulletin of the History of Medicine*, 51 (1977), 25–54; Anita Guerrini, 'The Varieties of Mechanical Medicine: Borelli, Malpighi, Bellini and Pitcairne', in *Marcello Malpighi*, ed. Domenico Bertoloni Meli (Florence, 1997), 111–28; Stephen Gaukroger, *Descartes' System of Natural Philosophy* (Cambridge, 2002), 180–214; Vincent Aucante, *La philosophie médicale de Descartes* (Paris, 2006); Domenico Bertoloni Meli, *Mechanism, Experiment, Disease: Marcello Malpighi and Seventeenth-Century Anatomy* (Baltimore, 2011); Fabio Zampieri, *Il Metodo Anatomo-Clinico Fra Meccanicismo Ed Empirismo: Marcello Malpighi, Antonio Maria Valsalva E Giovanni Battista Morgagni* (Rome, 2016).

[7] See 'The New Science' in Roy Porter, *The Greatest Benefit of Mankind: A Medical History of Humanity* (New York, 1999), 201–44; and Andrew Wear, 'Medicine in Early Modern Europe, 1500–1700' in Lawrence I. Conrad et al., *The Western Medical Tradition: 800 BC to AD 1800* (Cambridge, 1995), 250–361.

[8] Noël Chomel, *Algemeen huishoudelijk-, natuur-, zedekundig- en konstwoordenboek*, ed. Jacques Alexandre de Chalmot, 2nd ed. (Leiden and Leeuwarden, 1778), vol. 5, 2694. On Chomel, see Béatrice Camille Fink, 'Le Dictionnaire oeconomique de Noël Chomel: Un pot-pourri aigre-doux', in *Diderot, l'Encyclopédie et autres études*, ed. Jacques Proust, Marie Leca-Tsiomis, and Alain Sandrier (Ferney-Voltaire, 2010), 157–64. In revising Chomel's dictionary, De Chalmot collaborated with Petrus Camper, among others. See Arianne Baggerman, *Publishing Policies and Family Strategies: The Fortunes of a Dutch Publishing House in the 18th and Early 19th Centuries* (Leiden, 2014), 271–6.

be specifically applied to the field of medicine, explaining the conditions of life and health, also known as the doctrine of the animal oeconomy.[9] Boerhaave's students learned that

> By physiology we mean that part of medicine which explains the nature of the entire animal machine, consider'd in its natural state, when its operations are perform'd in greatest perfection; or in a state of health; which depends upon a just balance of the solids and fluids.[10]

De Chalmot might have mentioned any of a large number of physiologists, such as Friedrich Hoffmann (1660–1742) or Albrecht von Haller (1708–1777), but it was no accident that he chose to highlight Boerhaave. In De Chalmot's lifetime groundbreaking developments had occurred, originating with Boerhaave and his successors. They had developed new instruments and methods, such as thermometers and chemical laboratories that were able to measure and analyse bodily fluids to an extent never seen before. Because 'many of the most important parts in all the medical physiology are only to be known by chemistry'.[11]

The present study focuses on developments in eighteenth-century medicine largely passed over in historiography. Historians of science and medicine have often discussed the anatomical discovery of the circulation of blood and the consequent mechanisation of the body. The next momentous phase in the history of physiology has generally been identified as the experimental physiology of the nineteenth century. Yet this book develops a picture of physiology that revises this sequence of events by showing the establishment of a new physiology of fluids in the eighteenth century, which proved influential far into the nineteenth century. In doing so, this book aims to contribute to the current historiographical debates in medical and chemical history, the history of mechanist and vitalist theories, and the study of early modern material culture.

[9] Diderot, *Encyclopédie*, vol. 12, 537. For general overviews of physiology see Karl Eduard Rothschuh, *History of Physiology*, trans. Guenter B. Risse (Huntington, NY, 1973); Thomas S. Hall, *Ideas of Life and Matter: Studies in the History of General Physiology, 600 B.C.–1900 A.D.*, 2 vols (Chicago, 1969); E.M. Tansey, 'The Physiological Tradition', in *Companion Encyclopedia of the History of Medicine*, ed. W.F. Bynum and Roy Porter (London, 1993), 120–52.

[10] Boerhaave, *A New Method*, vol. 1, 194; idem, *Institutiones et experimenta chemiae*, vol. 1, 146–7.

[11] Herman Boerhaave, *A New Method of Chemistry: Including the History, Theory, and Practice of the Art*, trans. Peter Shaw, 2nd ed., 2 vols (London, 1741), vol. 1, 174.

The intricate relationship between physiology and anatomy has received much scholarly attention by historians of medicine. Anatomy and physiology, after all, appear to complement each other perfectly, because one needs to be intimately familiar with the structure of the internal organs before understanding their functions in the normal, living body. Anatomy has certainly been a popular research subject in recent years.[12] Tackling the complicated relationship between anatomy and physiology directly, Andrew Cunningham published an influential article in two parts titled 'The Pen and the Sword', in which he argued that physiology in the early modern period must not be interpreted anachronistically as a primitive predecessor of modern-day experimental physiology.[13] Furthermore, based on the case of Jean Fernel (1497–1558), the French physician who first used the term physiology, Cunningham demonstrates that early modern physiologists largely cared about the theoretical nature of the human body. Anatomists, by contrast, were occupied with manual investigations into the structure of human and animal body parts. Defining 'disciplinary boundaries' between the two, Cunningham corrects a number of historiographical misinterpretations: Fernel cannot be reduced to the role of the father of experimental physiology simply for using the word physiology, nor were early modern vivisectionists part of experimental physiology—they rather dealt with the manual art of anatomical observation. Even the case of Albrecht von Haller, who had famously performed innumerable vivisections and experiments on which he based his theory of sensibility and irritability, must not be confused with an experimental physiologist *avant la lettre*. Cunningham proposes that there was a clear divide between Von Haller's two roles, as the investigative anatomist and as the contemplating physiologist.

[12] See, for example Anita Guerrini, *The Courtiers' Anatomists: Animals and Humans in Louis XIV's Paris* (Chicago, 2015); Marieke M.A. Hendriksen, *Elegant Anatomy: The Eighteenth-Century Leiden Anatomical Collections* (Leiden, 2015); Andrew Cunningham, *The Anatomist Anatomis'd: An Experimental Discipline in Enlightenment Europe* (Farnham, 2010).

[13] Andrew Cunningham, 'The Pen and the Sword: Recovering the Disciplinary Identity of Physiology and Anatomy before 1800 – I: Old Physiology–the Pen', *Studies in History and Philosophy of Biological and Biomedical Sciences*, 33 (2002), 631–66; 'The Pen and the Sword: Recovering the Disciplinary Identity of Physiology and Anatomy before 1800 – II: Old Anatomy–the Sword', *Studies in History and Philosophy of Biological and Biomedical Sciences*, 34 (2003), 51–76.

While this may be the case, this book maintains that the binary division of anatomy and physiology was not the complete story. Take, for instance, the relatively short entry on physiology in Ephraim Chambers's *Cyclopaedia* (1728), published a year after Chambers's translation of Boerhaave's chemical lectures, *A New Method of Chemistry* (1727). The encyclopaedist indeed omitted anatomy from the entry, but instead referred to chemistry as the way to gain knowledge of the animal oeconomy: 'Physiology properly denotes only an internal reasoning or discoursing [...]. In this view Chymistry does not properly belong to Physiology, but is a kind of counter-part thereto, as imitating or mimicking Nature'.[14] As this quote suggests, more than the anatomical knife, the chemical flask was the instrument *par excellence* for imitating and observing changes in bodily fluids, and thereby supported physiological contemplation.

The role of chemical practices has not been absent from the history of medicine. Historians have written wonderful studies that paid specific attention to the role of medical chymical theories in the sixteenth and seventeenth centuries, largely analysing the works of Paracelsus, Jan Baptist van Helmont (1579–1644), Franciscus Sylvius (François de le Böe, 1614–1672), and their followers.[15] Medical chymistry, formerly known as iatrochemistry (from Greek *iatros*, 'physician'), denotes the Paracelsian school which sought to understand medicine and physiology in chymical terms. Practitioners working in Paracelsus's footsteps portrayed the body as a chemical laboratory, according to which, for example, food and drink were digested by means of acid fermentation in the stomach and intestines. Fermentation was caused by acid particles mechanically reacting with alkalis inside the body. Hence, Sylvius and his student Reinier de Graaf (1641–1673) went to great lengths to obtain the various intestinal fluids—saliva, stomach juice, bile, and pancreatic juice—to ascertain their acidity by the direct sensory experience of taste. As Evan Ragland has

[14] Ephraim Chambers, *Cyclopaedia, or, An Universal Dictionary of the Arts and Sciences*, 2 vols (London, 1728), vol. 2, 810.

[15] See, for example, Allen G. Debus, *Chemistry and Medical Debate: Van Helmont to Boerhaave* (Canton, MA, 2001); Amy Eisen Cislo, *Paracelsus's Theory of Embodiment: Conception and Gestation in Early Modern Europe* (London, 2010); Georgiana D. Hedesan, *An Alchemical Quest for Universal Knowledge: The 'Christian Philosophy' of Jan Baptist Van Helmont (1579–1644)* (London, 2016); Evan R. Ragland, 'Chymistry and Taste in the Seventeenth Century: Franciscus Dele Boë Sylvius as a Chymical Physician between Galenism and Cartesianism', *Ambix*, 59 (2012), 1–21; Harm Beukers, 'Acid Spirits and Alkaline Salts: The Iatrochemistry of Franciscus dele Boë, Sylvius', *Sartoniana*, 12 (1999), 39–59.

argued, this type of experiment not only showed the crucial role of sensory experience in experimentation, it was also an exemplary seventeenth-century approach to experiments derived from Harveian anatomical and Helmontian chymical traditions.[16]

How the relationship between physiology and chemistry continued during the eighteenth century, however, remains largely unexplored. In fact, many studies in the history of chemistry actually tend to pass over the early eighteenth century entirely.[17] This book, therefore, takes inspiration from those studies that are moving away from the traditional narrative of the chemical revolution in the late eighteenth century, and that focus on developments in chemistry in the years between Robert Boyle's (1627–1691) and Antoine Laurent Lavoisier's (1743–1794) active years instead.[18] A number of studies have already reinstated some important narratives in the history of chemistry from the *longue durée* of chemical knowledge.[19] This book adds to this growing body of work, first, by emphasising the diverse chemical practices that were developed in relation to medicine from the turn of the eighteenth century onwards. And second, it challenges the general argument that the academic institutionalisation of chemistry resulted in its taking on a more philosophical nature.[20] Seventeenth-century chymistry, as shown by William Newman and

[16] Evan R. Ragland, 'Experimenting with Chymical Bodies: Reinier De Graaf's Investigations of the Pancreas', *Early Science and Medicine*, 13 (2008), 615–64; idem, "Experimenting with Chemical Bodies: Science, Medicine, and Philosophy in the Long History of Reinier de Graaf's Experiments on Digestion, from Harvey and Descartes to Claude Bernard" (Doctoral thesis, University of Alabama, 2012).

[17] See, for example, Victor D. Boantza, *Matter and Method in the Long Chemical Revolution: Laws of Another Order* (Farnham, 2013).

[18] The fact that early eighteenth-century chemistry has been generally neglected is best expressed in Lawrence M. Principe, 'A Revolution Nobody Noticed? Changes in Early Eighteenth-Century Chymistry', in *New Narratives*, ed. Lawrence M. Principe (Dordrecht, 2007), 1–22. On the historiographic dominance of the chemical revolution, see Seymour Mauskopf, 'Reflections: "A Likely Story"', in *New Narratives*, ed. Lawrence M. Principe (Dordrecht, 2007), 177–93.

[19] See the special issue by Matthew D. Eddy, Seymour Mauskopf, and William R. Newman, eds., *Chemical Knowledge in the Early Modern World* (2014). See also Robert G.W. Anderson, ed. *Cradle of Chemistry: The Early Years of Chemistry at the University of Edinburgh* (Edinburgh, 2015); Hjalmar Fors, *The Limits of Matter: Chemistry, Mining, and Enlightenment* (Chicago, 2015).

[20] See, for example, Jonathan Simon, 'Pharmacy and Chemistry in the Eighteenth Century: What Lessons for the History of Science?', *Osiris*, 29 (2014), 283–97; John C. Powers, *Inventing Chemistry: Herman Boerhaave and the Reform of the Chemical Arts* (Chicago, 2012).

Lawrence Principe, denoted an artisanal practice involved in making things by chemical methods. Most seventeenth-century writers would not differentiate between chymical and alchemical practices.[21] In the early eighteenth century, however, academic scholars became increasingly aware of the discrepancy between commonly dispersed vernacular practices on the one hand, and the conscious attempt to make the practice of chemistry acceptable for academic purposes on the other. And indeed, learned men like Hieronymus David Gaubius (1705–1780) distinguished between seventeenth-century chymistry and the new chemistry, elevating chemistry as a demonstrative method, useful for the practice of natural philosophy and medicine.[22]

But this did not mean that chemistry lost any of its practical aspects. On the contrary, this book shows that the practical side of chemistry, too, gained a place in academic discourse: it became a practical foundation of academic medicine. The laborious nature of chemical experiments on bodily fluids thus significantly contributed to theoretical physiology and the development of a new medical system. According to Boerhaave's students, 'chemistry is no science form'd *à priori*; 'tis no production of the human mind, framed by reasoning and deduction'. Instead, truths in chemistry had to be 'collected *à posteriori*, from numerable experiments'.[23] At the same time in history, chemical operations comprised experiments on vegetal, animal, and mineral bodies by applying fire, water, earth, air, and *menstrua* (*i.e.*, solvents). By applying his new method of chemistry, Boerhaave formulated questions about the nature of bodily fluids, and painstakingly investigated blood, urine, and milk individually to understand how each of these animal materials changed *in vitro* (*i.e.*, in glass)

[21] William R. Newman and Lawrence M. Principe, 'Alchemy vs. Chemistry: The Etymological Origins of a Historiographic Mistake', *Early Science and Medicine*, 3 (1998), 32–65; William R. Newman, 'From Alchemy to 'Chymistry'', in *The Cambridge History of Science*, ed. Katharine Park and Lorraine Daston (Cambridge, 2006).

[22] Hieronymus David Gaubius, *Oratio inauguralis qua ostenditur chemiam artibus academicis jure esse inserendam* (Leiden, 1731); idem, *Oratio de vana vitae longae, a chemicis promissae, exspectatione* (Leiden, 1734); Abraham Kaau, *Declamatio academia de gaudiis alchemistarum* (Leiden, 1737). See also John C. Powers, 'Scrutinizing the Alchemists: Herman Boerhaave and the Testing of Chymistry', in *Chymists and Chymistry*, ed. Lawrence M. Principe (Sagamore Beach, 2007), 227–38; and Marieke M.A. Hendriksen, 'Criticizing Chrysopoeia? Alchemy, Chemistry, Academics and Satire in the Northern Netherlands, 1650–1750', *Isis*, 109 (2018), 235–53.

[23] Boerhaave, *A New Method*, vol. 1, vi. Boerhaave, *Institutiones et experimenta chemiae*, vol. 1, 2.

and how they worked *in vivo* (*i.e.*, in the living body). This physiological reasoning, based on experimentation, resulted in new insights, which proved essential to physiology and medicine, because 'many of the most important parts in all the medical physiology are only to be known by chemistry'.[24]

It is for this reason that this book prioritises the field of chemistry over anatomy in its investigation of early modern medicine and physiology. It takes its cue from studies in the history of medical chemistry, and applies the same to the often-neglected early eighteenth century. It will argue that chemical instruments and investigations of the bodily fluids in the eighteenth century played an essential role in the medical and physiological understanding of the body. As summarised by Boerhaave's successor Gaubius in 1731:

> Medicine, physiology, and chemistry cohere by an indissoluble bond, hence I take as the theme of my public lectures that most noble and most copious part of the entire human body, *the fluids of course*; because as the fluids are too subtle, the knife of the anatomist cannot reach them.[25]

While the aforementioned studies have addressed the entangled and complicated relationship between disciplines, a number of scholars have confronted the equally complex problem of two apparently competing physiological theories in the eighteenth century: mechanism and vitalism. Did the eighteenth-century study of bodily fluids using chemical experiments lead to a mechanist or vitalist understanding of physiology?

Inspired by William Harvey's (1578–1657) circulation of blood and Antoni van Leeuwenhoek's (1632–1723) blood corpuscles, mechanist theories explained the workings of the living body with reference to mechanical processes. The body could be seen as a piece of machinery, with external forces acting on a homogeneous matter. Boerhaave has often been cast as the pinnacle of mechanist medicine. Gerrit Lindeboom, for

[24] Boerhaave, *A New Method of Chemistry*, vol. 1, 174.

[25] Anonymous, 'Cl. Hier. Gaubii, in Academia Lugduno-Batava professoris extraordinarii et lectoris chemiae, praelectiones publicae chemicae annorum 1731, 1732 et 1733, sive examen chemicum humorum corporis humani, habitum in auditorio chemico', 2 vols, Bordeaux, Bibliothèque universitaire de lettres de l'Université Michel de Montaigne, Bdx.U3.Ms.4, 1731–1733, vol. 2, p. 313. Emphasis added. An exact copy of the first volume of these chemistry lectures on bodily fluids is kept at the Bibliothèque interuniversitaire, section médecine, Montpellier, Ms. H 166.

example, presented Boerhaave as a supporter of the iatromechanical or iatrophysical approach to medicine, claiming that Boerhaave 'went as far as a mechanistic way of thinking could lead him'.[26] Luyendijk-Elshout likewise argued that Boerhaave and his students kept their physiology strictly mechanistic, with Newtonian laws of gravity and attraction as the basic principle, and life understood as the circulation of fluids through the vessels.[27] Historians of medicine have argued that increased quantification and mechanisation gave rise to the idea that life itself, in fact, could be reduced to mechanical and physical phenomena.[28] This contributed to the understanding of mechanism and vitalism, associated with the supposition of a vital force or vital principle, as developing in strong oppositional reaction to each other.

In recent years, however, most historians have come to realise that this strict bifurcation of mechanist and vitalist theories of physiology is problematic. Detailed studies on Boerhaave, Von Haller, and the medical school of Montpellier all reveal the difficulties and limitations of keeping to a strict mechanist–vitalist opposition. Rina Knoeff's discussion of Boerhaave's textbook *Elementa chemiae* (1732) considered the theories of menstrua and of seminal principles.[29] A menstruum was a substance able to dissolve other substances over the course of a philosophical month (40 days). For Boerhaave, menstrua revealed many different powers, and he therefore considered them suited to discovering the vital powers of bodies, and also useful in the preparation of drugs. Furthermore, by intro-

[26] G.A. Lindeboom, 'Boerhaave's Concept of the Basic Structure of the Body', *Clio Medica*, 5 (1970), 203–8. Thomas Hall added that 'Boerhaave's causal interpretation [as] micromechanistic is, in itself, not new [but] he uses micromechanics more exhaustively and more consistently than any earlier biologist' in Thomas Hall, *Ideas of Life and Matter*, vol. 1, 390.

[27] A.M. Luyendijk-Elshout, 'Mechanicisme contra vitalisme: de school van Herman Boerhaave en de beginselen van het leven', *TGGNWT*, 5 (1982), 16–26.

[28] Erwin H. Ackerknecht, *A Short History of Medicine*, rev. ed. (Baltimore, London, 1982), 128–9. See also G.J. Goodfield, *The Growth of Scientific Physiology: Physiological Method and the Mechanist-Vitalist Controversy, Illustrated by the Problems of Respiration and Animal Heat* (London, 1960); Theodore M. Brown, 'From Mechanism to Vitalism in Eighteenth Century English Physiology', *Journal of the History of Biology*, 7 (1974), 179–216. See also students' rejection of Boerhaave's 'mechanistic learnings' in Frank Huisman, 'Medicine and Health Care in the Netherlands, 1500–1800', in *The History of Science in the Netherlands*, ed. Klaas van Berkel, Albert van Helden, and Lodewijk Palm (Leiden, 1999), 239–78, here 264–6.

[29] Rina Knoeff, *Herman Boerhaave (1668–1738): Calvinist Chemist and Physician* (Amsterdam, 2002), 125–7.

ducing seminal principles peculiar to every individual body, Boerhaave's approach to medicine was as mechanistic as it was vitalistic. Hubert Steinke has further pointed out the difficulty of positioning Albrecht von Haller in either the mechanist or vitalist school of thought once one realises that his contemporaries identified Von Haller with both and neither school at the same time.[30] Anita Guerrini has denied a sudden break between mechanist theories derived from Newtonian physics and vitalist theories. In the eighteenth century, she argued, '"vitalist" ideas were now revived by scientists who were dissatisfied with mechanical explanations. Yet there was not a dichotomy between mechanists and vitalists, and there were many gradations and syntheses between the two'.[31] Elizabeth Williams therefore decided to study vitalism in France neither as an abstract theory nor as a set of disembodied concepts, but rather as a broad discourse largely located in the medical school of Montpellier. According to Williams, Montpellier vitalism 'was not a unitary medical doctrine or program linked to a particular political or social philosophy but a protean, often fragmented discourse'.[32] Living plants, animals, and humans were thought to be endowed with life, and the Montpellier physicians searched for the unique laws governing the existence of living beings. They presupposed interconnected activities of the so-called animal oeconomy, which was empowered with a certain vital force through which humans could live. 'Thus physiology was the study of the interrelated, systematic, harmonious operations that simultaneously manifested and sustained the life of bodies that enjoyed vitality'.[33]

[30] Hubert Steinke, *Irritating Experiments: Haller's Concept and the European Controversy on Irritability and Sensibility, 1750–90* (Amsterdam, 2005), 175–229. Steinke also included animism in his discussion.

[31] Anita Guerrini, *Experimenting with Humans and Animals: From Galen to Animal Rights* (Baltimore, 2003), 60–1. In the case of Scottish physician George Cheyne (1671–1743), for example, Guerrini preferred to talk about 'a quasi-vitalistic physiology' which arose 'within the context of Newtonian aether theory'. See Anita Guerrini, 'James Keill, George Cheyne, and Newtonian Physiology, 1690–1740', *Journal of the History of Biology*, 18 (1985), 247–66, here 248.

[32] Elizabeth A. Williams, *The Physical and the Moral: Anthropology, Physiology, and Philosophical Medicine in France, 1750–1850* (Cambridge, 1994), 2.

[33] Ibid., 11; Elizabeth A. Williams, *A Cultural History of Medical Vitalism in Enlightenment Montpellier* (Aldershot, 2003); Rosalyne Rey, *Naissance et développement du vitalisme en France de la deuxième moitié du 18e siècle à la fin du Premier Empire* (Oxford, 2000). On vitalism, see also Charles T. Wolfe and Motoichi Terada, 'The Animal Economy as Object and Program in Montpellier Vitalism', *Science in Context*, 21 (2008), 537–79; Charles

This book builds on these studies and aims to exhibit a new perspective on the issue of mechanism and vitalism in eighteenth-century physiology. It approaches the theoretical binary between mechanism and vitalism through the lens of chemical *practices* and bodily fluids. Inspired by recent studies on mundane materials, such as ink, milk, and liquors, each chapter is devoted to a specific bodily fluid.[34] I analyse how each of these bodily fluids was handled and understood by physicians, chemists, and occasionally patients—from the bodily discharges of urine and sweat to the life-enforcing semen, milk, and blood. This will shed new light on early modern physiology and show the hybrid field of vitalist and mechanist physiologies. In tandem with this, I place a specific focus on a circle of physicians and chemists who were associated through their universities, kinship, friendship, intellectual beliefs, or a shared admiration of Herman Boerhaave.

The Boerhaave School

The notion that Boerhaave's students formed a school is a motif that permeates this book. The concept of a Boerhaave school is inspired by contemporary references to a medical community around Boerhaave. For example, the second hefty volume of Albrecht von Haller's *Commentaries* on Boerhaave's physiology textbook was dedicated to his friend and fellow physician António Sanches (1699–1783), because he was 'from the school of the great Boerhaave'. Volume five was dedicated to all disciples of Boerhaave.[35] Indeed, many eager students travelled to Leiden to study medicine and chemistry with the famous Boerhaave and his disciples. Many others who were unable to come at Leiden to meet the master nevertheless appropriated his teachings via the lecture notes and books that enjoyed a wide circulation at the time. Not only did all these men form a

T. Wolfe, 'From Substantival to Functional Vitalism and Beyond: Animas, Organisms and Attitudes', *Eidos*, 14 (2011), 212–35.

[34] On the history of materials see Ursula Klein and W. Lefèvre, *Materials in Eighteenth-Century Science: A Historical Ontology* (Cambridge, MA, 2007); Ursula Klein and E.C. Spary, eds., *Materials and Expertise in Early Modern Europe: Between Market and Laboratory* (Chicago, 2010); Helen Deutsch and Mary Terrall, eds., *Vital Matters: Eighteenth-Century Views of Conception, Life, and Death* (Toronto, 2012); E.C. Spary, *Eating the Enlightenment: Food and the Sciences in Paris* (Chicago, 2012).

[35] Albrecht von Haller, ed. *Praelectiones academicae in proprias institutiones rei medicae*, 6 vols (Göttingen, 1739–1744), vol. 2; vol. 5, pt. 2.

large group of physicians which have hitherto been little studied, their books and practices have also guided the selection of sources studied in this book. Boerhaave's students continued his work at other universities in the Dutch Republic and far beyond. The Boerhaave school, I argue, formed a large group of medical men across Europe who followed in Boerhaave's footsteps and helped establish new physiology based on the bodily fluids.

The term 'school' generally has two definitions: first, an institution or establishment at which instruction is given; and second, a group of people, usually artists or researchers, who share the same doctrine, have a common teacher or work in the same way.[36] Following the first meaning of school, I take Leiden University as the focal point for the Boerhaave school, yet also show the links to universities, academies, and courts in the Low Countries—Utrecht, Groningen, Franeker, and Harderwijk—as well as across Europe, such as Göttingen, Vienna, Edinburgh, and St Petersburg.[37] Like ordinary schools, the Boerhaave school denoted places for study, instruction, and learned conversation. Furthermore, I engage with the second meaning of school and allude to the phrase 'school of thought' to define the Boerhaave school as a group of intellectuals sharing certain ideas and methods. School, then, is comparable to the concept of movement in literary and art history: a group of writers or artists continue to work in the 'spirit' of a master or to whom they show a clear resemblance in method or opinion.

Historians have traditionally discussed 'Boerhaave's influence' from an institutional perspective. Gerrit Lindeboom, for example, has stated that Boerhaave 'greatly influenced the study and development of chemistry in Great-Britain', and Erwin Ackerknecht has claimed that Boerhaave made

[36] See *s.v.* 'school'. *OED Online*, December 2019, Oxford University Press. www.oed.com/view/Entry/172522 (accessed 21 February 2020); and *s.v.* 'school'. *Algemeen Nederlands Woordenboek*, Instituut voor de Nederlandse taal. http://anw.inl.nl/article/school (accessed 8 June 2020).

[37] For the Boerhaave school in Edinburgh, see Christopher Lawrence, 'Ornate Physicians and Learned Artisans: Edinburgh Medical Men, 1726–1776', in *William Hunter and the Eighteenth-Century Medical World*, ed. William Bynum and Roy Porter (Cambridge, 1985), 153–76; Andrew Cunningham, 'Medicine to Calm the Mind: Boerhaave's Medical System, and Why It Was Adopted in Edinburgh', in *The Medical Enlightenment*, ed. Andrew Cunningham and Roger French (Cambridge, 1990), 40–66; John C. Powers, 'Leiden Chemistry in Edinburgh: Herman Boerhaave, James Crawford and Andrew Plummer', in *Cradle of Chemistry*, ed. Robert G.W. Anderson (Edinburgh, 2015), 25–58.

Leiden 'the medical center of the world'.[38] Ashworth Underwood has meticulously investigated the life and works of Boerhaave's English-speaking students. Focusing on the students who enrolled at Leiden University during Boerhaave's tenure from 1701 to 1738, Underwood has identified 746 English-speaking students. Subsequently looking at the careers of these students, Underwood noted that 'there was hardly an institution associated with medicine [on the British Isles] where Boerhaave's men were not in positions of influence and authority'.[39] A new census of all Leiden medical dissertations in the eighteenth century has revealed that Boerhaave promoted a total of 182 (out of 702) doctoral students.[40] With an average of six dissertations completed each year, Boerhaave was the most popular doctoral advisor among his fellow professors, who included Johannes Rau (1668–1719) and Jacob le Mort (1650–1718) at the medical faculty.[41]

These quantitive insights are invaluable, as they give a good impression of the size and scope of the Boerhaave school. But the large diversity in students' career paths places limitations on a purely institutional interpretation of the Boerhaave school. The career of the Scotsman James Crauford (aka Crawford, *c.* 1685–1732) is a case in point. When Crauford matriculated at Leiden University on 31 May 1707, he had already studied in France. Furthermore, Crauford graduated as early as five weeks later, on 6

[38] G.A. Lindeboom, *Boerhaave and Great Britain: Three Lectures on Boerhaave with Particular Reference to His Relations with Great Britain* (Leiden, 1974). Ackerknecht, *A Short History of Medicine*, 130. Roy Porter also recognised the prestige and reputation enjoyed by Boerhaave across eighteenth-century Europe in Porter, *The Greatest Benefit*, 246.

[39] E. Ashworth Underwood, *Boerhaave's Men at Leyden and After* (Edinburgh, 1977). dust jacket. Willem Frijhoff has nuanced the statement that Boerhaave uniquely attracted many students from the British Isles, because there already existed a tradition of English and Scottish students at Leiden. See Frijhoff's *La société Néerlandaise et ses gradués, 1575–1814: Une recherche sérielle sur le statut des intellectuels* (Amsterdam, 1981), 103–7. Martine Zoeteman nevertheless underscores the importance of Boerhaave's reputation in attracting students from afar to Leiden in *De studentenpopulatie van de Leidse universiteit, 1575–1812: 'een volk op zyn Siams gekleet eenige mylen van Den Haag woonende'* (Leiden, 2011), 209, 62, 75–79.

[40] Earlier publications have recorded 178 doctoral students, such as J.E. Kroon, 'Boerhaave als hoogleeraar-promotor', *Nederlands Tijdschrift voor Geneeskunde*, 63 (1919), 87–99. This recount is based on the work of André Looijenga, complemented with Archieven van Senaat en Faculteiten, 1575–1877, Leiden University Library, UBL 042, MSS 348, 349, 350.

[41] It should be noted that Boerhaave was 'promotor' (doctoral advisor) in place of a colleague who was not able to attend for one reason or another on 27 occasions. Boerhaave himself was replaced by a fellow professor only 17 times.

July.[42] Boerhaave's reputation may have attracted Crauford to Leiden, but we can wonder whether Boerhaave's teaching was a major influence on Crauford's later achievements. And Crauford was only one of many. Indeed, Underwood's prosopographical study showed that the length of study differed significantly for the English-speaking students.[43] Some stayed for years, while others left after a few weeks; some took degrees, while others did not; some became renowned scholars, others worked in obscurity.[44] Numbers of graduates and doctorates, in other words, offer only a very limited understanding of the eighteenth-century Boerhaave school.

Moreover, statistics alone cannot explain Boerhaave's popularity. A new generation of historians has therefore studied the contents and methods of Boerhaave's teachings in greater detail, to answer the question *why* so many students were drawn to him. Andrew Cunningham examined Boerhaave's medical system and characterised it as 'irenic', 'eclectic', and based on sense experience and observation.[45] He has reconstructed Boerhaave's intellectual growth over the course of his professional life, from applying a method of 'prideful' reason to utilising experience, from an apparently mechanistic physiology of the body to one that was understood in chemical and vitalist terms. Additionally, Knoeff and John Powers have shown that Boerhaave's pedagogy was particularly unique and attractive.[46] Many students, such as Ulricus Wilhelmus Rhode (*fl.* 1720), mentioned their memories of Boerhaave's lectures in their dissertations.[47] Powers argued that the Leiden professor was not concerned with teaching

[42] Underwood, *Boerhaave's Men*, 99–101; Jacobus Crauford, *Disputatio medica inauguralis de scorbuto* (Leiden, 1707). More on Crauford see Powers, 'Leiden Chemistry in Edinburgh'. Powers suggests that Crauford modelled his chemistry lectures after Boerhaave's lecture course.

[43] Underwood, *Boerhaave's Men*, 57–9.

[44] Out of the 468 English-speaking students who ultimately obtained their MD, only 139 graduated at Leiden. Ibid., 110–1. This may have been largely motivated by financial reasons, as graduating at Rheims, for example, was far less expensive than graduating at Leiden.

[45] Cunningham, 'Medicine to Calm the Mind'; Harold J. Cook, 'Boerhaave and the Flight from Reason in Medicine', *Bulletin of the History of Medicine*, 74 (2000), 221–40, here 225; Knoeff, *Herman Boerhaave*, 34.

[46] Rina Knoeff, 'Herman Boerhaave at Leiden: Communis Europae praeceptor', in *Centres of Medical Excellence?*, ed. Ole Peter Grell, Andrew Cunningham, and Jon Arrizabalaga (Farnham, 2010), 269–86; Powers, *Inventing Chemistry*.

[47] Ulricus Wilhelmus Rhode, *Dissertatio inauguralis chirurgico-medica de nervi punctura* (Leiden, 1720), 4.

knowledge that was already established, but was rather invested in restructuring and reshaping various chemical practices for pedagogical purposes, and transforming these practices into an academic discipline with its own problems, parameters, and methods in the process. Boerhaave restructured and reinterpreted a variety of practices from diverse chemical traditions—didactic chemistry, transmutational alchemy, chymical medicine, and experimental natural philosophy—to develop his own methodology for chemical investigations.[48] Knoeff has added that, contrary to other medical professors, Boerhaave used a style that was less prescriptive and more pedagogical, stimulating students to think, observe, and experiment for themselves. Rather than presenting ready-made solutions, this style allowed for improvements, and proposing new hypotheses. In 1738, for example, Henry Pemberton (1694–1771), who had studied at Leiden in the 1710s and graduated in 1719 with 'supreme honours', remembered that when he pointed out mistakes in Boerhaave's ideas on vision, Boerhaave did not dismiss the remark, but instead praised him and treated him as an equal.[49] This was an exceptional and revolutionary teacher–student relationship.

But the success and longevity of the Boerhaave school, this book argues, has as much to do with Boerhaave's work and pedagogy, as with the consideration of students and their appropriation of knowledge.[50] Although Boerhaave passed away on 23 September 1738, he remained a much-revered physician well into the early nineteenth century. Indeed, in the early eighteenth century, Boerhaave soon became a household name in the field of medicine, and after his death, he turned into a cultural phenomenon. Take, for example, his successor Gaubius, who not only took on Boerhaave's teachings at Leiden but also owned a portrait of Boerhaave painted by Cornelis Troost (1696–1750), described as 'the original image of the famous Professor, Herman Boerhaave, in his professorial dress'.[51] Also Gaubius's student Matthias van Geuns (1735–1819),

[48] Powers, *Inventing Chemistry*, 37–62.

[49] Henry Pemberton, *Dissertatio physico-medica inauguralis de facultate oculi, qua ad diversas rerum conspectarum distantias se accommodat* (Leiden, 1719); Knoeff, 'Herman Boerhaave at Leiden', 283.

[50] Cf. Lissa Roberts, 'The Circulation of Knowledge in Early Modern Europe: Embodiment, Mobility, Learning and Knowing', *History of Technology*, 31 (2012), 47–68.

[51] Th.H. Lunsingh Scheurleer, C. Willemijn Fock, and A.J. van Dissel, *Het Rapenburg: Geschiedenis van een Leidse gracht*, 6 vols (Leiden, 1986–1992), vol. VIa, 401, 99. Extant copies are SK-A-2342, Amsterdam, Rijksmuseum; P02634, Leiden, Museum Boerhaave;

who had become a well-known professor of medicine in Harderwijk and Utrecht was, as it were, a second-generation Boerhaave student, for he had adorned his stately home with both Boerhaave's portrait and bust.[52]

As this book demonstrates, a defining characteristic of Boerhaave's disciples was their shared attention to bodily fluids. The physicians and chemists mainly considered in this book were those who found their inspiration in Boerhaave, appropriated his teachings and research practices, yet also made original contributions to the field and, in turn, passed on their knowledge to yet another generation of physicians. Most of these disciples had been former students of Boerhaave and, after gaining experience as practitioners, obtained permanent posts at universities across the Dutch Republic and beyond. They include, to name but a few, Hieronymus Gaubius in Leiden, who succeeded Boerhaave as the lecturer of chemistry and started his lectures with a focus on the chemistry of blood, urine, milk, and other bodily fluids; Johannes de Gorter in Harderwijk, who christened his third son Herman Boerhaave de Gorter (1732–1800) and researched the physiological phenomenon of sweat and insensible perspiration;[53] Paulus 's-Graeuwen (1715–1779) in Groningen, who gave lectures in chemistry and decorated his wallpaper with a depiction of Boerhaave's memorial monument; Albrecht von Haller in Göttingen, who commented and expanded on Boerhaave's physiology lectures; and Gerard van Swieten (1700–1772) in Vienna, who spent 30 years (from 1742 to 1772) to finalise all his notes and extensions on Boerhaave's *Aphorisms* on the knowledge and treatment of diseases, including one on urine and bladder stones. All these men, and many more, ensured the continuation and dissemination of Boerhaave's teachings at universities across Europe, and well into the early nineteenth century.

and in the Senate Room, Leiden University. On Troost's painting of Boerhaave's portrait, see G.A. Lindeboom, *Iconographia Boerhaavii* (Leiden, 1963), 7, 23.

[52] 'Bibliotheca Geunsiana, sive, Catalogus librorum quibus usus est vir celeberrimus Matthias van Geuns' (Utrecht, 1818), 326. On Matthias van Geuns see J.H. Sypkens Smit, *Leven en werken van Matthias van Geuns M.D., 1735–1817* (Assen, 1953).

[53] Wilhelm Michael von Richter, *Geschichte der Medicin in Russland*, 3 vols (Moscow, 1813–1817), vol. 3, 465. Johannes de Gorter was married to Susanna (1685–1758), and their other sons, David (1717–1784) and Theodorus (1720–1786), both also became physicians. Herman Boerhaave de Gorter also followed in his godfather's footsteps, and defended his doctoral dissertation on milk and lactation. See his *Disputatio medica inauguralis de lacte et lactatione* (Harderwijk, 1751).

But perhaps the most important feature of the Boerhaave school was its non-dogmatism. The Boerhaave school did not consist of disciples blindly following in their teacher's footsteps, but rather learned to think critically and make valuable contributions by filling in gaps or correcting mistakes. In 1775 Gaubius looked back at the history of the Boerhaave school when he delivered a two-hour speech to commemorate the bicentenary of the founding of Leiden University. Reflecting on the history of medicine, he observed that 'people have eventually seen the rise of the Boerhaavian school, not a new school, but one very old, based on the Hippocratic one'.[54] This school of physicians, Gaubius argued, was imbued with the spirit and system of Boerhaave, grounded in indisputable discoveries and pure reasoning. He admitted that 'it is far from me, audience, not to recognise that also this Sun had its spots', because the very fact that Boerhaave had his shortcomings was the strength of the Boerhaave school: throughout the century it constantly improved.[55] And indeed, many had done so. Gaubius replaced Boerhaave's textbook with his own system of medicine in 1758. And when his friend Sanches only dared to anonymously publish his book because he disagreed with Boerhaave, Gaubius interfered and republished the work with Sanches's name on the title page.[56]

More than once Gaubius took on the role of the respected patriarch, which guaranteed the continuation and dissemination of thought from the Boerhaave school. Naturally, the common interest in bodily fluids and the shared dedication to improve physiology and pathology was of crucial importance. But this enterprise would not have been possible without institutional support. In 1753, when the curator of the University of Groningen, Anthony Adriaan van Iddekinge (1711–1789), sought to appoint a new professor of medicine, he consulted Gaubius. In the correspondence that unfolded over the next six months, Gaubius first of all recommended the appointment of not one but two professors: one in medicine and anatomy, the other in chemistry and botany. Second, alongside providing a list of potential candidates, Gaubius did not hesitate to

[54] Hieronymus David Gaubius, *Oratio panegyrica in auspicium seculi terti Academiae Batavae* (Leiden, 1775), 53; idem, *Feestrede van Hieronymus David Gaubius by den heuglyken aanvang der derde eeuwe van Hollands Hooge Schoole te Leyden, den agsten van Sprokkelmaand 1775*, trans. Pieter van den Bosch (Leiden, 1775), 89–93.

[55] Ibid., 94–6.

[56] Gaubius to Sanches, 2 February 1777 in Sophia W. Hamers-van Duynen, *Hieronymus David Gaubius (1705–1780): Zijn correspondentie met Antonio Nunes Ribeiro Sanches en andere tijdgenoten* (Assen and Amsterdam, 1978), 163–6.

point out his most promising students; to offer his apologies on behalf of his own nephew Johannes David Hahn's (1729–1784), who had just accepted a position at Utrecht University; and to cast doubt on those who were suspected to follow the Lutheran faith.[57] Gaubius's powers of persuasion paid off when the experienced scholar Tiberius Lambergen (1717–1763) and the 24-year-old Wouter van Doeveren, both Leiden alumni, held their inaugural lectures in Groningen in 1754. While this is just a single example for the dissemination of the Boerhaave school, in the course of the eighteenth century all Dutch medical faculties—Leiden, Franeker, Groningen, Utrecht, and Harderwijk—saw the appointments of former students of Boerhaave and Gaubius.

This book's focus on the Boerhaave school, in sum, builds on previous work on Leiden University as an institution and on the teachings of Boerhaave. But it also characterises the Boerhaave school as a (trans) national community of medical men working in Boerhaave's footsteps, either directly or indirectly. The Boerhaave school was a social phenomenon expressed in visual and material culture, and based on its followers' dedication to a research programme to further knowledge of physiology and pathology based on the knowledge of bodily fluids.

Sources and Structure

The Boerhaave school functions as a guiding principle for this book's corpus of study. I employ an assemblage of textual, visual, and material sources, all of which shed light on Boerhaavian physicians' engagement with bodily fluids. Personal letters and unauthorised publications are considered together with printed textbooks and official documents. Other written sources include dissertations, journal contributions, Latin and vernacular treatises, manuscripts, and students' lecture notes. Visual and material sources, such as engravings and paintings, chemical and medical instruments, provide a unique insight into the work of chemists and physicians of the Boerhaave school. They also demonstrate technological innovation. Boerhaave, for example, transformed the little stove ordinarily used as a foot warmer into a scientific instrument by placing a glass flask

[57] Hieronymus Gaubius to Anthony Adriaan van Iddekinge, Leiden, 8 February, 15 May, 24 August, and 2 September 1753, The Hague, Royal Library, 424 B 2, nr. 14, fol. 1–11. On van Iddekinge and the University of Groningen see Klaas van Berkel, *Universiteit van het noorden: Vier eeuwen academisch leven in Groningen* (Hilversum, 2014).

on top of it, making it a highly effective and safe furnace for research in and the teaching of chemistry and botany. This instrument came to be called the 'Boerhaave little furnace', grew popular among students as a safe and easy-to-use instrument, and remained in use throughout the eighteenth century.[58]

A large part of this study involves the analysis of university textbooks written by Boerhaave, Gaubius, and many other of Boerhaave's disciples. Textbooks offer many advantages for this investigation since, to a large extent, they reveal contemporary medical perceptions of bodily fluids. The historian of science Thomas Kuhn has argued that textbooks were essential to education, because they functioned as pedagogical instruments, establishing a community of scientists in agreement with the prevalent paradigm. Textbooks described 'exemplars', not only for the purpose of instructing university students in primary methodologies for solving problems but also for setting the standard for what could be considered the 'normal science' view.[59] Many early modern academic books started as a series of lectures delivered in front of students. Interestingly, these lectures were then collected in a coherent and concise *vade mecum*, often at the students' request. Textbooks, therefore, prove most useful for providing insights into rational and learned ideas about the physiological functions in which the fluids play an essential role, for example, in circulation, digestion, procreation, perspiration, lactation, and menstruation. Indeed, the textbooks by Boerhaave and Gaubius went through multiple editions and translations, suggesting that there was a large demand for this kind of medical information. The wide circulation and dissemination of the textbooks may, therefore, have struck a chord with their readers. Boerhaave's and Gaubius's works were arguably among the most widely read and frequently reprinted medical texts in the eighteenth century.[60]

[58] Boerhaavian furnace, V25790, Rijksmuseum Boerhaave, Leiden. Herman Boerhaave, *Elementa chemiae, quae anniversario labore docuit in publicis, privatisque scholis*, 2 vols (Leiden, 1732), vol. 1, 886–9; Boerhaave, *A New Method of Chemistry*, vol. 1, 589–90. Boerhaave wrote that he had invented this little furnace for his own investigations around 1692. Marieke M.A. Hendriksen and Ruben E. Verwaal, 'Boerhaave's Furnace: Exploring Early Modern Chemistry Through Working Models', *Berichte zur Wissenschaftsgeschichte*, 43 (2020), 385–411.

[59] Thomas S. Kuhn, *The Structure of Scientific Revolutions*, 2nd ed. (Chicago, 1974).

[60] G.A. Lindeboom, *Bibliographia Boerhaaviana: List of Publications Written or Provided by H. Boerhaave or Based upon his Works and Teaching* (Leiden, 1959). See this book's Chap. 7 on semen for the publication history of Gaubius's *Institutiones pathologiae medicinales* (1758).

Their textbooks provided an intellectual core for the Boerhaave school, for the instruction of its students and its furthering of the formation of a new physiology based on the fluids.

To tell the story of new medical perceptions relating to the bodily fluids, I integrate analyses of published sources with ego-documents and physicians' correspondence. These letters reveal much about the medical practice in the eighteenth century, because the authors refer to problematic cases, diagnoses of illnesses, and the prescription of medicines. Long after graduating from university, local practitioners continued to correspond with and requested advice from their former university teachers.[61]

Dispersed in libraries across Europe and North America today, students' lecture notes form another unique source material for the content of university lectures on bodily fluids.[62] A comparison of several of these manuscripts shows that some unique copies, written in a quick and illegible hand, were produced as a professor was lecturing. Most surviving lecture notes, for example, those by the Schaffhausen student Georg Michael Wepfer (1692–1774), show a very clear and neat hand, and were probably fair copies of hasty, perhaps abbreviated notes taken during the lecture.[63] Some student lecture notes reveal transfer across countries and reuse by others. Anton Gabriel Meder (b. 1686) from Norden in East Frisia, for example, studied in Leiden in 1710 and took notes in Boerhaave's chemistry lectures for three consecutive years.[64] Graduating in 1714, he returned to the north and worked as a physician. In around 1729 these notes came into the possession of one Rudolph Pabus (1712–1769), who studied theology in Groningen, close to East Frisia. Pabus started adding an index of subjects to the manuscript which, however, he never finished. When Pabus reached old age in 1764 he decided to donate his personal library,

[61] Many letters used here are published in G.A. Lindeboom, ed. *Boerhaave's Correspondence*, 3 vols (Leiden, 1962–1979); and Hamers-van Duynen, *Hieronymus David Gaubius.*

[62] On the distinctive characteristics of student lecture notes, see Georgette Taylor, 'Plummer to Cullen: Novelty in William Cullen's Chemical Pedagogy', in *Cradle of Chemistry*, ed. Robert G.W. Anderson (Edinburgh, 2015), 59–84.

[63] Georg Michael Wepfer, 'Collegium Chymicum experimentale', Schaffhausen, 1714. Uppsala, University Library, Waller MS cod-00068. Another example of neatly written out lecture notes is [anonymous], 'Descripta quaedam ab Ore Hermanni Boerhaavii inter pronunciandum lectiones ejus de Chemia in 2 tomis. Lugd. Bat. Anno 1719', 2 vols, Leiden, c. 1719–1730. British Library, London, Sloane MS 3183–3184.

[64] Anton Gabriel Meder, 'Tractatus collegii chymici, anno 1710 hab., nunc vero 1713 a Dr. Meder descript. & collect'. Leiden, 1713. Groningen, University Library, HS 98.

including the lecture notes, to the university, to benefit the Groningen students.[65]

In short, the Boerhaave school was formed by a fascinating and important group of medical men sharing similar ideas on physiology and methods of research that demand closer study. At the same time, for the core material of this book, the Boerhaave school has guided the assemblage of a coherent collection of sources and materials. The allegorical frontispiece to Herman Boerhaave's medical textbook *Kortbondige spreuken* (1741), a Dutch translation to his *Aphorisms* (first published in 1709), beautifully sums up the set-up and argument of this book (see Fig. 1.1).[66] Commissioned by bookseller Johannes Gysius, the engraving by an unknown artist shows the medical professor directing his student to the steps ideally undertaken in order to become a physician. Guarded by the female figures of reason and experience, the bottom steps of the stairs are philosophy and botany. The student then proceeds along with two doctors of medicine, understanding the structure of the body through anatomy. At the pinnacle, however, is chemistry, where Asclepius, the god of healing, is sat on a throne to crown the student as a medical doctor. The skills of observation and practice are symbolised by two gates leading to a botanical garden on the left and, of course, a chemical laboratory on the right. At the turn of the eighteenth century, in other words, both Boerhaave and his students found a new appreciation for the chemistry of the fluids. The use of chemical instruments and methods to bodily fluids led to an extraordinary set of changes, which established a new physiology of the fluids.

Each chapter in this book is dedicated to one bodily fluid, in order to follow its transformations in the eighteenth-century Boerhaave school, and to discover the ways in which the latter contributed to the new physiology of the fluids. Chapter 2, 'Savouring Alchemy', asks who Boerhaave's main precursors were in the study of the fluids by means of chemistry. Focusing on the case of saliva and digestion, this chapter examines how Boerhaave rejected medical chymistry to solidify his own chemistry of medicine. Although Boerhaave was less dismissive of chymistry than he

[65] Rudolph Pabus to Anthony Adriaan van Iddekinge, Farnsum, 27 January 1764 in National Library, The Hague, 424 B 2 no.14. The *Catalogus librorum*, which lists Pabus's contributions to the university library, also lists this manuscript.

[66] Herman Boerhaave, *Kortbondige spreuken wegens de ziektens, te kennen en te geneezen*, trans. Kornelis Love Jakobsz (Amsterdam, 1741).

Fig. 1.1 Frontispiece to Herman Boerhaave, *Kortbondige spreuken wegens de ziektens, te kennen en te geneezen* (Amsterdam, 1741). Utrecht University Library, MAG: O oct 5068

publicly led on, ultimately he was able to develop an irenic and inclusive model of physiology, establishing it as the standard for decades to come.

Chapter 3, 'The Nature of Blood', analyses the changing perceptions of blood, both by Boerhaave's disciplines and by his rivals. It discusses physicians' analyses of the peculiar properties of blood and how this gave an impetus to develop new understandings of physiology. Furthermore, it demonstrates that the new method of chemistry did not remain uncontested. Some grew deeply sceptical about chemistry, convinced that blood *in vitro* and blood *in vivo* were drastically different. To overcome this problem, other instruments were put forward with which living blood could be examined. Ultimately the debate went beyond the problem of methodology and was directly linked to the question of the essential yet disputed question surrounding blood: was blood alive?

Boerhaavian experiments on blood demonstrate the important role of chemistry in the reconceptualisation of the fluids and the physiology of the body. Chapter 4, 'Piss Prophets and Urine Matters', delves deeper into the role of the 'philosophy by fire' in this process by demonstrating the shift from chymical uroscopy to urine chemistry at the turn of the eighteenth century. By means of distilling urine, some practitioners believed to be able to diagnose patients more precisely than ever before. But for physicians like Gaubius, the chemistry of urine was not so much a diagnostic instrument applicable to any disease; rather, it was a means for investigating the changing properties of urine in the course of a patient's condition. These doctors explained specific urinary disorders like bladder stones in terms of the chemical elements in urine and their cohesive powers. The method and practice of urine chemistry, then, had very different aims and applications, moving from diagnostics to pathology. These new goals and methods would prove to be paramount for the development of modern medicine at the turn of the nineteenth century.

While the first three chapters on saliva, blood, and urine mainly focus on the changing perceptions of the fluids among academics, Chap. 5, 'Crying over Spilt Milk', reveals that the renewed interest in and reconsideration of the fluids also occurred outside academia. The practice of breastfeeding was never self-evident nor static, because the assistance of wet nurses and hand-feeding was popular practice across early modern Europe, for reasons of necessity or convenience. This chapter discusses how physicians published moralising texts to point mothers into the direction of maternal breastfeeding. I argue that in the eighteenth century, chemical experiments on milk supplied physicians with new arguments for emphasising the exceptional nutritious qualities of milk in general, and the importance of mother's milk for the health of both mother and child. Furthermore, stimulated by the programmes of learned societies, physicians, surgeons, and apothecaries were increasingly involved in the evaluation of galactagogues, the substances stimulating and inducing lactation. Instrument-makers even introduced lactation technologies such as breast pumps to the market. Yet mothers, it appears, continued to make up their own minds about the nursing of their children.

The fluids of saliva, blood, urine, and milk were easily collected and subjected to chemical experimentation. Yet sweat's ephemeral and volatile properties posed serious challenges for investigators. And still, perspiration, too, underwent a reconceptualisation into the new physiology, as much as its more congealed counterparts. Chapter 6, 'Sweat It Out', focuses on the concept of insensible perspiration and the sudorific drug of *sal ammoniac*. Looking at the Dutch reception of Santorio Santori's (1561–1636) weighing chair, this chapter argues that the works of Johannes de Gorter reveal a drastic shift in thought about insensible perspiration. Instead of recognising the role of balance of the humours, De Gorter paid particular attention to the role of microscopic nerves in the body and to the nature of individual and invisible bodily fluids, such as the so-called nervous juice. De Gorter's study on perspiration incorporated neurological descriptions into its attempt to develop a more detailed theory of the internal physiology of perspiration. Furthermore, it allowed him to describe the pathology of diseases like catarrh, and to justify the efficacy of his preferred treatment—*sal ammoniac*—which would make his patients sweat profusely and, literally, sweat out the disease.

All chapters touch upon fluids in relation to patients' illnesses and their treatments, but the final chapter tackles the concept of disease head-on. Chapter 7, 'Semen in Flux', demonstrates the development of eighteenth-century physicians' notions of disease. It argues that investigations relating to semen helped formalise pathology as a separate discipline. While observations made in nosology and morbid anatomy focused on the effects and end results of disease, they provided no indication as to the possible cause of a disease. Gaubius's *Institutiones pathologiae medicinalis* (1758) presented a third type of disease theory, namely one based on the chemistry of fluids. By exploring semen and gonorrhoea in relation to pathology, I argue that chemistry helped explain the causes of disease, and hence furthered the establishment of a new pathology. This was not one which was quickly forgotten, but one that fostered a new understanding of disease by dissemination across Europe.

Ultimately, the conclusion highlights what came of the redefinition of the nature of bodily fluids and the development of a new medical system in the eighteenth century. Granted, laboratory medicine and hospital medicine were established at the turn of the nineteenth century, allowing for statistical and more detailed quantity-based research. But this book argues that many of these nineteenth-century advancements in medicine and science could never have taken place, if it were not for the important work on bodily fluids undertaken in the early eighteenth century by members of the Boerhaave school.

CHAPTER 2

Savouring Alchemy

Early modern physicians drew salutary lessons from patients' retention and secretion of saliva. Consider the example of a 12-year-old patient from Amsterdam at the turn of the eighteenth century who suffered from a painful ulcer in her mouth. It left an open sore on her lower lip, causing saliva to continuously dribble down her chin. Over time, the girl struggled with digestion, grew increasingly thin, and remained relatively small for her age. The Amsterdam surgeon Adriaan Verduyn (fl. 1694–1732) was able to clean and stitch the wound, which then healed within a week and stopped the drooling. The girl soon recovered, as she could now retain her saliva and continued to grow.[1] Drawing conclusions from such observations, physicians and surgeons were convinced that saliva played an important part in digestion and hence was of the essence to one's growth and health. Herman Boerhaave concurred with this view. He taught his students that 'when it [saliva] is spit away too profusely there follows a loss of appetite, a bad digestion, and a wasting of the whole

[1] This case is shared in Frederik Ruysch, *Alle de ontleed- genees- en heelkundige werken*, trans. Ysbrand Gysbert Arlebout, 3 vols (Amsterdam, 1744), vol. 3, 1007–9. Verduyn was one of the overseers of the Amsterdam surgeons' guild from 1729 to 1731. See Abraham Titsingh, *Diana, ontdekkende het geheim der dwaazen die zig vroedmeester noemen: ter eeren van chirurgia geschreeven* (Amsterdam, 1750), 66, 9.

R. E. Verwaal, *Bodily Fluids, Chemistry and Medicine in the Eighteenth-Century Boerhaave School*, Palgrave Studies in Medieval and Early Modern Medicine,
https://doi.org/10.1007/978-3-030-51541-6_2

body'.[2] The medical professor explained that the lack of saliva would hinder tasting because one tastes nothing in a dry mouth. Saliva deficiency would also complicate chewing and obstruct swallowing. Purely from an observational perspective, then, saliva was essential to digestion.

Yet these observations did not explain how saliva supported the vital function of digestion inside the human body. Did it help heat up nourishment and support the cooking or concoction of the food in the stomach? Or did saliva perhaps cause some fermentation to break down the food? Early modern physicians and chemists such as Jan Baptista van Helmont (1577–1644), Franciscus Sylvius (François de le Boë, 1614–1672), and Herman Boerhaave proposed that only chemistry could reveal the digestive properties of the bodily fluids. Despite this common understanding, however, their research methods, results, and theories were painfully at odds.

This chapter focuses on medical theories of digestion and the practice of chemistry through the lens of saliva. Academic lectures, textbooks, and students' lecture notes reveal that deep into the eighteenth century, saliva played a key role in a discussion on the healthy functions of ingestion and digestion. Using saliva as a case study, I also demonstrate that Boerhaave transformed the medical chymistry of his forerunners into a discipline of chemistry to support medicine.[3]

Historical scholarship on digestion has paid little attention to saliva and its properties. Historians have mainly discussed the seventeenth-century intellectual debates on digestion and the shift towards vitalist perceptions. Allen Debus, for example, perceived a dichotomy between chymists who proposed digestion and absorption by acids, and mechanists who resorted to triturable operations of grinding, mincing, chewing, and churning.[4] Antonio Clericuzio has nuanced this view, showing that not all seventeenth-

[2] Albrecht von Haller, ed. *Dr. Boerhaave's Academical Lectures on the Theory of Physic: Being a Genuine Translation of his Institutes and Explanatory Comment*, 6 vols (London, 1742–1746), vol. 1, 134, vol. 5, 446.

[3] 'Chymistry' denotes all chemical practices prior to 1700 when the words alchemy and chemistry were used interchangeably. This includes chrysopoeia (*i.e.*, transmutation of base metals into noble ones), spagyria (*i.e.*, purification of substances), and iatrochemistry or chemiatria (*i.e.*, the school of medicine explaining disease in terms of chemical doctrines and the use of metals and minerals in treatment). See William R. Newman and Lawrence M. Principe, 'Alchemy vs. Chemistry: The Etymological Origins of a Historiographic Mistake', *Early Science and Medicine*, 3 (1998), 32–65.

[4] Allen G. Debus, *Chemistry and Medical Debate: Van Helmont to Boerhaave* (Canton, MA, 2001).

century physicians were convinced by the mechanist approach and mainly researched chemical operations.[5] Focusing on the second half of the eighteenth century, Elizabeth Williams has uncovered a view of the digestive process that was neither chemical nor mechanical. She argues that French physicians developed a vitalist conception of digestion that entailed the entire *animal oeconomy*, that is, physiology including emotional well-being.[6] This chapter suggests that the study of saliva bridges these apparently competing approaches of chemical and vitalist digestion: Boerhaave's physiology held that the body's life and vital functions originated in the actions and properties of the most subtle fluids, explored by his new method of chemistry. I will first discuss how Boerhaave's chemistry of digestion built on, yet also differed from the medical chymistry of Paracelsus (c. 1493–1541), Van Helmont, and Sylvius. I will then turn to the central role of saliva in the teachings of Boerhaave and his star pupil Albrecht von Haller (1708–1777). Close examination of their works reveals that their chemical work on saliva led to an understanding of vital powers peculiar to the living body.

A Salivary Theory of Digestion

Upon accepting the chair of chemistry at Leiden University in 1718, Herman Boerhaave held an inaugural lecture in which he looked back at the discipline's development. Reflecting on the works of his predecessors, he quickly singled out Franciscus Sylvius.[7] From his pulpit in the university auditorium (see Fig. 2.1), Boerhaave lectured that 'Sylvius, Professor of medicine at this University, recommended chemical science on all occasions with immoderate praise; and being wholly engrossed in it, he extolled the great benefits of this method before a large audience with the persuasive force of his authority, eloquence, and example'. As a result, many physicians had adopted Sylvius's 'idea that Nature acted through chemical

[5] Antonio Clericuzio, 'Chemical and Mechanical Theories of Digestion in Early Modern Medicine', *Studies in History and Philosophy of Biological and Biomedical Sciences*, 43 (2012), 329–37.

[6] Elizabeth A. Williams, 'Sciences of Appetite in the Enlightenment, 1750–1800', *Studies in History and Philosophy of Biological and Biomedical Sciences*, 43 (2012), 392–404; ead. 'Food and Feeling: 'Digestive Force' and the Nature of Morbidity in Vitalist Medicine', in *Vital Matters*, ed. Helen Deutsch and Mary Terrall (Toronto, 2012), 203–21.

[7] Herman Boerhaave, *Sermo academicus de chemia suos errores expurgante* (Leiden, 1718).

Fig. 2.1 Herman Boerhaave gives an inaugural speech in a packed lecture hall. Frontispiece to Boerhaave, *Sermo academicus de comparando certo in physicis* (Leiden, 1715). (Credit: Wellcome Collection. CC BY)

means and that human life was determined by them'.[8] At first glance, then, Boerhaave unequivocally celebrated the late professor's dedication to the art of chemistry and its importance to the field of medicine.

On a more personal level, too, Boerhaave and Sylvius shared many characteristics. Sylvius had been born into a French Protestant family of merchants, just as Boerhaave was a faithful Calvinist.[9] Both men had also studied medicine and were appointed professors at the same university. Furthermore, they argued in favour of the use of chemistry in medicine and developed medicines in their laboratories. In 1730 Boerhaave even bought the same stately home at the Rapenburg canal in Leiden in which Sylvius had lived from 1658 until his death in 1672.[10]

Yet on closer inspection Boerhaave distanced himself from Sylvius's ideas. In his lecture he acknowledged the rise of medical chymistry, of which Sylvius become the pinnacle, but he sharply criticised its central tenets. Boerhaave dismissed its central theory of the human body, explained in terms of acids and alkalis, because he claimed it could be learned in a single hour. Besides criticising the simplicity of the theory, Boerhaave was abhorred by its speculative nature and implications. In the chymical theory of digestion, for example, a 'ridiculous equation' was made between the 'acetous force of biting substances' that corroded metals and 'the substance that dissolves the food in the stomach'.[11] According to Sylvius, saliva played a key role in generating fermentation in the stomach. He argued that together with pancreatic juice and bile, an acid-alkali reaction took place in the digestive system that helped to break down the ingested food. Boerhaave thought this was a ludicrous idea. If this salivary theory of digestion was taken seriously, it meant that the human stomach was no longer an internal organ, but a 'Hermetical crucible'.[12]

[8] As translated in E. Kegel-Brinkgreve and A.M. Luyendijk-Elshout, eds., *Boerhaave's Orations* (Leiden, 1983), 207.

[9] E.D. Baumann, *François dele Boe Sylvius* (Leiden, 1949); Rina Knoeff, *Herman Boerhaave (1668–1738): Calvinist Chemist and Physician* (Amsterdam, 2002).

[10] Pamela H. Smith, *The Body of the Artisan: Art and Experience in the Scientific Revolution* (Chicago, 2004), 298–9.

[11] Kegel-Brinkgreve and Luyendijk-Elshout, *Boerhaave's Orations*, 207–8.

[12] Ibid. For a detailed discussion on the idea of the human body as chemical laboratory, see Antonio Clericuzio, 'The Internal Laboratory: The Chemical Reinterpretation of Medical Spirits in England (1650–1680)', in *Alchemy and Chemistry*, ed. Piyo Rattansi and Antonio Clericuzio (Dordrecht, 1994).

It is, of course, likely that claims about Sylvius and medical chymistry were exaggerated by skilful rhetoric in Boerhaave's lecture, aptly titled *On Chemistry Purging Itself of its Own Errors*. Yet his oration also reveals competing perceptions of the body at the turn of the eighteenth century that require further study. If Sylvius and Boerhaave both excelled in chemical experimentation and were dedicated physicians, why did Boerhaave dismiss the chymical concepts of his predecessor? To understand why Boerhaave distanced himself from Sylvius's theories, we must first explore Sylvius's career and his chymical study of human physiology and digestion in more detail.

Franciscus Sylvius was appointed to the chair of practical medicine at Leiden University in 1658, and under his wing medical teaching flourished. He had gained the necessary experience as a professor of anatomy at Leiden between 1638 and 1642, while he had also practised as a physician in Amsterdam. During his professorship, the number of medical students matriculating at Leiden continued to grow. Student Reinier de Graaf (1641–1673) was particularly fond of his master: 'It had pleased us so much to be around him, that we never attended his lectures, in public or especially at his house, without such heartfelt joy, as when we together with other students of his gentleman were encouraged to delve further into the truth and put our hands to work'.[13] Students defended their doctoral theses under Sylvius's supervision, further developing the chymical model of physiology. Abraham Quina (b. *c.* 1635), for example, discussed the fermentation of food in the stomach in 1659.[14]

Sylvius was passionately devoted to chymistry and worked in the laboratory. When still practising medicine in Amsterdam, he had befriended the apothecary Johann Rudolf Glauber (1604–1670), who owned an elaborate chemical laboratory. Once in Leiden, Sylvius had his house equipped with a personal laboratory with a separate distilling room, furnished with iron and copper furnaces, oil pots, glass bottles, copper kettles, iron tongs, brass mortar, copper bowls, and other instruments.[15]

[13] Reinier de Graaf, *Alle de wercken, so in de ontleed-kunde, als andere deelen der medicyne* (Amsterdam, 1686), 499.

[14] Sylvius collected the dissertations of Quina and nine other students on various physiological topics in Franciscus Sylvius, *Disputationum medicarum: pars prima primarias corporis humani functiones naturales ex anatomicis, practicis & chymicis experimentis deductas complectens* (Amsterdam, 1663).

[15] For the complete inventory, see Harm Beukers, 'Het laboratorium van Sylvius', *TGGNWT*, 3 (1980), 28–36.

Here, Sylvius tested descriptions sent by Glauber and had his students perform experiments. The sisters Catharina and Johanna Agache, who had moved in after the death of Sylvius's second wife in 1667, were also skilled at manufacturing medicaments, such as the so-called Sylvius's salt (*sal volatile*). Yet Sylvius believed chemistry should also become part of the official university curriculum. In 1664 he petitioned the university governors to establish a separate chair in chemistry and a chemical laboratory, but they refused. Only when Sylvius threatened to leave two years later did they reconsider. Sylvius's persistence was rewarded with the appointment of Carel de Maets (*c.* 1640–1690) as the first professor of chemistry at Leiden and the establishment of a chemical laboratory in the botanical garden in 1669.[16] That same year, Giovanni Battista Gornia (1633–1684), private physician to Grand Duke Cosimo de' Medici of Tuscany, visited Sylvius, describing him as a 'vain and self-assured man, though a skilled chemist and an expert of chronic diseases'.[17]

Chymistry dominated Sylvius's study of human physiology.[18] Sylvius advocated a medical chymistry which explained all phenomena of the body in terms of a series of chemical reactions and the interaction between acids and alkalis. In gastric digestion, food was changed by a process 'we incline to call *Fermentation*'.[19] During this fermentation, those parts of the food which would nourish the body separated as chyle; the rest continued down as excrement. In Sylvius's words, 'the tender, fluid, and whitish Part we call, by way of excellency, Chyle, which goes forward through the spungy Crust of the Guts, the Lacteal Veins, and Thoracic Passage to the Heart, to get the Form of Blood; whilst the thick, solid and diversly colour'd part

[16] Baumann, *Sylvius*, 67–9; Ronald Sluijter, *'Tot ciraet, vermeerderinge ende heerlyckmaeckinge der universiteyt': Bestuur, instellingen, personeel en financiën van de Leidse universiteit, 1575–1812* (Hilversum, 2004), 79. On De Maets, see John C. Powers, *Inventing Chemistry: Herman Boerhaave and the Reform of the Chemical Arts* (Chicago, 2012), 49–55.

[17] G.J. Hoogewerff, ed. *De twee reizen van Cosimo de' Medici, prins van Toscane, door de Nederlanden (1667–1669): Journalen en documenten* (Amsterdam, 1919), 311–2.

[18] On Sylvius's chymical medicine, see Evan R. Ragland, 'Chymistry and Taste in the Seventeenth Century: Franciscus Dele Boë Sylvius as a Chymical Physician between Galenism and Cartesianism', *Ambix*, 59 (2012), 1–21; Harm Beukers, 'Acid Spirits and Alkaline Salts: The Iatrochemistry of Franciscus dele Boë, Sylvius', *Sartoniana*, 12 (1999), 39–59.

[19] Franciscus Sylvius, *A New Idea of the Practice of Physic*, trans. Richard Gower (London, 1675), 28. This translation was based on Franciscus Sylvius, *Praxeos medicae idea nova* (Leiden, 1671).

of Food is thrust down to the great Guts, in which it gets the name and form of Excrements'.[20]

The essential fluid to start this fermentation in the stomach was saliva. Whereas humans consumed food at set intervals during the day, saliva was continuously produced and swallowed. Sylvius reasoned that spittle must accumulate in the stomach, ready to start digesting food: 'To Spittle we ascribe, that by its promoting the Fermentation of Food more and more, and by diminishing the viscidity thereof, it makes it all more fit for its desir'd Separation'.[21] In chymical terms, saliva consisted of 'much Water, a little Volatile Spirit, and mixed and tempered with a very little lixivial Salt, with Oil and very little of Spirit of acid'.[22] During gastric digestion, the water of saliva was supposed to dissolve salts, while the spirituous parts promoted fermentation of food, just as grape juice fermented into wine.

Besides saliva, Sylvius identified two other fluids that played an important role in digestion. Once food and saliva had fermented in the stomach, the mass continued into the duodenum (the first part of the small intestines), where it was mixed with bile and pancreatic juice. The confluence of these three acidic and alkaline fluids would bring about an effervescence, separating the most pure and nutritious fluids from the impure and thick solids. The bile, which Sylvius believed to be alkaline in nature, promoted separation, 'in as much as it mildly cuts the viscous parts of Food by its volatil [*sic*] and oily Salt, and frees the Fluid parts from them, and makes these more fluid, yea flowing'. The acid spirit in the pancreatic juice 'does more potently loosen the same obstinate Viscidity of Food, and gives occasion also by compelling its most viscous parts of flowing together with the more fluid parts, and presently to flow through'. As Sylvius summed up, the so-called Triumvirate of saliva, bile, and pancreatic juice swayed the animal oeconomy of the body.[23]

The question still remained whether pancreatic juice was really acidic in nature. Sylvius's student Reinier de Graaf famously corroborated Sylvius's theory by means of *in vivo* experiments. Resorting to vivisection, he cut open a living dog and attached a fistula to the pancreas in order to tap the pancreatic juice. After collecting, observing, and even tasting the dog's fluids, De Graaf held a public lecture under the supervision of Sylvius,

[20] Sylvius, *Practice of Physic*, 50–1.

[21] Ibid., 51.

[22] Sylvius, *Disputationum medicarum decas*, vol. 5, §36.

[23] Sylvius, *Practice of Physic*, 51–2.

defending his anatomical-medical thesis and ascribing acidic properties to the pancreatic juice.[24] Ultimately, Sylvius had shown the strength of chymistry as an explanatory model for physiology by pointing at the acidity of the digestive fluids. Sylvius's physiology of the body had become not unlike a chemical laboratory in which intricate series of fermenting and effervescing processes occurred. Disease too, Sylvius presumed, must spring from an acidic-alkaline imbalance. Treatment was successful when balance could be restored by neutralising the excess acid or base.

During his oration, however, Boerhaave denied the validity of medical chymistry, both on grounds of research methodology and its speculative nature. Boerhaave taught his students that a chemist should go about formulating hypotheses and performing experiments, and he emphasised what kind of knowledge the chemist produced: 'For chemistry is no science form'd *à priori*; 'tis no production of the human mind, framed by reasoning and deduction'. Instead, Boerhaave argued that truths of chemistry are 'collected *à posteriori*, from numerable experiments'.[25] Ursula Klein has rightly pointed out that Boerhaave was strongly committed to Baconian empiricism and the latter's 'experimental history', that is, the study of animals, plants, and minerals by means of experiment, as distinct from natural history, which was mainly occupied with observation. Experimental history differed from experimental philosophy because the latter assumed a theoretical framework and methodology, whereas experimental history was merely concerned with fact-producing by means of repeating additional operations on various bodies.[26] Hypothesising that

[24] Reinier de Graaf, *De succi pancreatici natura et usu exercitatio anatomico-medica* (Leiden, 1664). For a more detailed discussion on De Graaf's experiment, see Evan R. Ragland, 'Experimenting with Chymical Bodies: Reinier De Graaf's Investigations of the Pancreas', *Early Science and Medicine*, 13 (2008), 615–64. On De Graaf, see Hans L. Houtzager, ed. *Reinier de Graaf 1641–1673: In sijn leven nauwkeurig ontleder en gelukkig geneesheer tot Delft* (Rotterdam, 1991); Mart J. van Lieburg, 'Reinier de Graaf en zijn plaats in het fysiologisch onderzoek van de zeventiende eeuw', *Gewina*, 15 (1992), 73–84.

[25] Herman Boerhaave, *A New Method of Chemistry: Including the Theory and Practice of that Art: Laid down on Mechanical Principles, and Accommodated to the Uses of Life*, trans. Peter Shaw and Ephraim Chambers, 2 vols (London, 1727), vol. 1, vi; Herman Boerhaave, *Institutiones et experimenta chemiae*, 2 vols ('Paris', 1724), vol. 1, 2. See also Powers, *Inventing Chemistry*, 112.

[26] Ursula Klein, 'Experimental History and Herman Boerhaave's Chemistry of Plants', *Studies in History and Philosophy of Biological and Biomedical Sciences*, 34 (2003), 533–67, here 538–41.

pancreatic juice was acidic before having determined it was doing science the wrong way round.

Boerhaave also warned against the extrapolation of experiments on a single material into a general law applicable to other materials. He objected to Sylvius's inference that if the mixture of acid and alkaline fluids in a glass flask caused effervescence, it would necessarily mean the same occurred inside the intestines. Even if a general rule may present itself after thousands of experiments, caution is required:

> Nor is it allowable to extend even such rule, however true it may prove; but only to apply it to single cases, where the same common reason is found to obtain. For certain bodies have *peculiar powers*, from whence effects arise, that do not come within the compass of any general theorem; but depend on a certain constitution.[27]

In other words, Boerhaave stressed the idiosyncrasy of different matters and taught his students to perform extensive experiments with chemical instruments, solvents, and processes—not within a predetermined theoretical or philosophical framework, but rather as a scientific method to support fact-finding on particular materials.

Sylvius had indeed been less strict when it came to making sweeping statements based on his laboratory work. It was not uncommon for him to make inferences without supporting evidence. This was no idleness, but merely his way of knowing, summarised by the motto 'reason is the leader accompanied by experience'.[28] Whereas Boerhaave was strict about the extent to which a rule could be generalised, Sylvius recognised the validity of comparable experience. And even though he supported his theory with sensory experience and experiment, it was specifically the method of analogy to which Boerhaave objected: 'If the acetous force of biting substances corroded metals', Boerhaave said, 'then the substance that dissolves the food in the stomach was called an acid too' by the chymists.[29] Comparing the body to a chemical laboratory was an interesting perspective, but Boerhaave protested against Sylvius's lack of experimental evidence to

[27] Herman Boerhaave, *A New Method of Chemistry: Including the History, Theory, and Practice of the Art*, trans. Peter Shaw, 2nd ed., 2 vols (London, 1741), vol. 1, 2–3. Emphasis added.

[28] Disputationum medicarum II in Franciscus Sylvius, *Opera medica* (Amsterdam, 1679), 13. The original dispute was held by Ludovicus Meyer in 1659.

[29] Kegel-Brinkgreve and Luyendijk-Elshout, *Boerhaave's Orations*, 207–8.

support such claims. Where was the proof of the existence of a 'spirituous part' of saliva?

Yet Boerhaave did not reject his findings out of hand. Arguing for diligence, he took it upon himself to ascertain whether the bodily fluids indeed had acidic or alkaline properties. He took quantities of urine, milk, blood, and other bodily fluids to test their acidity. In the case of milk, Boerhaave held two glass flasks with a warm quantity of the white fluid (see Fig. I in Fig. 2.2). To the first, he added an alkaline, such as the oil of tartar, and to the second he added an acid, like the spirit of nitre. If milk was acidic or alkaline, one would expect a reaction to take place in one of the two vessels. The students, however, observed that in neither flasks the milk showed any motion or ebullition. Then Boerhaave combined the two mixtures, which immediately caused a great effervescence. With this process Boerhaave confirmed that milk contained neither acid nor alkali, 'for if either of these were latent in it, the process ought to make the discovery, as being generally allow'd the proper criterion'.[30] Boerhaave then distilled milk in a glass alembic over a fire of 150 °F (see Fig. V in Fig. 2.2). This procedure conveyed 'an aqueous liquor' and left behind a 'yellow, thick, unctuous mass'. Boerhaave performed the same operation as before, adding two different reagents, one acidic and one alkaline. Again, the liquor nor the mass gave 'the least appearance of containing any thing acid, alcaline, or saline, upon all the trials made to discover it'.[31] The process was repeated with blood, the urine of a healthy and strong man, as well as the morning urine from a man who had taken down five pints of wine at once the evening before. According to Sylvius's theory of digestion, one would expect the kidneys to purify the body from impure excesses and hence leave acids, alkalis, or fermented spirits in the urine. But the results were exactly the same: none of the mixtures showed any traces of ebullition, and thus contained no acid nor alkali. Boerhaave therefore concluded that the body, too, contained none. On the basis of numerous experiments, then, he had proved Sylvius's chymical theory of the body in terms of acids and alkalis to be false.

As we have seen, the question to what extent Boerhaave built on Sylvius's medical chymistry is a complex one. On the one hand he venerated Franciscus Sylvius for insisting chemistry was crucial to understand phenomena of life like digestion. He also paid tribute to Sylvius's

[30] Boerhaave, *A New Method*, vol. 2, 182–3.
[31] Ibid., vol. 2, 186–7.

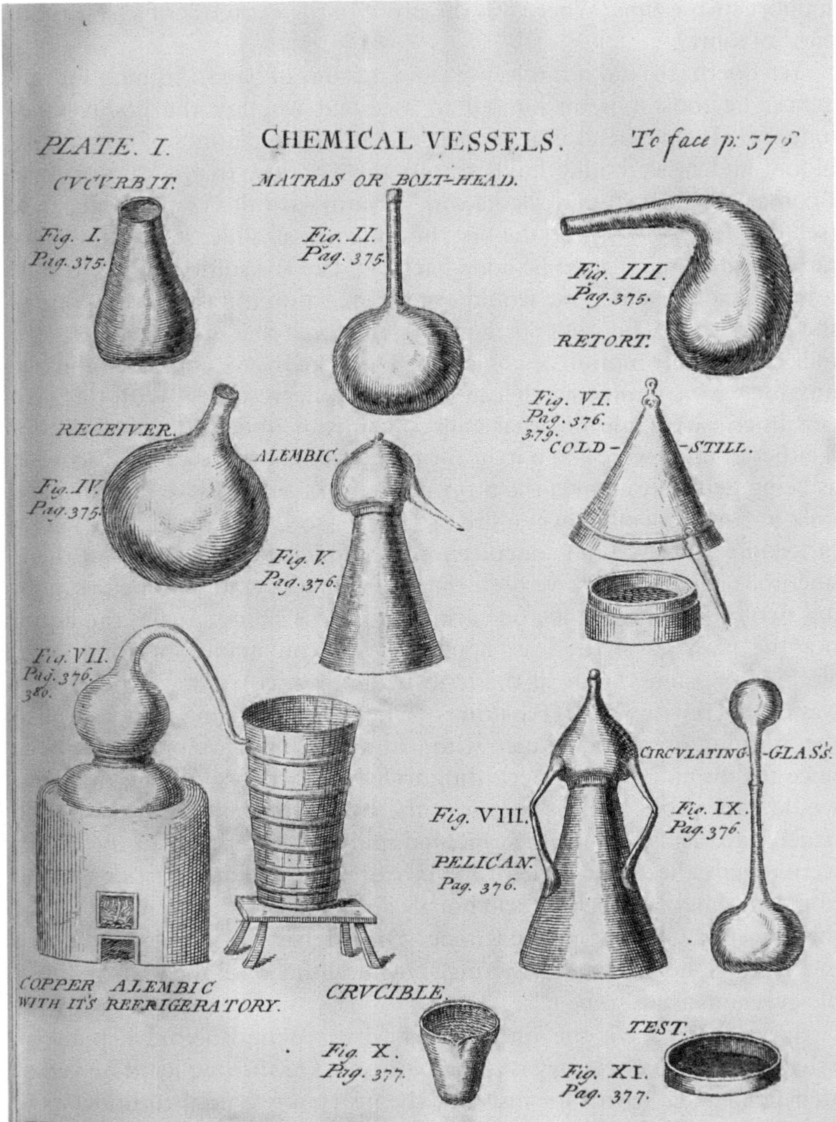

Fig. 2.2 Chemical vessels. Plate I in Herman Boerhaave, *A New Method of Chemistry* (London, 1727). (Credit: Wellcome Library. CC PDM 1.0)

campaign for the foundation of a chemical laboratory at Leiden University and implementing the practice of chemistry in the medical curriculum. There was also no denying that Sylvius's influence had been significant, both in the Low Countries and across Europe. Yet in his 1718 oration and in his chemistry lectures, Boerhaave dismissed the analogous reasoning inherent in Sylvius's theory, as well as the acid-alkali interaction and Sylvius's notion of the body as a 'chemical laboratory, an arena in which the chemists could stage their games'.[32] Boerhaave stressed that medicine should not be made to depend on the practice and speculation of chemistry: *in vitro* experiments in the laboratory should not be equated to *in vivo* processes in the body. Chemists should not seek verification to physiological and pathological theories, nor make grand generalisations. Instead, they ought to remain modest and limit themselves to performing experiments and demonstrating the effects of chemical reactions.

Heat and Acid

Although Boerhaave distanced himself from Sylvius, he seems to have admired the strongest advocates of medical chymistry, Paracelsus and Jan Baptist van Helmont. He often referred to Van Helmont favourably in his correspondence.[33] Boerhaave's former student and biographer William Burton (1703–1756) also recalled that Boerhaave 'had read over carefully Paracelsus four, and Helmontius seven times'. The Leiden professor 'could even repeat some whole chapters of Van Helmont almost verbatim'.[34] In addition, we know that Boerhaave shared his fascination with Paracelsus and Van Helmont with his students during his lectures. In his course on the theory and practice of chemistry, Boerhaave evaluated the works of a multitude of chemists and alchemists, from ancient times until his present

[32] Kegel-Brinkgreve and Luyendijk-Elshout, *Boerhaave's Orations*, 206.

[33] See, for example, Boerhaave to Joannes Baptista Bassand, 23 July 1733, in G.A. Lindeboom, ed. *Boerhaave's Correspondence*, 3 vols (Leiden, 1962–1979), vol. 2, 320–1. Knoeff, *Herman Boerhaave*, 181.

[34] William Burton, *An Account of the Life and Writings of Herman Boerhaave* (London, 1743), 67, 158. Boerhaave owned editions of their main works: Paracelsus, *Opera medico-chemica*, 3 vols (Frankfurt, 1603); idem, *Opera: Bücher und Schrifften* (Strassburg, 1616); idem, *Opera omnia medico-chemico-chirurgica*, 3 vols (Geneva, 1658); and Jan Baptist van Helmont, *Ortus medicinae*, ed. Franciscus Mercurius van Helmont (Amsterdam, 1652); idem, *Opera omnia* (Frankfurt, 1707). See 'Bibliotheca Boerhaaviana sive catalogus librorum instructissimae bibliothecae viri summi D. Hermanni Boerhaave' (Leiden, 1739).

day, with particular emphasis on the lives and works of Paracelsus and Van Helmont.

Why Paracelsus and Van Helmont merited a more thorough discussion than any other chemist is not immediately obvious. For if Boerhaave had so harshly criticised Sylvius, whose theory of salivary digestion was in fact based on these two protagonists of medical chymistry, why did not the same fate befall Paracelsus and Van Helmont? To answer this question, this section explores Boerhaave's lectures on Paracelsus and Van Helmont and evaluates the extent to which Boerhaave adopted them as a model. As will become clear, Boerhaave was less dismissive of medical chymistry than he had claimed in his 1718 oration. He had in fact much admiration for the pioneering work of medical chymists, but it was their tainted reputation and questionable claims that made it necessary to criticise them in front of his colleagues.

Boerhaave started giving public lectures in chemistry at Leiden in 1718.[35] In the absence of a much-desired textbook, however, some eager students took matters into their own hands. As was customary at early modern universities, they took notes during lectures, but they then proceeded to collect them and combined the various versions into a single manuscript for publication. In 1724, their unauthorised chemistry textbook appeared as *Institutiones et experimenta chemiae*, ascribed to the 'honourable Boerhaave' and allegedly printed in Paris. The book carried a disclaimer about the quality of the work. Despite the minor mistakes, the editors recommended the work for its new experiments and its overall usefulness. The 'illustrious author', they argued, had given his course on chemistry a very clear organisation and shown the strengths of chemical products.[36] The book quickly became a bestseller. Its small octavo size made it affordable for students. The Venetian bookseller Sebastian Coleti reprinted the book in 1726; the Englishmen Peter Shaw (1694–1763) and Ephraim Chambers (*c.* 1680–1740) translated the work and published it as *A New Method of Chemistry* in 1727.[37]

These lecture notes demonstrate Boerhaave's profound appreciation of medical chymistry. In the first part on the history of chemistry, Boerhaave

[35] Boerhaave had offered private lectures in chemistry since 1702. See Powers, *Inventing Chemistry*, 68, 84, 9.

[36] Boerhaave, *Institutiones et experimenta chemiae*, [*2].

[37] Herman Boerhaave, *Institutiones et experimenta chemiae*, 2 vols (Venice, 1726); J.R.R. Christie, 'Historiography of Chemistry in the Eighteenth Century: Hermann Boerhaave and William Cullen', *Ambix*, 41 (1994), 4–19.

referred to the main alchemists and chemists throughout history, who had worked on metals and alchemy, such as Roger Bacon (*c.* 1214–1294), Arnoldus de Villa Nova (*c.* 1240–1311), and George Ripley (*c.* 1415–1490).[38] Attention quickly shifted to those chemists 'who have cultivated the art with a view to Medicine; whether in order to a discovery of an universal remedy for all Diseases, or of particular ones in the ordinary way of physic'.[39] In other words, Boerhaave highlighted those chemists who had paid particular attention to applying chemistry to medicine, either as the preparation of medicaments or as the use of stones, metals, and minerals in therapy.[40] He first considered Basil Valentine (fl. fifteenth century), who had become famous for his remedies and was later judged the 'founder of the chemical pharmacy'.[41] Boerhaave paid most attention to Paracelsus and Van Helmont: besides their production of medicaments, their propagation of a chymical doctrine profoundly changed the medical theory and practice.

The reason why Paracelsus had blended alchemy and medicine was his training and experience as a miner and physician. Boerhaave elaborately lectured on the life of Paracelsus to drive home this point. Born as Theophrastus Phillipus Aureolus Bombastus von Hohenheim in Switzerland (see Fig. 2.3), Paracelsus was the son of a physician and was trained in both medicine and alchemy at a young age. He travelled throughout the German lands and Hungary, working as an apprentice in the mines and acquiring much of his practical knowledge of mineralogy and metallurgy. After further travels to Russia and Turkey and experiences in both surgery and medicine—two fields ordinarily exercised by separate practitioners—Paracelsus 'assumed the title of ***utriusque medicinae doctor***; or doctor both of internal, and external medicine'.[42] His break with Galenic medicine started during a syphilis outbreak. Paracelsus applied to

[38] More on these and other alchemists, see Lawrence M. Principe, *The Secrets of Alchemy* (Chicago, 2012).

[39] Boerhaave, *Institutiones et experimenta chemiae*, vol. 1, 21.

[40] For lapidary and mineral medicine in this context, see Marieke M.A. Hendriksen, 'Boerhaave's Mineral Chemistry and Its Influence on Eighteenth-Century Pharmacy in the Netherlands and England', *Ambix*, 65 (2018), 303–23; ead. 'The Repudiation and Persistence of Lapidary Medicine in Eighteenth-Century Dutch Medicine and Pharmacy', in *Gems in the Early Modern World*, ed. Michael Bycroft and Sven Dupré (Cham, 2019), 197–220.

[41] Boerhaave, *Institutiones et experimenta chemiae*, vol. 1, 22.

[42] Boerhaave, *A New Method*, vol. 1, 24. Emphasis in original.

Fig. 2.3 Portrait of Paracelsus by Romeyn de Hooghe, in Godfried Arnold, *Historie der kerken en ketteren* (Amsterdam 1701–1729). (Credit: Wellcome Collection. CC BY)

bleeding, purging, and vomiting, but when these Galenic remedies proved ineffective he turned to mercury as a more efficacious treatment. The application of quicksilver caused copious salivation in the mouth, a sign that syphilis was being cured.[43] Some physicians were already familiar with the excessive salivation induced by mercury, but Paracelsus gained great renown because of his innovative preparations. The opium and mercurial remedies Paracelsus produced were shaped in the form of pills, allowing him to heal in a 'gentler manner'. It was also said that 'he cured the itch [scabies], lepra [leprosy], ulcers, *Neapolitan* disease, and even gout'.[44] Paracelsus rejected the traditional framework of medicine, instead of basing his medical knowledge and practice on experience and alchemy.[45] His definitive departure from Galenic medicine was prompted by a rather explosive and controversial event, which Boerhaave shared with the students. Upon accepting a professorship in medicine and philosophy at the University of Basel, Paracelsus 'procured a fire to be brought in a brazen vessel, into the middle of the school; where, after casting in sulphur and nitre, in a solemn manner, he burnt the writings of *Galen*, and *Avicenna*: alledging [*sic*], that he had held a dispute with them in the gates of hell, and had fairly routed and overcome them'. From then on, 'physicians should all follow him; and no longer call themselves *Galenists*, but *Paracelsists*'.[46]

At the turn of the seventeenth century, a similar shift from Galenic medicine to medical chymistry occurred in the work of Jan Baptista van Helmont. A physician from Brussels, Van Helmont was the second major chymist whom Boerhaave introduced to his students. After his studies Van Helmont grew increasingly disappointed by the lack of success in his practice of medicine and began to doubt the authority of Galen and Hippocrates. Following an accidental meeting with 'a *Paracelsian*

[43] Throughout the seventeenth century, mercury was praised for its penetrating and cleansing qualities. See Lawrence M. Principe, *The Aspiring Adept: Robert Boyle and His Alchemical Quest: Including Boyle's "Lost" Dialogue on the Transmutation of Metals* (Princeton, 1998), 138–80; and Marieke M.A. Hendriksen, 'Anatomical Mercury: Changing Understandings of Quicksilver, Blood, and the Lymphatic System, 1650–1800', *Journal of the History of Medicine and Allied Sciences*, 70 (2015), 516–48.

[44] Boerhaave, *A New Method*, vol. 1, 24. Emphasis in original.

[45] Andrew Weeks, *Paracelsus: Speculative Theory and the Crisis of the Early Reformation* (Albany, NY, 1997); Walter Pagel, *Paracelsus: An Introduction to Philosophical Medicine in the Era of the Renaissance* (Basel, 1958).

[46] Boerhaave, *A New Method*, vol. 1, 25.

chemist' and the curing of his itch by means of sulphur, Van Helmont 'began now to take heart again: this new light soon alter'd the course of his thoughts; and to chemistry he falls with all his might'.[47] Judging by the lecture notes, Boerhaave favoured Van Helmont over Paracelsus, who since his professorship at Basle had indulged himself in excessive drinking. Van Helmont, in contrast, 'was very sober, and regular in his way of living; of a firm body, and a healthy constitution'.[48]

Paracelsus and Van Helmont had viewed alchemy not simply as the production of medicaments, but as the core of the world and the human body. Paracelsus was convinced that the key to understanding the entire world was alchemy. In order to uncover the hidden virtues and workings of nature, one had to submit the world's objects to chemical analysis, break down substances to reveal their quintessential qualities and refine the pure from the impure to improve upon nature. Paracelsus argued that God was the supreme alchemist: since He had divided heaven from earth, was Creation not a continuous separation of matter to create all parts of the natural world?[49] All those who performed manual labour in nature—such as artisans and craftsmen—gained knowledge of nature and as such worship God. Through the application of one's body and senses as instruments, humans could manifest the invisibility of God's revelation.[50] And just as the natural world could be reduced to chemical processes, Van Helmont believed that the human body could also be perceived as a complex of chemical processes. He argued that every organ was governed by a so-called 'archaeus', a vital spirit working as an 'inner chemist' to constantly separate pure from impure, and thus keep the body healthy. The chief archaeus resided in the stomach and was the principle of digestion.[51] In Boerhaave's words, 'if now any heterogeneous body happen to be present to the *archaeus*; it rises into a fervour, endeavours to expel the hostile matter'. Illness was the result of a failing archaeus and 'to cure any disease, therefore, is to pacify, and compose this *archaeus*'.[52] Where then could one find this archaeus? Van Helmont claimed that it had originally grown in

[47] Ibid., vol. 1, 31–2.

[48] Ibid.

[49] Allen G. Debus, *The Chemical Philosophy: Paracelsian Science and Medicine in the Sixteenth and Seventeenth Centuries* (New York, 1977), 45–57.

[50] Coined as the 'artisanal epistemology' in Smith, *The Body of the Artisan*, 82–93.

[51] Lester S. King, *The Road to Medical Enlightenment, 1650–1695* (London, 1970), 43–5.

[52] Boerhaave, *A New Method*, vol. 1, 33–6.

the tree of life, but that God had hidden it inside metals after the expulsion of Adam and Eve from Paradise.[53]

The case of digestion demonstrates how Van Helmont extended and improved upon the ideas of Paracelsus, rejecting some while generalising others. Paracelsus had argued that digestion occurred in the mouth as much as it did in the stomach, because 'in the mouth there is the heat of digestion'.[54] In the stomach, he believed, this heat of digestion 'efficiently seethes and cooks, not unlike the fire outside'.[55] Van Helmont disagreed, though he based his arguments on similar analogical reasoning, stating that 'heat does not digest efficiently, but excitingly'.[56] He compared it with the effect of heat when cooking meat: the basic structure of the meat remained unchanged.

Besides heat, Paracelsus thought that acid could play a role in the stomach, particularly in the case of disease or in certain animals. In his books *On the Origin and Cause of Diseases* (1531) and *On the Tartarus Diseases* (1537–1538), he explained how food and acid in the stomach could form pathological deposits called 'tartar'. This abnormal accretion prevented normal digestion and could lead to various diseases of the belly, such as colic, constipation, and diarrhoea. Paracelsus compared this process to the curdling of milk, which also required heat and the accession of acid.[57] Some powerful acids called *Acetosa Esurina* ('hungry vinegars') could be found in certain animals, Paracelsus argued, enabling them to digest rigid and poisonous materials such as metal, bone, and venomous spiders, frogs, and snakes. Van Helmont generalised the notion of acid in the stomach, but instead of using it to explain the disease, Van Helmont saw the role of acid as the normal digestive process common in all animals and humans. Van Helmont studied and described the nature of this digestive acid in

[53] Debus, *The Chemical Philosophy*, 104–17. See also Georgiana D. Hedesan, *An Alchemical Quest for Universal Knowledge: The 'Christian Philosophy' of Jan Baptist Van Helmont (1579–1644)* (London, 2016).

[54] Paracelsus, *De morborum origine et causa* (1531) as quoted in Andrew Weeks, ed. *Paracelsus (Theophrastus Bombastus von Hohenheim, 1493–1541): Essential Theoretical Writings* (Leiden, 2008), 531.

[55] Paracelsus, *Opus Paramirum* (1531), lib. 3, as quoted in Walter Pagel, 'J. B. Van Helmont's Reformation of the Galenic Doctrine of Digestion—and Paracelsus', *Bulletin of the History of Medicine*, 29 (1955), 563–8, here 565.

[56] Van Helmont, *Ortus medicinae* (Amsterdam, 1648) as quoted in Walter Pagel, 'Van Helmont's Ideas on Gastric Digestion and the Gastric Acid', *Bulletin of the History of Medicine*, 30 (1956), 524–36, here 524.

[57] Pagel, *Paracelsus*, 153–61.

chemical terms. Digestion, then, came to mean fermentation—a description later adopted by Sylvius.

Having discussed the lives and works of Paracelsus and Van Helmont, Boerhaave then assessed their merits for the study of medicine. Their alchemical quests for *chrysopoeia*—the transmutation of base metals into noble ones—and for a universal medicine to cure all diseases and promote long life were both dismissed as 'empty smoke, and idle ostentation'. The death of Van Helmont, his wife, and sons only showed that there was no obvious evidence of a panacea to cure all diseases or prevent premature death. 'It appears, then, contrary to the unanimous voice of the whole tribe of chemists, that *Helmont* had not the universal medicine'.[58] Yet at first he had not rejected the notion. As he confessed to his students:

> After reading the writings of the two authors abovemention'd; I was at once discouraged, and cast off from all thought of ever practising physic more. For, as they assert it needless to enquire into the cause, nature, seat, symptoms, effect, &c. of a disease; or to trouble one's-self with regimen, diet, &c. since with one little, simple medicine, all diseases may be cured, alike; and death shut out from every door.[59]

Boerhaave was surprisingly forthcoming in sharing his feelings of 'inquietude', perusing books day and night trying to find concrete references to the universal medicine. Ultimately, an acquaintance who had personally known Van Helmont and his son disabused Boerhaave of the fanciful notion that a universal medicine was anything more but the result of imagination and enthusiasm.

Despite this deception, the student edition reveals that Boerhaave recognised Paracelsus's and Van Helmont's important work and contribution to the field of medicine. Paracelsus's fame was rightly based on his skill as a surgeon, his knowledge of medicine, the art of making medicaments, the beneficial cures by means of opium, and lastly his acquaintance with the virtues of mercury. Although Boerhaave admitted that some 'represent him [Paracelsus] as one of the most vile, flagitious, and worthless of the race of men', he endorsed his overall work for having 'reform'd and alter'd the face of medicine, and turn'd it altogether into the vein of chemistry'.[60] He also praised Van Helmont, because since his time the field of chemistry

[58] Boerhaave, *A New Method*, vol. 1, 36; Debus, *The Chemical Philosophy*, 104–17.
[59] Boerhaave, *A New Method*, vol. 1, 36.
[60] Ibid., vol. 1, 22.

had grown immensely—producing more books in the 70 years that had passed than in all the ages before.[61]

But Boerhaave's lectures, which revealed his close affinity with the medical chymistry of Paracelsus and Van Helmont, were supposed to never leave Leiden's lecture hall. When Boerhaave heard that an edition of his chemical lectures was published without his consent, he was filled with rage. Those students even took the unauthorised textbook with them to lectures, so as to compare the text with his utterances, infuriated Boerhaave even more.[62] He maintained that the book should never have seen the light of day because it cheated the public and only lined the pockets of greedy booksellers. Yet Boerhaave was just as concerned about his own reputation as a chemist and physician. In a local newspaper, he declared 'I own none such for my works, being fraudulently publish'd without my knowledge, contrary to my will, being and full of such gross and dangerous mistakes, as tend equally to my discredit, and the reader's prejudice, who relies on them'.[63] In addition, Boerhaave took legal action to prevent future mishaps. He submitted a formal request to the States of Holland to have the reprinting and pirating of both his own works and those of his fellow professors forbidden unless they, their heirs, or the university gave permission. After long discussions with the guilds, these rights were granted in 1728.[64]

But Boerhaave knew that the best defence was to mount a counterattack. Although he had resigned from the chair of chemistry in 1729, he took up his quill to produce his own chemical textbook, which was published in 1732.[65] Comparing Boerhaave's *Elementa chemiae* to the unauthorised student edition reveals that entire sections on the history of chemistry were now excised. 'I will not rehearse the many false, ridiculous and absurd things there attributed to me in every page: a detail too nauseous to entertain the reader with', Boerhaave noted in the preface.[66] While his students had heard a professor who freely expressed his thoughts

[61] Ibid., vol. 1, 36–7.

[62] Boerhaave, *A New Method of Chemistry*, vol. 1, vi.

[63] As quoted in Burton, *An Account*, 166–7.

[64] Robert Verhoogt and Chris Schriks, 'Reflecting Media: A Cultural History of Copyright and the Media', in *The History of Information Security*, ed. Karl de Leeuw and Jan Bergstra (Amsterdam, 2007), 83–119, here 88.

[65] Herman Boerhaave, *Elementa chemiae, quae anniversario labore docuit in publicis, privatisque scholis*, 2 vols (Leiden, 1732).

[66] Boerhaave, *A New Method of Chemistry*, vol. 1, v.

and experiences about medical chymistry, the *Elementa chemiae* displayed a far more cautious Boerhaave. In particular the sections on the lives and works of Paracelsus, Van Helmont, and other alchemists and chymists were significantly shorter, to the point of being not much more than a bibliography. This clearly suggests that Boerhaave was not just concerned about typographical errors, but that he was also trying to rescue his reputation and legacy. There had to be no doubt that Boerhaave had set up a new method of chemistry for the use of medicine, both in physiology and pathology, which was distinctly different from the medical chymistry as propagated by seventeenth-century chymists.

The Chemical Analysis of Saliva

As this section demonstrates, Boerhaave's own medical writings on saliva demonstrate the importance he ascribed to his chemical method for a better understanding of physiology. They can be found in Boerhaave's *Institutiones medicae*, first published in 1708 but updated and extended by his student Albrecht von Haller following Boerhaave's death in 1738.[67] Appointed as professor of anatomy, botany, and medicine at the newly founded Hanoverian University of Göttingen, Von Haller had quickly established his name as a physiologist.[68] Some factors of eighteenth-century chemistry for medicine expose an adoption of certain features of medical chymistry: Boerhaave, for example, did not abstain from using the same terminology, nor did he completely abandon theorising about physiological processes on the basis of chemical analyses. Yet in contrast to medical chymistry and analogical reasoning, Boerhaave's texts reveal an empirical use of chemistry, based on numerous experiments with saliva. Furthermore, they reveal a bounded notion of the chemistry of bodily fluids, one that did not make chemistry the all-encompassing explanatory model but corroborated mechanist and vitalist theories of digestion. Finally, the chemical method as practised by Boerhaave and Von Haller was not aimed at prescribing medicines but primarily sought to

[67] Herman Boerhaave, *Institutiones medicae, in usus annuae exercitationis domesticos digestae* (Leiden, 1708); and the last edition appearing during Boerhaave's lifetime, *Institutiones medicae in usus annuae exercitationis domesticos digestae*, 5th ed. (Leiden and Rotterdam, 1734); Albrecht von Haller, ed. *Praelectiones academicae in proprias institutiones rei medicae*, 6 vols (Göttingen, 1739–1744).

[68] Hubert Steinke, Urs Boschung, and Wolfgang Proß, eds., *Albrecht von Haller: Leben – Werk – Epoche* (Göttingen, 2008).

understand physiology—a feat that was to have a lasting impact on eighteenth-century medicine.

In the *Institutiones*, chemical tests with saliva assisted in confirming and falsifying claims made about its origin: in other words, chemistry corroborated anatomical descriptions of the salivary glands. Boerhaave summed up the various salivary glands where blood was transformed into saliva (see Fig. 2.4): (1) the parotid glands situated at the root of the ear; (2) the two large submaxillary glands in the lower jaw; (3) the sublingual glands situated under the tongue; and (4) the 'lenticular and miliary Glandules' perforating the tongue and discharging 'a much thinner Saliva'; and finally small glands in the back part of the palate, which discharge 'a more thick or mucous Saliva'.[69] Hence, different types of salivary glands produced different types of saliva. In the 1720s, salivary glands became the subject of heated debate, as the young Albrecht questioned the discovery of a new salivary duct by Georg Daniel Coschwitz (1679–1729), professor of medicine and anatomy at the University in Halle.[70] Haller exposed the falsity of Coschwitz's claim by multiple dissections and injections. Furthermore, using chemical analysis he rejected the hypothesis that lymphatic fluid might be added to the saliva coming from the parotid glands. Heated lymphatic fluid coagulated like egg white, whereas saliva always completely evaporated. Clearly, saliva as secreted in the mouth did not contain any lymphatic fluid.[71]

Besides anatomy, the essential factor in Boerhaave's chemistry of medicine was his detailed analysis of the fluids. As Boerhaave explained in his textbook *Elementa chemiae*, 'in each part of an animal, we find humours of a peculiar kind, which always appear specifically different from another', so they should be studied separately.[72] Instead of discussing the various branches of arteries, veins, and nerves leading to and from the glands, the fluid of saliva and its properties were his main subject of study. In conducting chemical experiments on saliva, chemists chose the saliva of a healthy

[69] Von Haller, *Academical Lectures*, vol. 1, 135–6. On Boerhaave's conception of secretion, see Rina Knoeff, 'Chemistry, Mechanics and the Making of Anatomical Knowledge: Boerhaave Vs. Ruysch on the Nature of the Glands', *Ambix*, 53 (2006), 201–19.

[70] Albrecht von Haller, *Dissertatio inauguralis sistens experimenta et dubia circa ductum salivalem novum Coschwizianum* (Leiden, 1727). See also Rainer Godel, 'Controversy as the Impetus of Enlightened Practice of Knowledge', in *Scholars in Action*, ed. André Holenstein, Hubert Steinke, and Martin Stuber (Leiden, 2013), 413–31, here 416–21.

[71] Von Haller, *Academical Lectures*, vol. 1, 137, 41.

[72] Boerhaave, *A New Method of Chemistry*, vol. 1, 153.

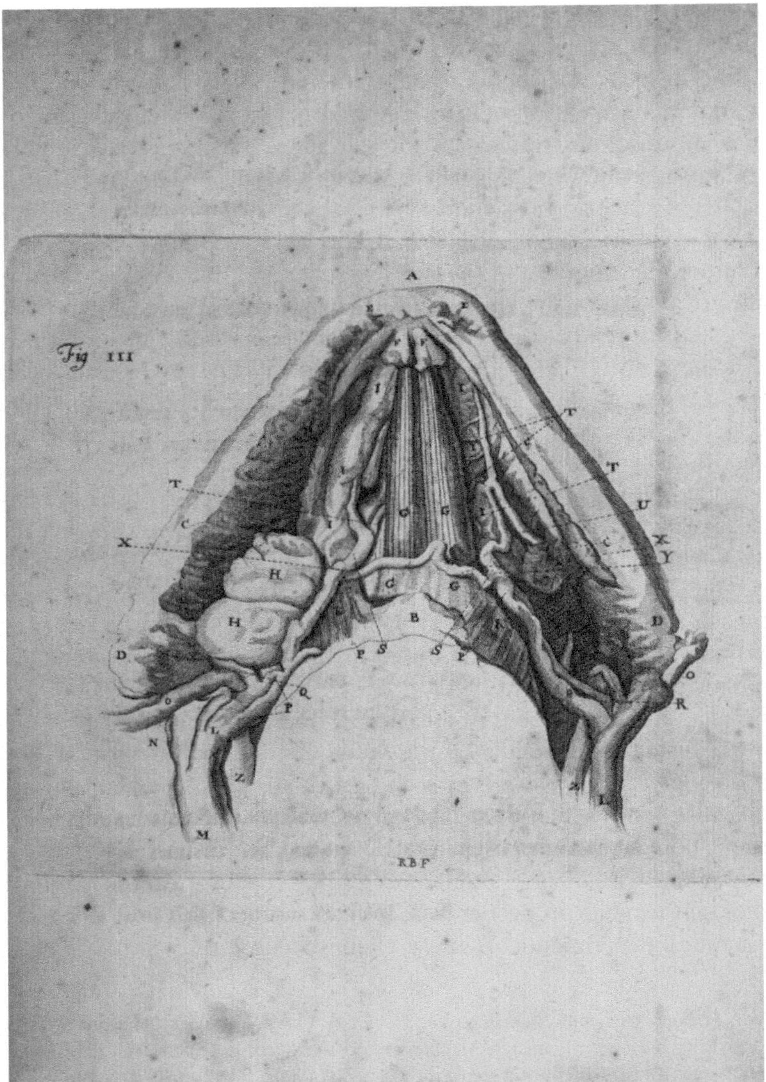

Fig. 2.4 Anatomical drawing of the lower jaw by R.B.F., in Albrecht Haller, *Dissertatio inauguralis sistens experimenta et dubia circa ductum salivalem novum Coschwizianum* (Leiden, 1727). 'HH' refers to the submaxillary glands; 'III' are the sublingual glands. (Credit: Wellcome Collection. CC PDM 1.0)

and young man after he had washed his mouth. Initial examination of this spittle revealed that the fluid was thin, transparent, and devoid of smell and taste; when heated it turned into vapour, and when stirred formed a ropy froth. This froth could complicate further chemical analysis, as Von Haller explained: during a chemical operation like distillation, too much froth could rise up and obstruct the long neck of a retort, potentially causing it to break (see Fig. III in Fig. 2.2).[73] Von Haller therefore stressed that in the examination of saliva, one should apply a gentle heat, just slightly warmer than body temperature. Doing so revealed the composition of saliva:

> Twenty Ounces of Saliva will afford nineteen of simple Water, like common Water, and there remains about an Ounce of a gritty or tartarous Substance. If that tophaceous residuum be distill'd with a strong Fire, it then affords a little volatile and foetid Salt, being a Mixture of both Oil and Salt, leaving black Faeces behind, which also contain some Oil; it indeed contained no Spirit, if by Spirit you mean one that is inflammable, and capable of mixing as well with Oil as Water; but it contains a little Salt, which is neither of an acid nor alcaline Nature; and such is the Composition of healthy *Saliva*.[74]

Chemistry, in short, revealed the nature of saliva as a compound of water, salt, and oil. The analysis once again refuted Sylvius's theory about the acid-alkali dichotomy of the body and the proposed 'spirit' in the saliva.

This meticulous attention to the chemical properties of saliva was useful when examining saliva during abnormal conditions. Fasting, extreme feelings of hunger, and the licking of wounds among animals all revealed a quality or effect of saliva, which were expressed in chemical terms. According to Boerhaave, saliva 'flows more plentifully, fluid and sharp into the Mouths of hungry People; but after long fasting is extremely *acrimonious*, *deterging*, penetrating and *dissolving*'.[75] Observations supported these claims. First, the acrimony of saliva in those who were fasting could be sensed by the foetid smell of their breath and spittle, as well as by the observation of their sore and painful gums. Second, the effective, deterring quality of saliva was commonly seen among dogs, who by licking

[73] Von Haller, *Academical Lectures*, vol. 1, 140–5.

[74] Ibid., vol. 1, 144–5. According to Boerhaave, distillation of saliva yielded 'a water of a somewhat unpleasant, indeed, but not a fetid odour': Boerhaave, *A New Method*, vol. 2, p. 188.

[75] Von Haller, *Academical Lectures*, vol. 1, 140.

cleansed a wound and promoted the healing process.[76] And finally, the dissolving or 'saponaceous' quality of saliva in those who had not eaten ensued the sensation of appetite: the dissolving power of saliva affected the inner coats of the stomach and thereby occasion the feeling of hunger.[77] It was this dissolving property of saliva that was explored experimentally by eighteenth-century naturalists and physicians.

Chemical analyses of saliva allowed Boerhaave and Von Haller to distinguish true from false saliva. It is yet another example of how Boerhaave's chemistry of medicine informed physiological and pathological understanding. For one of the questions that remained was whether salivation caused by the application of mercury to cure venereal disease was also a form of saliva. Based on observation and experiment, it was known that ordinary saliva consisted predominantly of water and evaporated when heated. But when the morbid fluid was collected and heated, it instead 'shoots into Crystals, almost like Nitre, which arise from the acid Salts of the Mercurius Dulcis, and are of a quite different Kind from the natural Salts of the Saliva'.[78] Hence applying the very same chemical operation using this kind of saliva had very contrasting effects. By means of chemical analysis, the copious salivation caused by mercurial treatments were no longer interpreted as the secretion of ordinary saliva.

Whereas the chemical analysis of saliva differed from Sylvius's understanding of the chymical composition of saliva, Boerhaave's discussion on the role of saliva in digestion still closely resembled Sylvius's. For example, the main function of saliva was to change and assimilate food to help nourish the body. In the process of thoroughly chewing and grinding aliments in the mouth, Boerhaave believed that saliva would 'form an intimate Mixture of their [the food's] oily and aqueous Parts [...] dissolve their saline Parts [and] excite a *Fermentation*'.[79] Apparently Boerhaave was not opposed to using chemical words and concepts propagated by chymists, in particular fermentation. Indeed, we have seen how Boerhaave freely used terms such as acid, alkali, and saponaceous to describe properties of bodily

[76] On the therapeutic properties of saliva in wounds, itching, and skin peeling were discussed in Martinus Houttuyn, *Natuurlyke historie of uitvoerige beschryving der dieren, planten en mineraalen, volgens het samenstel van den heer Linnaeus*, 37 vols (Amsterdam, 1761–1785), vol. 1, 313–4.

[77] Von Haller, *Academical Lectures*, vol. 1, 142–3.

[78] Ibid., vol. 1, 145.

[79] Ibid.

fluids and their role in physiology.[80] The process of fermentation, on which the entire process of digestion rested according to Van Helmont and Sylvius, also occurred in some fashion in Boerhaave's view of digestion. Although he had objected to equating the body to a chemical laboratory and opposed speculative approaches, he nonetheless utilised chemical processes like fermentation to enlighten physiology and pathology.

The fact that Boerhaave's teachings on saliva and digestion were a departure from chymistry, even though they were flavoured with chymical terminology, was a point quickly elucidated by Boerhaave and Von Haller. As the latter explained, chymists and mechanists had proposed dissimilar theories of digestion, but they erroneously assumed their one-cause understanding of physiology counted for all bodily functions:

> Fermentation is drawn in by the Chemists to account for every Operation in Nature, while their Adversaries as strenuously exclude it from having the least Share in any of her Appearances; but in this, both of them over-shoot the Mark widely; for whenever a fermentable Substance is excited by Heat, Moisture, and a free Admission of the Air, there must inevitably arise a Fermentation.[81]

Indeed, food in combination with saliva, air, and body heat provided the perfect condition for a fermentable process to occur. The ordinary occurrence of belching, Von Haller argued, could only but substantiate this line of thinking. Boerhaave also touched upon digestion in his chemical lectures, suggesting the process was more akin to putrefaction than fermentation. Even if a person only ate vegetables, fruit, and other acidic foods, no part of the body or excrements would show any acidity or fixed salt after distillation, but only a volatile alkali. Hence, digestion more closely resembled putrefaction than fermentation.[82]

Whether food putrified or fermented, Boerhaave maintained that chemistry could only support a physiological understanding of the body, but could never become the overarching explanation. He therefore recognised the phenomenon of digestion as a complex blend of chemical, mechanical, and vital causes, but he refrained from providing a definitive explanation. Once the food had been masticated by the teeth, diluted by saliva, and swallowed down the throat, it would enter the stomach, where

[80] Powers, *Inventing Chemistry*, 93.
[81] Von Haller, *Academical Lectures*, vol. 1, 147.
[82] Boerhaave, *A New Method*, vol. 2, p. 177.

it underwent both chemical and mechanical action: the aliments would mix with and dissolve in gastric fluids and additional saliva, as well as being acted upon by the muscular coat of the stomach.[83] Indeed, in addition to chemical actions, Boerhaave identified other possible factors that could be involved in the digestive tract, such as heat, pulsations, vibrations of the arteries and organs, the force of the nervous juices, and internal agitations. Ultimately, however, he asked whether physiologists could ever satisfactorily explain how digestion worked. Boerhaave admitted that 'the causes already explained may indeed seem insufficient thus to digest and change the Food in the Stomach'.[84] Nevertheless, the observations and *in vitro* experiments in the chemical laboratory supported the existence of vital principles. According to Boerhaave, the transformation from food to body—or from its vegetal to animal nature—could only be explained in reference to some latent power: of the things we take in, 'we call alimental, if it proves so yielding that its nature may be changed into ours, *by the force residing in the body*'.[85] Although the chemistry of saliva could not satisfactorily explain the entirety of digestion and transformation of food into chyle, it had nevertheless resulted in the observation of some vital principle at work: 'All the acids of the aliment are therefore subdued by *the vital powers of animals*, and converted into volatile salts of an alkaline nature; yet without a real or actual putrefaction, tho by an operation that nearly approaches to it'.[86]

Boerhaave's understanding of vital powers peculiar to the living body was not limited to saliva and the process of digestion. Indeed, Rina Knoeff has argued that Boerhaave considered chemistry the best way to discover the latent powers in bodies and hence the best way to study physiology in general. This did not result in the establishment of general laws everybody abided, but rather a meticulous study of all the particular properties of fluids, vegetables, and materials.[87] The employment of chemistry enabled

[83] Von Haller, *Academical Lectures*, vol. 1, 184–98.

[84] Ibid., vol. 1, 212.

[85] Boerhaave, *A New Method*, vol. 1, p. 345. Emphasis added. Cf. Boerhaave, *A New Method of Chemistry*, vol. 1, 148–53; Knoeff, *Herman Boerhaave*, 200. See also Boerhaave's reference to 'the united Force' in Von Haller, *Academical Lectures*, vol. 1, 219.

[86] Boerhaave, *A New Method*, vol. 2, p. 177. Emphasis added. Cf. Boerhaave, *A New Method of Chemistry*, vol. 2, 189.

[87] Knoeff, *Herman Boerhaave*, 193–201; ead. 'Practising Chemistry 'after the Hippocratical Manner': Hippocrates and the Importance of Chemistry for Boerhaave's Medicine', in *New Narratives*, ed. Lawrence M. Principe (Dordrecht, 2007), 63–76.

Boerhaave to delineate all matters alimentary, medicinal, and poisonous. As Ursula Klein has shown, Boerhaave examined the chemical properties of lists of different plants to ascertain their nourishing and medicinal uses.[88] Also the chemistry of milk, both its production as mother's milk and its effects in the body, informed therapeutic and dietetic advice to mothers and wet nurses in eighteenth-century society. But the previously unexplored case of saliva has shown the intricate ways in which Boerhaave tried to negotiate the tainted legacy of chymistry, in which saliva had played a significant part. Furthermore, Boerhaave's approach to digestion proved to be more irenic rather than divisive, for he had found a way to have chemical and mechanical investigations into digestion work together rather than against each other.

Boerhaave's chemical analysis of saliva in digestion and his understanding of latent powers had a lasting impact, furthering research and vitalist thought in the course of the eighteenth century. Von Haller, first of all, significantly expanded Boerhaave's dense *Institutiones* by means of his authoritative, six-volume *Praelectiones academicae*. Although Von Haller researched motion and sensation more so than digestion, his innovative physiological research in Göttingen led to the development of the concepts of irritability and sensibility, which gave a strong impetus to new vitalist theories.[89] In France, the salience of Boerhaave's new physiology had a considerable impact on Denis Diderot's *Encyclopédie* in general, most of all thanks to the work of physician Louis de Jaucourt (1704–1779). A medical *philosophe par excellence* and disciple of Boerhaave, De Jaucourt contributed thousands of entries to the *Encyclopédie*, the majority of which were summaries of or extracts from the works of his former teacher.[90] His entry on saliva summarised Boerhaave's analysis of saliva, its nature and chemical properties, its relation to taste and hunger, and of course its digestive role.[91]

In the entry on digestion, chemist and physician Gabriel-François Venel (1723–1775) summarised Boerhaave's physiology as 'the explanation of

[88] Klein, 'Experimental History'.

[89] Hubert Steinke, *Irritating Experiments: Haller's Concept and the European Controversy on Irritability and Sensibility, 1750–90* (Amsterdam, 2005), 175–229.

[90] Elizabeth A. Williams, *A Cultural History of Medical Vitalism in Enlightenment Montpellier* (Aldershot, 2003), 123; Richard N. Schwab, 'The History of Medicine in Diderot's *Encyclopédie*', *Bulletin of the History of Medicine*, 32 (1958), 216–23.

[91] Louis de Jaucourt, 'Salive', in Denis Diderot and Jean le Rond d'Alembert, eds., *Encyclopédie, ou dictionnaire raisonné des sciences, des arts et des métiers*, 17 vols (Paris, 1751–1765), vol. 14, 572–3.

the modern physiologists', yet relegating it as 'a kind of concordance of all systems'. Venel wanted to present digestion as something that went beyond discrete processes in the viscera and instead was linked to feelings of hunger and fatigue, daily activities and mental well-being—in other words, how it had 'a general and essential influence on the whole animal oeconomy'.[92] Venel and the Montpellier vitalists located hunger at the back of the mouth and in the stomach, exactly where saliva was secreted and collected. And the dissolving and hunger-instilling property of saliva was the cause of much research as late the 1770s.

At Pavia, the Italian abbot and natural historian Lazzaro Spallanzani (1729–1799) once again asked whether the process of digestion was by fermentation, putrefaction, solvent, or trituration, 'or whether, according to the opinion of the great Boerhaave, it rather depends upon all these causes operating in conjunction'.[93] To answer this question, Spallanzani became a noted expert on matters vomitory. Spallanzani had birds, frogs, dogs, and other animals swallow and regurgitate little containers with food to see the effect of gastric fluids. Spallanzani even self-experimented by gulping down linen bags with food to be brought up again and analysed. He vomited on an empty stomach to obtain samples of gastric fluid and have them subject to chemical analysis. Spallanzani concluded that digestion was effected by fermentation nor putrefaction, but by 'the solvent power' of gastric fluid and saliva. Throughout his book Spallanzani was in direct conversation with Boerhaave, citing passages from the *Praelectiones academicae* and critically comparing them with his own findings.[94] Though modest this reception history may be, it shows how profound Boerhaave's influence was on Enlightenment science and medicine. Since Boerhaave's chemistry and physiology was palatable to all schools of medicine, he had set a new standard for further investigations into saliva and the essential process of digestion.

* * *

[92] Gabriel-François Venel, 'Digestion', in Ibid., vol. 4, 999–1003; Williams, 'Food and Feeling', 207–8.

[93] Lazzaro Spallanzani, *Dissertations Relative to the Natural History of Animals and Vegetables*, trans. Thomas Beddoes (London, 1784), vi.

[94] Lazzaro Spallanzani, *Dissertazioni di Fisica Animale, e Vegetabile*, 2 vols (Modena, 1780); Evan R. Ragland, "Experimenting with Chemical Bodies: Science, Medicine, and Philosophy in the Long History of Reinier de Graaf's Experiments on Digestion, from Harvey and Descartes to Claude Bernard" (Doctoral thesis, University of Alabama, 2012), 455–8.

This chapter has explored the essential role of saliva in early modern theories of digestion. Boerhaave transformed the field of medical chymistry into a chemistry for medicine at the turn of the eighteenth century. This chemistry was not the sole methodology by which physicians gathered knowledge, such as had often been the case among chymists. Rather, Boerhaave's physiology can be considered irenic and more inclusive, because it allowed for multiple causes and approaches to come together in a unified understanding of physiological processes like digestion. Having evaluated textbooks and students' lecture notes, this chapter has nuanced the dichotomy between Boerhaave and his predecessors. Boerhaave performed an academic balancing act, paying tribute to those alchemists who had made chemistry suitable for medicine, while at the same time distancing himself from them so as to prevent close association.

On the one hand, Boerhaave introduced a chemistry for medicine that relied on a new research method. In his attempt to understand the composition of saliva, Boerhaave rejected speculative approaches and instead advocated a chemistry that was empirically grounded. His 'Hippocratic method' entailed a Baconian empiricism: a good chemist ought to perform numerous observations and experiments on individual bodily substances. All options had to be considered carefully, rather than made up *a priori*. The body was no chemical laboratory governed by the balance between acids and alkalis but was firmly constituted out of fluids, each with their particular properties, motions, and variabilities depending on health or sickness. Chemistry was crucial in uncovering the nature of these fluids, which would lead to a better understanding of digestion and physiology.

On the other hand, some parts of the Boerhaave's chemistry of medicine clearly echoed previous theories of medical chymistry. Although Boerhaave rejected the more extravagant claims and speculative theories of his predecessors, he appreciated that Paracelsus and Van Helmont had outgrown the pharmaceutical preparation of drugs. Van Helmont in particular had played a crucial part in Boerhaave's thinking about the role of chemistry in the study of physiology and medicine. In fact, in his last oration held in 1731 Boerhaave explicitly recommended the works of Van Helmont, because he had acknowledged nature as being the generative force setting everything in motion. Boerhaave's theory of digestion came to include latent powers of the living body, which to some extent were inspired by Van Helmont's seminal principles. Although for Boerhaave these peculiar powers were not to be a source of speculation, they could be studied by chemistry. Paradoxically, then, the study of the material properties of saliva exposed the hidden forces bringing about digestion.

CHAPTER 3

The Nature of Blood

Blood, that warm, red fluid running through arteries and veins to every part of the body, was a subject of considerable debate in the Dutch Republic. Although William Harvey (1578–1657) had already demonstrated in 1628 that blood circulated in the human body, even a century later, anatomical studies of the circulatory system still could not explain the nature and properties of blood itself. In their attempts to understand the nature of blood medical men in the Dutch Republic at the turn of the eighteenth century increasingly relied on chemistry, but as we shall see, their approach was not universally accepted. Medical professors at Leiden University such as Hieronymus Gaubius were confident that chemistry could indeed uncover blood's true nature, arguing that chemical investigations would reveal its constituent parts. But others, most notably Thomas Schwencke, a city physician from The Hague, argued that there existed a fundamental difference between blood *in vitro* (in glass) and blood *in vivo* (in the living body) and, therefore, seriously doubted the merits of a chemical approach. This chapter argues that the disagreement about how to research the nature of blood allows us to understand important issues in early eighteenth-century medicine, in particular, the definition of bodily fluids, the increasingly important role of chemistry in

R. E. Verwaal, *Bodily Fluids, Chemistry and Medicine in the Eighteenth-Century Boerhaave School*, Palgrave Studies in Medieval and Early Modern Medicine,
https://doi.org/10.1007/978-3-030-51541-6_3

medicine, and the apparent rivalry between mechanist and vitalist approaches to physiology.[1]

The notion of blood as the most important bodily fluid existed for many centuries. Yet it was only at the turn of the eighteenth century that the actual nature and function of blood in the so-called animal oeconomy became the subject of extensive chemical and empirical study. Blood raised a multitude of questions that were avidly debated by medical men: what was the relationship between the movement of blood and its unique ability to coagulate? What were the differences between blood in sick and in healthy bodies? And, most importantly: was blood alive? Studying blood thus allowed Dutch physicians to approach some of the biggest questions concerning the mechanisms of life.

The contemplation of blood brought into contact theories of physiology and pathology with practices from experimental and observational traditions. Blood provides a particularly fruitful case for historians aiming to explore questions of methodology in eighteenth-century medicine, specifically the relationship between medical theory and experimental practices since it required medical researchers to seek out new ways in their attempt to understand its workings. Gaubius and Schwencke observed that blood flowing through the arteries and veins of the living body remained firmly fluid and homogeneous but began to clot once gushing out of a ruptured vessel. Recognising that the anatomical studies of the circulatory system could not explain the nature and properties of blood itself, they subjected blood to measuring instruments and chemical apparatuses. Schwencke's and Gaubius's studies of blood serve as a basis for the argument in this present chapter that their debate on the nature of blood, in effect, spurred innovative interactions between medical theory and experimental research.

Scholars of the most recent decades have shifted their attention away from the circulation of blood towards blood itself, asking how physicians

[1] On the purported dichotomy between mechanism and vitalism, see Anita Guerrini, 'James Keill, George Cheyne, and Newtonian Physiology, 1690–1740', *Journal of the History of Biology*, 18 (1985), 247–66. On modern misconceptions of vitalism, see Charles T. Wolfe, 'From Substantival to Functional Vitalism and Beyond: Animas, Organisms and Attitudes', *Eidos*, 14 (2011), 212–35. Wolfe and Terada encourage scholars to instead focus on animal economy or physiology in Charles T. Wolfe and Motoichi Terada, 'The Animal Economy as Object and Program in Montpellier Vitalism', *Science in Context*, 21 (2008), 537–79.

understood its properties throughout history.[2] Domenico Bertoloni Meli, for example, has traced the problems facing seventeenth-century physicians debating the value of colour indications as evidence of the essential properties of blood.[3] Andrew Cunningham has shown that in England blood did not become a serious subject for research until the second half of the seventeenth century.[4] Although Harvey never intended to research the nature of blood itself he granted it precedence over the heart as the source of bodily activity, warmth, and life in his *Exercitationes de generatione animalium* (1651).[5] Both Bertoloni Meli and Cunningham demonstrate that increasing attention was paid to blood and its physical properties in the seventeenth century, but they do not discuss the developments following the turn of the century. By contrast, Noel Coley's work reveals the advances made with the help of new instruments in clinical chemistry and haematology in late eighteenth-century England and France. Coley argues that these studies were inspired by the Dutch professor Herman Boerhaave, who 'prepared the ground for the rise of a new form of humoralism, one in which the chemical composition of body fluids in health and disease would figure prominently'.[6] However, Coley considered Boerhaave in isolation from other Dutch medical men, implying that nothing else of importance happened in the Dutch Republic at the time.[7]

This chapter argues that the Dutch Republic was, in fact, crucial for the development of innovative blood research. Focusing on the physicians Thomas Schwencke and Hieronymus Gaubius, this chapter examines their

[2] Maxwell M. Wintrobe, ed. *Blood, Pure and Eloquent: A Story of Discovery, of People, and of Ideas* (New York, 1980); Douglas Starr, *Blood: An Epic History of Medicine and Commerce* (London, 1999). On circulation, see Roger French, 'Harvey in Holland: Circulation and the Calvinists', in *The Medical Revolution*, ed. Roger French and Andrew Wear (Cambridge, 1989), 46–86; Roger French, *William Harvey's Natural Philosophy* (Cambridge, 1994).

[3] Domenico Bertoloni Meli, 'The Color of Blood: Between Sensory Experience and Epistemic Significance', in *Histories of Scientific Observation*, ed. Lorraine Daston and Elizabeth Lunbeck (Chicago, 2011), 117–34.

[4] Andrew Cunningham, 'The Principality of Blood: William Harvey, the Blood, and the Early Transfusion Experiments', in *Blood – Symbol – Liquid*, ed. Catrien Santing and Jetze Touber (Leuven, 2012), 193–205.

[5] William Harvey, *Exercitationes de generatione animalium* (Amsterdam, 1651); idem, *Anatomical Exercitations Concerning the Generation of Living Creatures* (London, 1653).

[6] Noel G. Coley, 'Early Blood Chemistry in Britain and France', *Clinical Chemistry*, 47 (2001), 2166–78, here 2167.

[7] Lawrence M. Principe, 'A Revolution Nobody Noticed? Changes in Early Eighteenth-Century Chymistry', in *New Narratives*, ed. Lawrence M. Principe (Dordrecht, 2007), 1–22.

proposals for blood chemistry and haematology, respectively. Schwencke's haematology made detailed observations and measured properties of blood with thermometers and hydrometers. But since his results from these measurements differed significantly from Gaubius's chemical analyses Schwencke expressed strong criticism against blood chemistry. This chapter concludes with the argument that this debate about the methodology of studying blood was linked directly to the ontological question of its essential yet disputed quality: was blood alive?

Lifeblood

Early modern physicians believed that blood was the most important bodily fluid, embodying notions of life and soul. Blood was seen as the nourishment for the body, the sustenance of health, and the source of all other bodily fluids. Medical professors repeated this notion in their lectures and textbooks often: Albrecht von Haller commented on Boerhaave's lectures, 'the red fluid we call blood [...] contains the materials of all the other juices found in every part of the body'.[8] In his chemical lectures Boerhaave himself hailed blood as the 'perfect animal juice' and the 'utmost universal humour', not just because blood raced to all corners of the human body, but also because, in his view, all other bodily fluids were derived from it.[9] Schwencke and Gaubius—both of whom had studied medicine at Leiden under Boerhaave in the 1710s and 1720s—represented the mainstream thinking on blood when they, too, observed that fluids flow 'from the general ocean of blood', which was, as it were, 'the common mother of them all'.[10]

The nourishing and life-giving property of blood was particularly located in the female body. Pregnancy provided evidence for the benign

[8] Albrecht von Haller, ed. *Praelectiones academicae in proprias institutiones rei medicae*, 6 vols (Göttingen, 1739–1744), vol. 2, 325; idem, *Dr. Boerhaave's Academical Lectures on the Theory of Physic: Being a Genuine Translation of his Institutes and Explanatory Comment*, 6 vols (London, 1742–1746), vol. 2, 175.

[9] Herman Boerhaave, *A New Method of Chemistry: Including the Theory and Practice of that Art: Laid down on Mechanical Principles, and Accommodated to the Uses of Life*, trans. Peter Shaw and Ephraim Chambers, 2 vols (London, 1727), vol. 1, 167.

[10] Thomas Schwencke, *Schets van de heelmiddelen en haar uytwerkingen op het lichaam* (The Hague, 1745), 3. Hieronymus David Gaubius, *Institutiones pathologiae medicinalis* (Leiden, 1758), 155; idem, *The Institutions of Medicinal Pathology*, trans. Charles Erskine (Edinburgh, 1778), 103.

quality of blood: the absence of menstrual blood during gestation was explained in reference to the menstrual blood as foetal nourishment. Physicians calculated that nine months' worth of menstruation was of sufficient quantity to nourish a foetus during that period.[11]

The primacy of blood was further supported by the belief that it directly influenced the passions (see Fig. 3.1). Blood played a pivotal role in the interaction between body and soul since mental disturbances were linked to bodily imbalances. Schwencke, for example, observed that 'in anger the pulse becomes large and rapid, and as a result the heat increases, as the Thermometer demonstrates; but that there is a speedy and small pulse, accompanied with coldness, when one is frightened'.[12] When a woman's 'excessive sexual desires' were out of control, Gaubius similarly advised that bloodletting would 'weaken' and 'cool' her, so that there would be no difficulty to 'restrain her from her unlawful habits'.[13] Specific motions of blood were thus a direct manifestation of strong feelings and desires—and through the treatment of blood, physicians were able to modify certain faculties of the soul directly.

These two notions of blood's primacy, namely, as the 'mother of them all' and as the reflection of emotions, had ancient origins. In humoral theory, health and disease were understood to correspond to a balance or imbalance of the four essential humours: blood, yellow bile, black bile, and phlegm. In this system, the humours corresponded to the four Aristotelian elements (air, water, fire, and earth), the four qualities (hot, cold, dry, and wet), and the four temperaments (sanguine, phlegmatic, choleric, and melancholic). Blood literally was the most sanguine of all four humours, allowing humans to be characterised by optimism, patience, and thoughtfulness—as opposed to being irritable or melancholic in the case of yellow or black bile.[14] But the notions of ancient blood and early modern blood were quite different: in the early modern period, the humours were related to, but not directly identified with specific fluids.

[11] Cathy McClive, *Menstruation and Procreation in Early Modern France* (Farnham, 2015).

[12] Thomas Schwencke, *Haematologia, sive Sanguinis historia, experimentis passim superstructa* (The Hague, 1743), 68; idem, *Haematologia, ofte Verhandeling van het bloed*, trans. Abraham Westerhoff (The Hague, 1744), 109–10.

[13] Hieronymus David Gaubius, *Sermo academicus de regimine mentis quod medicorum est* (Leiden, 1747), 113; idem, *On the Passions: or a Philosophical Discourse Concerning the Duty and Office of Physicians in the Management and Cure of the Disorders of the Mind*, trans. J. Taprell (London, 1760?), 116.

[14] Noga Arikha, *Passions and Tempers: A History of the Humours* (New York, 2007).

Fig. 3.1 Lady Caliste bleeds to death after physician's fatal mistake. Engraving by Noach van der Meer Jr., after Jacobus Buys, in C.F. Gellert, *Fabelen en vertelsels* (Amsterdam, 1777). (Credit: Rijksmuseum, Amsterdam. CC0 1.0)

The humoral theory was, above all, a rational and logical construct, and not necessarily grounded in observation and experience.[15] In the course of the early modern period, however, physicians shifted their focus from humoral blood to blood as bodily fluid.

Finally, the primacy of blood appears in early modern physicians' notions of life. When Gaubius showed how blood clotted, he noted that 'consistency of life and health therefore especially depends on an unimpeded transfer of blood, for blood can easily cross from a fluid state to a solid state'.[16] Schwencke and Abraham Westerhoff were equally explicit in the discussion on the 'juice of life'.[17] Schwencke argued that, when blood flowed out of a dog's body during vivisection, the functions of the soul—such as the will to move—also seeped away. Blood not only sustained life, but 'it is said that the soul and the life are *in the blood*'.[18] These statements and experiments echoed Harvey's claim that blood was the agent of life, since if the blood stopped flowing to parts of the body, life in the entire body, or the relevant parts, would stop.[19] In blending medical and metaphysical notions of blood, then, these academic medical texts clearly show how blood was considered to be the primary bodily fluid and to reflect one's passions, life, and health.[20]

Despite the wide-ranging consensus over the primacy of blood, however, eighteenth-century physicians still faced intriguing questions concerning the nature and production of blood. What exactly was blood made of? And where and when did food loose its vegetal nature and transform

[15] Lawrence I. Conrad et al., *The Western Medical Tradition: 800 BC to AD 1800* (Cambridge, 1995).

[16] 'Praelectiones publicae chemicae', vol. 2, p. 45.

[17] Schwencke, *Verhandeling van het bloed*, xxii. In the original Latin, Schwencke used the more generic term *humor*, but Westerhoff decided to translate it as 'juice of life'. Westerhoff came from Gouda had also studied at Leiden, promoting in 1738.

[18] Schwencke, *Haematologia*, 9. Emphasis added.

[19] Harvey, *De generatione*, 304. C. Webster, 'Harvey's *De Generatione*: Its Origins and Relevance to the Theory of Circulation', *British Journal for the History of Science*, 3 (1967), 262–74, here 272. On Harvey's notion of the soul residing in the blood, see Christopher Hill, 'William Harvey and the Idea of Monarchy', *Past and Present*, 27 (1964), 54–72; Ann Thomson, *Bodies of Thought: Science, Religion, and the Soul in the Early Enlightenment* (Oxford, 2008), 68–74.

[20] Scottish anatomist John Hunter (1728–1793) continued to reiterate the notion of blood as the indispensable factor of life when he stated that 'Blood is not only alive itself, but it is the support of life in every part of the body' in John Hunter, *A Treatise on the Blood, Inflammation and Gun-Shot Wounds* (London, 1794), 85.

into a liquid of an animal nature? The latter question was particularly complex—since antiquity natural philosophers had generally divided the natural world into the vegetable, animal, and mineral kingdoms, thus problematising the transformation from vegetal nourishment to the animal body. In the eighteenth century, however, these separate realms were increasingly being questioned, as explorers discovered living creatures who exhibited characteristics of both animals and plants, such as freshwater polyps and coral.[21] Consequently, physicians like Gaubius adopted chemistry as a method of investigation, hoping that it would help them to resolve the physiological problem of the nature of blood.

At the turn of the eighteenth century, experiments and hands-on observations to better understand the nature and production of blood were, indeed, incorporated in medical curricula at Dutch universities (see Fig. 3.2). In Leiden, for example, medical students learned that blood developed from chyle, the milky fluid draining from the lacteals of the small intestine into the lymphatic system. As Boerhaave explained, chyle was extracted from vegetables during digestion, and therefore vegetal in nature. Yet when this chyle 'is further prepared in the intestines, and received into the lacteals, [it] approaches more to the nature of an animal juice, and undergoing divers circulations therein, it becomes a perfect animal juice, and is called *blood*'.[22]

Chemistry supported the notion that vegetal chyle was transformed into animal blood. Boerhaave had demonstrated that, when animal fluids were burnt, the ashes were found to be insipid, and the remaining salts volatile, whereas vegetal juices left fixed salts in all their ashes. Moreover, none of the animal fluids contained acid, whereas vegetal juices did.[23] The Irishman John Ellis (c. 1710–1776) applied exactly this chemical process when he argued that vegetal-appearing corallines were, in fact, animals. He posited that corallines and plants differ in chemical production, 'For Sea-Plants ... afford in Distillation little or no Traces of a volatile Salt:

[21] Susannah Gibson, *Animal, Vegetable, Mineral? How Eighteenth-Century Science Disrupted the Natural Order* (Oxford, 2015).

[22] Boerhaave, *A New Method*, vol. 1, 167. Emphasis in original. Barbara Orland, 'The Fluid Mechanics of Nutrition: Herman Boerhaave's Synthesis of Seventeenth-Century Circulation Physiology', *Studies in History and Philosophy of Biological and Biomedical Sciences*, 43 (2012), 357–69.

[23] Boerhaave, *A New Method*, vol. 1, 168.

Fig. 3.2 Students in the laboratory, in Herman Boerhaave, *Institutiones et experimenta chemiae* ('Paris', 1724). Ghent University Library

Whereas all the Corallines afford a considerable Quantity'.[24] The chemical process of combustion thus provided a new method to properly distinguish between animal and vegetal materials. Chemistry also revealed close similarities between different fluids. When Gaubius analysed blood and milk in his lectures, he concluded that 'when we compare the serum of the blood with the whey of milk, red blood with the cream, and the fibre with the curd, we find many things in common to both'.[25] Chemistry, in other words, allowed physicians to reveal constituent parts of different bodily fluids, and to differentiate more clearly between animal and vegetal bodies.

But not everyone was convinced of the merits of chemistry for furthering knowledge on the nature and the origins of blood. One of its fiercest critics was Thomas Schwencke. In the preface to his *Haematologia* (1743), he stated that 'we have also shortly addressed the Chemical examination of the blood, but at the same time we reported the reason why a Physician by this way can never understand the natural character or condition of blood'.[26] The nature of materials obtained from blood by distillation, he argued, was different from that of the living blood running though the veins. The difference of opinion between Schwencke and Gaubius manifested perhaps most clearly in their respective publications and lectures. Despite their different approaches to studying the nature of blood, Gaubius and Schwencke avoided direct confrontation and did not reference each other directly in their arguments. Rather, Schwencke talked about 'chemists', in general, or safely referred to 'Boerhaave in his chemical work' who, by then, had passed away.[27] But it was clear that their writings were directed as one another. In the 1740s Antonius de Haen (1704–1776), who practised medicine in The Hague and like Schwencke was a member of the local *collegium medicum*, complained that he and his fellow Boerhaavians ('*Boerhaavii* filios') were often picked on by the older generation of physicians. According to De Haen, Schwencke would at first act pleasant and polite in the company of the young physicians. But the moment they had diagnosed and prescribed treatment to a patient, he would intervene, prescribe a different medicament, and claim the honours

[24] John Ellis, *An Essay towards a Natural History of the Corallines, and other Marine Productions of the Like Kind, Commonly Found on the Coasts of Great Britain and Ireland* (London, 1755), 2.

[25] Gaubius, *Institutiones*, 160; idem, *Institutions*, 106.

[26] Schwencke, *Haematologia*, xi–xii; idem, *Verhandeling van het bloed*, xxi–xxii.

[27] Ibid., xviii, 34, 82, 139, 252.

in case of a successful recovery.[28] It appears, then, that Schwencke not only suspected blood chemistry but sceptically approached Boerhaave and his school in general. Although an overtly hostile language characteristic of polemics was carefully avoided, the authors' difference of opinion is nonetheless clearly apparent, as both Schwencke and Gaubius formulate their ideas and methodology very precisely.

Blood Chemistry

When Hieronymus Gaubius (see Fig. 3.3) started his chemical lectures at Leiden University in 1731 as Boerhaave's successor, he affirmed that 'blood is the liquid of animals, which proceeds from the heart into the arteries', yet hastened to add that 'we cannot give a definition of its [blood's] own *particular nature* before we have investigated it by chemistry'.[29] Blood chemistry evolved in the course of the eighteenth century, and with its methods, chemists like Gaubius envisioned blood as a complex blend of rudimentary substances and principles. Chemical textbooks and lecture notes from this period all attest to the frequent experiments on blood that took place in the Leiden *Laboratorium Chymicum*, a modest space located in the botanical garden of the university (see Fig. 3.2).[30] These sources reveal that professors like Gaubius approached blood with chemical methods, teaching students how to experiment on blood, how to handle it, and how to determine its properties using instruments as well as their own senses.[31]

In the course of his own studies at Leiden between 1723 and 1725, Gaubius had learned about the use of chemistry in medicine from Herman Boerhaave, the renowned professor of medicine who had delivered regular

[28] Antonius de Haen to Gerard van Swieten, The Hague, 1743–1751, in Antonius de Haen, *Opuscula quaedam inedita*, ed. Joseph Eyerel, 2 vols (Vienna, 1795), vol. 1, 48. Cf. Jacob Boersma, *Antonius de Haen, 1704–1776: Leven en werk* (Assen, 1963), 17.

[29] 'Praelectiones publicae chemicae', vol. 2, p. 37. Emphasis added.

[30] Besides Boerhaave's official chemical textbook *Elementa chemiae* (Leiden, 1732), students' lecture notes circulated widely, such as 'Viri Clarissimi Hieronymi Davidis Gaubii [...], Dictata in Chemiam', 3 vols, Leiden, *c.* 1750. University Library, Leiden, MS BPL 1477.

[31] Rina Knoeff, *Herman Boerhaave (1668–1738): Calvinist Chemist and Physician* (Amsterdam, 2002); Ursula Klein, 'Experimental History and Herman Boerhaave's Chemistry of Plants', *Studies in History and Philosophy of Biological and Biomedical Sciences*, 34 (2003), 533–67; John C. Powers, *Inventing Chemistry: Herman Boerhaave and the Reform of the Chemical Arts* (Chicago, 2012).

Fig. 3.3 Hieronymus David Gaubius. Line engraving by Jan Houbraken, 1744, after Hieronymus van der Mij, 1741. (Credit: Wellcome Collection. CC BY)

lectures on the history, theory, and practice of chemistry until 1729. Boerhaave argued that the method of chemistry was essential in understanding physiology because '[n]o-body but a chemist could say what kind of liquor the blood is'. Medical students needed to extend beyond the anatomical structure of the body and explore the very materials of which it was made because blood 'exerts a force of its own; which chemistry explains, by shewing [*sic*] it possessed of certain active principles, as salts, spirits, oils, &c'. Once these active principles were known, 'a chemist will nicely understand, and advantageously distinguish between the signs of health and sickness'.[32] Boerhaave's conception sharply contrasted Georg Ernst Stahl's opinion on the role of chemistry in medicine. While Stahl (1660–1734), a German chemist and physician, contributed significantly to the field of chemistry, and his work was widely read,[33] he played a minor role in the blood chemistry of Boerhaave and Gaubius: his approach to medicine favoured the role of *anima* (soul) in health and disease over chemical principles of body parts.[34] Boerhaave and his medical students in Leiden embraced the work of Stahl's rival, Friedrich Hoffmann (1660–1742), instead, who had not only enriched the field of chemistry but also recognised the service of chemistry to medicine.[35] Hoffmann's best-known chemical work concerned the analyses of mineral waters, in which he identified aether, water, earthy, and saline components within

[32] Quotations are taken from Boerhaave, *A New Method*, vol. 1, 194. This exposition was translated – with poetic licence – from idem, *Institutiones et experimenta chemiae*, 2 vols ('Paris', 1724), vol. 1, 146–7. For his authorised version of the usefulness of chemistry to medicine, see Herman Boerhaave, *Elementa chemiae, quae anniversario labore docuit in publicis, privatisque scholis*, 2 vols (Leiden, 1732), vol. 1, 83–4; idem, *Elements of Chemistry: Being the Annual Lectures*, trans. Timothy Dallowe, 2 vols (London, 1735), vol. 1, 52–3.

[33] Both Boerhaave and Gaubius owned a copy of Stahl's *Fundamenta chymiae dogmaticae et experimentalis* (Nuremberg, 1723). See the auction catalogues 'Bibliotheca Boerhaaviana sive catalogus librorum instructissimae bibliothecae viri summi D. Hermanni Boerhaave' (Leiden, 1739), 56; 'Bibliotheca Gaubiana sive Catalogus librorum viri celeberrimi Hieronymi Davidis Gaubii' (Leiden, 1783), 65. For the sums of money for which Boerhaave's 3334 titles were sold, see 'Bibliotheca Boerhaaviana', Leeuwarden, Tresoar, s 66 AW.

[34] On Stahl countering iatrochemical perspectives, see for example Ku-ming Chang, 'Fermentation, Phlogiston and Matter Theory: Chemistry and Natural Philosophy in Georg Ernst Stahl's Zymotechnia Fundamentalis', *Early Science and Medicine*, 7 (2002), 31–63.

[35] Boerhaave, *Elementa chemiae*, vol. 1, 29; idem, *Elements of Chemistry*, vol. 1, 18. Here Boerhaave specifically suggested his students to read Hoffmann's *Observationum physico-chymicarum* (Halle, 1722).

the waters.[36] Dutch medical professors and their students turned their attention to the body itself and recognised that in order to explain phenomena in the human body one required chemistry, in particular chemistry of blood.

Gaubius was no exception. He performed many experiments on blood, first and foremost to uncover what he called the constituent parts. In this demonstration, Gaubius warned his students that appearances could be deceiving. Although to the naked eye blood seemed homogeneous, chemical experiments revealed it was, in fact, made up of different substances.[37] Gaubius showed students how blood drawn from a healthy person and poured in a bowl was of a 'uniform dark and red colour, unctuous and glutinous to the touch'. Left to rest, however, it would start to coagulate and split into three parts: one ruby-coloured, hard part, known as *crassamentum* (*i.e.*, thickness), another white and watery part called *serum* (whey), and finally a fibre 'of a membraneous texture'. The apparent homogeneous fluid of blood was, thus, in fact, an amalgamation of parts, which could be identified according to the process of decomposition. As Gaubius summarised in his medical lectures, 'it is evident that, besides the *spirit* [*i.e.*, the smell] of the blood, there are three different constituent parts, the serum, the red blood, and the fibre'.[38]

Experimentation did not stop there. In another experiment, Gaubius collected a quantity of healthy blood in a clean glass vessel—so that the transformation might be clearly observed—and subjected it to distillation at different temperatures. This process produced four basic elements: water, bitter oil, volatile salt, and black earth.[39] Gaubius assigned great significance to this experiment because it not only revealed the hidden elements in blood but also informed about the elements that constitute the body as a whole. He argued that the human body consisted of fluids and solids, all constituted by four elements: water, salt, earth, and phlogiston—a substance chemists believed to exist in all combustible bodies,

[36] Victor D. Boantza and Leslie Tomory, 'The "subtile Aereal Spirit of Fountains": Mineral Waters and the History of Pneumatic Chemistry', *Early Science and Medicine*, 21 (2016), 303–31, here 319–22.

[37] In 1731, the Leiden *Ordines lectionum* listed Gaubius's lectures as 'naturam Humorum corporis humani experimentis Chemicis' in P.C. Molhuysen, *Bronnen tot de geschiedenis der Leidsche Universiteit*, 7 vols (The Hague, 1913–1924), vol. 5, 39.

[38] Gaubius, *Institutiones*, 200–312; idem, *Institutions*, 106.

[39] Ibid., 103–6.

which was released through burning.[40] The proportion of the principle of water in relation to the other elements imparted fluidity to bodily liquids and flexibility to solid parts. Similarly, the principle of earth assigned the quality of thickness to the fluids, and of firmness to the solids. Bones, for example, were considered to consist almost exclusively of earth and to be devoid of water. The serum was mostly water, the red *crassamentum* mostly phlogiston, and the fibre mostly earth and salt. Gaubius believed, then, that he could understand the general physiology of the human body through the analysis of blood.

Gaubius's understanding of the chemical make-up of the animal body was essentially principlist (as opposed to compositionist). Historian of chemistry Robert Siegfried has shown that compositionist explanations of matter were based on the presumption of immutable and basic chemical substances. These chemical substances could be combined and separated, yet they remained intact in themselves and did not change their nature in the process.[41] Principlism, on the other hand, was a system formed around the concept of principles, namely, fundamental substances that imparted characteristic properties to other substances. According to Hasok Chang, principlism had three defining epistemic activities: first, classifying substances according to observable properties; second, explaining the properties of substances by reference to principles; and lastly, effecting transformations of substances by the application (or withdrawal) of principles.[42] Gaubius's perception of blood and the human body was clearly based on the notion of four principles which actively modified bodily fluids and solids. For example, he referred to phlogiston as 'the inflammable part, the principle of heat and light', which gave tenacity to solids and fluids.[43] The principle of water imparted fluidity to blood, whereas the principle of earth lent its solidity. In a unique constellation, then, the principles accounted for blood and its properties. Gaubius spoke of blood's

[40] The term phlogiston was derived from the Greek *phlogistós*, 'set on fire'. For a good reappraisal of the phlogiston theory, see Hasok Chang, *Is Water H2O? Evidence, Realism and Pluralism* (Dordrecht, 2012), 1–70.

[41] Robert Siegfried, *From Elements to Atoms: A History of Chemical Composition* (Philadelphia, 2002).

[42] Hasok Chang, 'Compositionism as a Dominant Way of Knowing in Modern Chemistry', *History of Science*, 49 (2011), 247–68; idem, *Is Water H2O?*, 37–42. Chang prefers to use the term 'principlist' over 'principalist', because it refers to principles, not principals.

[43] Gaubius, *Institutiones*, 57; idem, *Institutions*, 37.

'own particular nature' or *propria natura*.[44] The Latin *proprius* ('one's own' or 'particular') provided the root for the concept of property as an attribute or characteristic of a substance.[45] Gaubius described the chemical properties typical of blood, such as viscosity: the state of having a thick and sticky consistency between solid and fluid. In describing blood, Gaubius referenced the principles of water, phlogiston, salt, and earth, which all imparted characteristic properties to the red fluid.[46]

Gaubius's principlist chemistry proved highly successful in explaining pathological phenomena, which he structured in a coherent system. Gaubius was convinced that solidity, or the force of cohesion, was intrinsic to the principles of water and earth, which was also responsible for the strength of cohesion and the scale of fluidity. This force of cohesion differed from one bodily fluid to another and was influenced by a person's age, sex, temperament, and way of life. For example, blood was different from milk, because by circulation and air 'an acescent [*i.e.*, acid] salt is turned into an alcalescent [alkaline], a fermentible [*sic*] into a putrescible; the phlogiston is more evolved, the earth condensed'.[47] In some cases, however, the force of cohesion could be excessive (causing viscidity and toughness) or deficient (producing tenuity and thinness). If someone had consumed too much glutinous food, the level of *crassamentum* would rise in disproportion to the serum, and even cohere to earth elements. The blood would circulate more slowly, stagnate, obstruct, and potentially cause a tumour. By contrast, if a person drank too much, the resulting superfluous water in the serum could lead to haemorrhages.[48]

The rise of chemistry in eighteenth-century medicine radically transformed perceptions of bodily fluids such as blood. Thinking about their constituent parts and principles not only allowed Gaubius to investigate blood more closely, but it also enabled him to theorise about blood in its varying physiological and pathological states. Thus, Boerhaave's and Gaubius's experiments on blood reflected a growing optimism among physicians about the use of new methods in the investigation of the human body.[49]

[44] 'Praelectiones publicae chemicae', vol. 2, p. 37.
[45] Nalini Bhushan, 'What is a Chemical Property?', *Synthese*, 155 (2007), 293–305.
[46] Gaubius, *Institutions*, 37–8, 85–6.
[47] Gaubius, *Institutiones*, 161; idem, *Institutions*, 107.
[48] Ibid., 83–9.
[49] Conrad et al., *The Western Medical Tradition*, 371–6.

Haematology

Blood chemistry did not immediately win the hearts and minds of the entire medical profession but was confronted with collegial scepticism and rivalling methods. The fact that chemical analyses changed the material appearance of blood led some physicians to doubt the validity of the experimental results. The most outspoken of these critics was Thomas Schwencke (see Fig. 3.4), who published a systematic study of blood titled *Haematologia, sive sanguinis historia* (1743).[50] Schwencke agreed that physicians ought to study blood because it played a crucial role in the assessment of patients' medical conditions. Schwencke's haematology was, therefore, not only aimed towards the acquisition of new knowledge about blood but was also directly and actively involved in the treatment of blood, including the advantages and disadvantages of phlebotomy and the medicines to treat wounds and stop the bleeding. Haematology, in other words, was supposed to be applicable in medical practice. In his studies, Schwencke closely observed the colours and smells during coagulation, using hydrometers and thermometers to measure the density and temperature of the blood in different places (arteries and veins) and circumstances. But as we shall see, measuring was also crucial to Schwencke's haematology, because it allowed him to question the usefulness of chemistry for the analysis of blood.

The reason why Schwencke began to doubt the chemical approach was because his medical career had taken him into very different directions from those Boerhaave and Gaubius had followed. His was a practical, hands-on career, not the kind of 'armchair medicine' often pursued by university professors. Born in the city of Maastricht, Schwencke had begun his medical studies in surgery and pharmacy before he matriculated at Leiden University in 1712. After defending his dissertation on saliva in 1715 Schwencke began working as a physician in The Hague. His practice in medicine and obstetrics soon earned him a fine reputation, and in 1723 the town council appointed him professor of anatomy and surgery. His duties included the teaching of anatomy to local midwives and surgeons. In 1737 he was also installed as court physician to Prince Karl Christian and Princess Carolina van Nassau-Weilburg, brother-in-law to and sister of

[50] Schwencke, *Haematologia*. Schwencke first coined the term haematology according to Harry F.P. Hillen, 'De eerste Nederlandse hematoloog: Thomas Schwencke (1694–1767)', *Nederlands Tijdschrift voor Hematologie*, 7 (2010), 17–20.

Fig. 3.4 Hendrik Pothoven, after a painting by Verheyden, portrait of Thomas Schwencke, 1782. Collection Haags Gemeentearchief, The Hague

Stadtholder William V (1745–1808), respectively.[51] Another of Schwencke's famous patients was Adrienne Marguerite Huguetan, Countess de Nassau La Lecq: in 1752 he paid her thirteen visits and held 38 consultations, for which he received 84 guilders.[52]

The contrast between Schwencke and Gaubius could not have been greater. Whereas both had studied medicine at Leiden and trained to become physicians, Gaubius hated the practice of medicine. After three years of working as a physician in Deventer and Amsterdam, Gaubius was glad to be appointed at Leiden University in 1731 because, as he confessed to a colleague, it relieved him of his 'aversion of the very tiresome practice'.[53] To his close friend António Sanches he confessed that he disliked medical practice and public service because he felt that it tied him down.[54] Indeed, when he was appointed court physician to William V in 1761, Gaubius admitted that 'I would have liked to be excused at the beginning of old age, to be able to only attend to my peaceful profession joined with a practice much less bristled with thorns'.[55] While Gaubius sought to explain the pathological theory to students, Schwencke devoted most of his energy to surgical practice. Based on his experiences, for example, Schwencke presented six different ways to stem bleeding, namely, by fire, alcohol, stitches, various bandages, and the external application of agaric, a tree mushroom.[56] Schwencke's hands-on approach to medicine meant that his focus was more directed towards successful treatment than Gaubius's, who preferred developing and teaching theories of pathology.

Central to Schwencke's haematology were his detailed descriptions of blood, in particular, the many weight and density measurements he took. Schwencke closely observed how the healthy blood of a strong man, once

[51] Eduard Sandifort, 'Levensbeschryving van den Hooggeleerden Heere Thomas Schwencke', in idem, *Natuur- en genees-kundige bibliotheek*, 11 vols (The Hague, 1765–1775), vol. 3, 429–36.

[52] Eduard van Biema, *Les Huguetan de Mercier et de Vrijhoeven: histoire d'une famille de financiers huguenots de la fin du XVIIe jusqu'à la moitié du XVIII siècle* (The Hague, 1918), 58.

[53] Gaubius to Albrecht von Haller, 23 March 1743 in Sophia W. Hamers-van Duynen, *Hieronymus David Gaubius (1705–1780): Zijn correspondentie met Antonio Nunes Ribeiro Sanches en andere tijdgenoten* (Assen and Amsterdam, 1978), 181–3.

[54] Gaubius to António Sanches, 9 March 1748. Ibid., 95–7.

[55] Gaubius to Sanches, 23 September 1761. Ibid., 123–5.

[56] Thomas Schwencke, 'Aanmerkingen over verscheide manieren van bloedstelpen, en de voornaamste bloedstelpende middelen in de heelkunde', *Verhandelingen van de Hollandse Maatschappy der Weetenschappen*, 2 (1755), 225–50.

collected in a bowl, naturally separated in two parts serum and one part *cruor* (*i.e.*, red blood).[57] Schwencke also compared the respective ratios of density in blood and water, using a hydrometer built by the Amsterdam instrument-maker Daniel Fahrenheit of thermometer fame (1686–1736). Fahrenheit's hydrometer consisted of two glass spheres connected by a cylinder (see Fig. 3.5). The lower sphere was filled with quicksilver. Schwencke then lowered the instrument into a large cylindrical vessel filled with blood and observed how deep it sank, repeating the experiment with the individual constituent parts of blood. He discovered that water weighed 1110 grains, serum of blood 1142 grains (1/34 heavier than water), red blood 1204 grains (1/12 heavier than water), and ordinary blood 1173 grains. When he measured the densities again using different blood samples, he discovered that these ratios differed with the donor's age and sex.[58]

Microscopy also formed an important part of Schwencke's haematology. He incorporated Antoni van Leeuwenhoek's findings on the microscopic size and shape of red blood globules into his *Haematologia* (see Fig. 3.6). Van Leeuwenhoek had found that red blood consisted of heavy, elastic, and round globules of red, purple, or black colour. Schwencke argued that one red globule consisted of a cluster of six serum globules, together with gaining weight and darkness but preserving the same size. Furthermore, he believed that every serum globule was composed of six chyle globules. One red blood globule therefore consisted of 36 chyle globules, stuck together by an inherent, elastic force emanating from each individual globule.[59]

Finally, Schwencke gained new knowledge about blood and coagulation by means of measuring temperature. His efforts were part of a wider development: Hasok Chang has shown how natural philosophers all faced the problem establishing fixed points in thermometry and debated the use and applicability of thermometers.[60] John Powers has shown that Boerhaave integrated Fahrenheit's thermometer in his chemical lectures, and that other chemistry instructors followed suit.[61] We know indeed

[57] Schwencke, *Verhandeling van het bloed*, 141–9.

[58] Ibid., 200–9.

[59] Ibid., 176–80.

[60] Hasok Chang, *Inventing Temperature: Measurement and Scientific Progress* (Oxford, 2004).

[61] John C. Powers, 'Measuring Fire: Herman Boerhaave and the Introduction of Thermometry into Chemistry', *Osiris*, 29 (2014), 158–77.

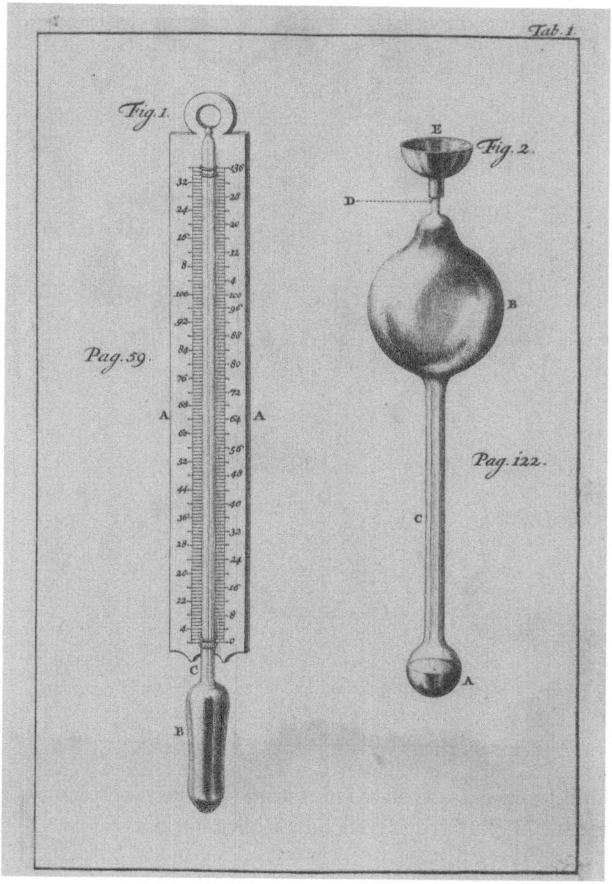

Fig. 3.5 Thermometer (left) and hydrometer (right) in Thomas Schwencke, *Haematologia* (The Hague, 1743). (Credit: Wellcome Library. CC PDM 1.0)

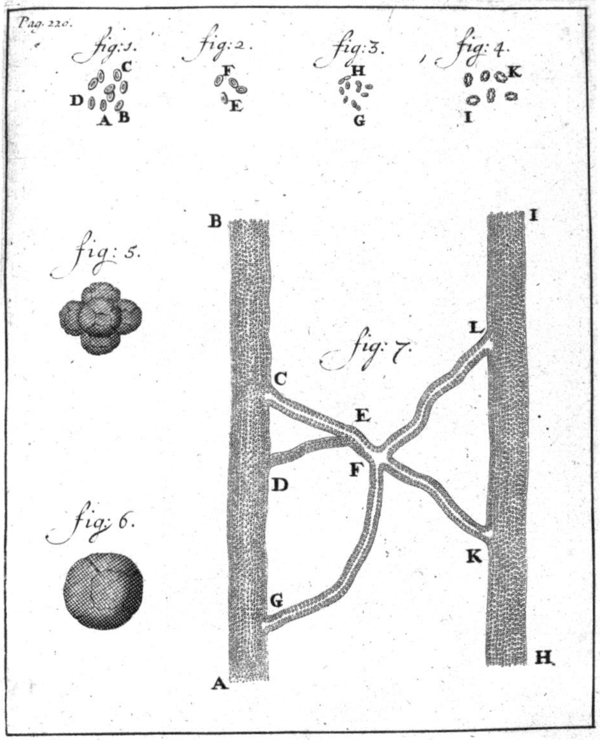

Fig. 3.6 Blood corpuscles depicted in Antoni van Leeuwenhoek, *Arcana natura detecta* (Leiden, 1719). (Credit: Wellcome Collection. CC BY)

from Gaubius's post-mortem inventory that he owned a number of thermometers, one of which was graduated in Fahrenheit, Wachendorff, and Réaumur.[62] He also used the Fahrenheit thermometer in experiments and demonstrations, as student lecture notes attest.[63] Gaubius observed, for example, that blood serum 'by the heat of 150 °F thermometer, is

[62] 'Musei Gaubiani pars sive Catalogus partis supellectilis, qua usus est vir celeberrimus H. D. Gaubius' (Leiden, 1783), 31. Besides Fahrenheit, these scales of temperature were named after Evert Jacob van Wachendorff (1703–1758), professor of medicine, botany and chemistry at Utrecht University, and French naturalist René-Antoine Ferchault de Réaumur (1683–1757).

[63] 'Praelectiones publicae chemicae', vol. 1, pp. 38, 39, 101, 102, etc.

suddenly converted, without much dissipation, into a whitish scissile mass, like the white of an egg'.[64]

Schwencke, too, argued that the only way in which degrees of heat could be properly measured was with a thermometer. He regarded the quicksilver thermometer as the 'true indicator of the heat', but he used it in quite a different manner.[65] In 1731, Schwencke performed a number of experiments on the blood of oxen and cows with the thermometer. He took the temperature simultaneously in the carotid artery and jugular vein, measuring 97 °F and 94 °F, respectively, and concluded that the body's heat was dependent upon the movement and friction of blood. Excitement caused the pulse to throb more rapidly as the heat increased.[66] Schwencke established that a healthy person had a temperature of 96 °F and he reasoned that a correlation existed between body temperature, mental state, and disease. Future physicians, Schwencke argued, ought to systematically measure the temperature of patients in all diseases, which would greatly improve diagnostics.[67] And indeed, Boerhaavian physician Antonius de Haen took up this call. Having access to hundreds of patients in the Citizen's Hospital in Vienna—where orphans, the pour, and the elderly were accommodated—De Haen made the first attempt to test body temperatures in health and disease by the thermometer. De Haen's numerous observations and systematic investigations of blood temperature in patients falsified some ancient notions of blood. Whereas Hippocrates had famously stated that young kids had a higher body temperature than adults, and the elderly even colder body temperatures, De Haen provided the evidence disproving that claim.[68] Introducing the term of haematology, therefore, Schwencke squarely placed this study of blood within the context of what has been called 'the quantifying spirit'.[69] The extensive measurements of blood density and temperature, and microscopic observations of blood globules, exemplify the early modern passion to order and systematise, to measure, and calculate.

[64] Gaubius, *Institutiones*, 157; idem, *Institutions*, 104.

[65] Schwencke, *Haematologia*, xi; idem, *Verhandeling van het bloed*, xix.

[66] Ibid., 49.

[67] Ibid., xix–xx.

[68] Boersma, *Antonius de Haen*, 78–85. Antonius de Haen, *Ratio medendi in nosocomio practico* (Vienna, 1756–1773).

[69] Tore Frängsmyr, John L. Heilbron, and Robin E. Rider, eds., *The Quantifying Spirit in the 18th Century* (Berkeley, 1990).

Chemistry Contested

But measurements also led Thomas Schwencke to question the chemical approach to the study of blood. As part of his study on the material disposition of blood, Schwencke reproduced Gaubius's chemical experiments. Just like Gaubius had done, Schwencke drew six ounces of blood from a healthy person and distilled it in a glass alembic, indeed producing water, a bitter oil, a volatile salt, a black and malodorous oil, and black earth.[70] Yet when he weighed the various elements, he discovered that water made up a staggering 7/8 parts of the blood, instead of the two-thirds obtained after natural coagulation. Struck by the high degree of variability that chemistry had produced *in vitro* Schwencke began to doubt the applicability of chemistry *in vivo*. How did the proportion of two-thirds serum and one-third *cruor* by hydrometry align with the very different division obtained by distillation? From this Schwencke argued that chemical operation by fire not only divided some of the blood's elements but also produced new elements not found in the living body. No matter how carefully chemists attempted to reassemble these ingredients, he claimed, they would never be able to recombine them to form human blood, not in colour, smell, or any other property.

Schwencke also discredited chemistry by pointing out the innumerable differences between chemistry and nature. In procreation, nature can produce all body parts of an infant, solely on the basis of blood and mother's milk. This process was yet to be explained by chemistry. Furthermore, physicians commonly agreed that blood was the mother of all bodily fluids. Yet chemists could produce not one of them. Surely, Schwencke argued, the chemist cannot afford to use only fire for experimentation but would require the use of a thousand more processes, as many as there were parts in the body. As he warned his readers, 'although we have presented here the chemical analysis of the blood, no one will believe that we have therefore demonstrated and argued the nature and quality of the same'.[71]

The debate between advocates for and opponents to blood chemistry can be explained with the help of the contrasting contemporary principlist and compositionist understandings of matter. I argue that Schwencke's opposition to blood chemistry stemmed from his preference for compositionism. In Schwencke's view, chemical processes were mere processes of

[70] Schwencke, *Verhandeling van het bloed*, 246–7.

[71] Schwencke, *Haematologia*, 153; idem, *Verhandeling van het bloed*, 252.

re-grouping basic components—a view also emerging in his understanding of the blood globules. The impossibility to reconstruct blood, or any other bodily fluid, conflicted with the notion of his building block or compositionist understanding of matter. This ideological difference between Gaubius and Schwencke was, however, not as clear-cut as it may seem, especially because Gaubius's principlist blood chemistry included (for lack of a better word) some compositionist elements. Gaubius's analysis of blood as a substance consisting of constituent parts did not directly explain the notion of principles of water, phlogiston, salt, and earth. As Albrecht von Haller later commented, 'these are not the natural but factitious principles of the blood, produced or made by the intensity of fire'.[72] The debate between principlist and compositionist understandings of nature continued throughout the century, ultimately leading to compositionist chemistry at the turn of the nineteenth century.

Schwencke's concerns about the essential difference between blood *in vitro* and *in vivo* also found echoes outside the Dutch Republic. One of the most notable critics of blood chemistry was the French physician Théophile de Bordeu (1722–1776), who opined that blood in alembics bore no relation to living blood. Bordeu eventually settled in Paris, where court physician Louis de Lacaze (1703–1765) mobilised his help for the completion of a treatise on the theoretical foundations of medicine.[73] This work sharply criticised those who obtained knowledge about diseases with the help of the dissection of corpses. If readers really wished to understand 'the essential character of a disease', they required information about its course and tested methods of treatment. In other words, the 'mechanism of life' could only be understood by the 'observation of the living body'.[74] It was the rhythmical throbbing of the arteries that won Bordeu general acclaim. In 1756 he published a book on the pulse, which received positive reviews. The study of the many species of the pulse was the observation of living blood and recognition of the signs of diseases. Looking for

[72] Von Haller, *Academical Lectures*, vol. 2, 167–8. In the original, Von Haller made a distinction between those particles brought about by natural body heat, and those particles made by the intensity of fire. *Praelectiones*, vol. 2, 312.

[73] Although ascribed to Bordeu, it appeared as Louis de Lacaze, *Idée de l'homme physique et moral, pour servir d'introduction à un traité de médecine* (Paris, 1755); Elizabeth A. Williams, *A Cultural History of Medical Vitalism in Enlightenment Montpellier* (Aldershot, 2003); Anne C. Vila, *Enlightenment and Pathology: Sensibility in the Literature and Medicine of Eighteenth-Century France* (Baltimore, 1998).

[74] Lacaze, *Idée de l'homme*, 9–10.

ways of diagnosing and treating patients, then, both Schwencke and Bordeu focused their attention on the symptoms and circumstances of living blood.[75]

Bordeu soon recognised that feeling the pulse produced quite different results from distilling blood serum. Whereas Dutch chemists had shown that serum putrefied under moderate heat—a phenomenon Boerhaave considered the cause of many diseases[76]—Bordeu observed that 'the blood in the living body never shows any signs of putridity or acrimony, and that it is even incapable of any fermentation inside it. This completely precludes all notions about the bitterness of liquids contained in their vessels, and at the same time eliminates all the applications that supposedly can be made of the changes observed in dead liquids on the various changes that are consequently believed to happen in various stages of an illness, and on the liquids of the living body'.[77] For Bordeu, then, all changes observed in blood through chemistry were useless, because they had no relation to the properties of living blood, whether diseased or healthy. Some physicians' practical approach to medicine led them to doubt the necessity of chemistry for an understanding of bodily fluids *in vivo*.

In the eyes of Gaubius, however, blood was not either dead or alive. Rather, he perceived latent powers inherent in bodily materials like blood. Besides recognising the existence of an immaterial and eternal soul, Gaubius developed a material theory of life in which the constituent parts of the body acted on and explained physiological phenomena. The principles of water, phlogiston, salt, and earth explained how it was possible for blood to cohere, separate, and coagulate, and hence, to nurse and nourish. These principles within blood, in other words, explained the life of the body. A similar threefold explanation applied to the relation between body and soul: Gaubius envisioned an *enormôn* in the nerves in order to bridge the dichotomy between body and soul, since this principle, or agent of arousal, was responsible for movements of the body, as well as for transmitting bodily changes via the senses to the mind.[78] In short, Gaubius saw no disjunction between blood as a material that underwent changes in

[75] Théophile de Bordeu, *Recherches sur le pouls, par rapport aux crises* (Paris, 1756); idem, *Inquiries Concerning the Varieties of the Pulse: And the Particular Crisis Each More Especially Indicates* (London, 1764); Vila, *Enlightenment and Pathology*, 59.

[76] Boerhaave, *A New Method*, vol. 2, 211–2.

[77] Lacaze, *Idée de l'homme*, 44.

[78] Lelland J. Rather, *Mind and Body in Eighteenth-Century Medicine: A Study Based on Jerome Gaub's* De regimine mentis (Berkeley and Los Angeles, 1965), 59–65.

various pathological circumstances, and blood as a fluid embodying vital functions of the body. In his chemical lectures, for example, he told his students that

> You have seen, I hope, that the experiments with blood have taught you its parts, properties, nature, mutability, origin, and forces. This was the matter of more than forty lectures, and no wonder, for blood is the universal liquid of our microcosm; from which Moses has said that in it is the living soul, and he forbade the use of blood.[79]

Yet such statements carried no small degree of risk: locating life in the blood came dangerously close to materialism and atheism. If the soul was indeed in the blood, how could this spiritual part be regarded as immortal? The career of the French physician Julien Offray de La Mettrie (1709–1751) aptly illustrates the dangers of locating the soul in the blood. La Mettrie visited Leiden in the 1730s and 1740s, attended lectures by Boerhaave and Gaubius, and disseminated Boerhaave's ideas with the publication and translation of several of his works. Inspired by Gaubius's lecture *De regimine mentis* (1747), on the harmful and beneficial effects of the body on the soul, La Mettrie published his book *L'homme machine* (1748), in which he described the human body as 'a clock, but huge and built with so much artifice and ability'. Discussing the phenomenon of sleeping, he wrote that 'if the blood circulates too fast, the soul cannot sleep. If the Soul is too agitated, the Blood cannot calm down; it gallops through the veins with such a sound one can hear'.[80] Although La Mettrie did not eliminate the soul altogether by reducing life to matter alone, he did present it as a secondary quality of the body, largely dependent on physiological processes. In this philosophy the soul still existed but was no longer a metaphysical entity that would survive the body after its death.[81] For many, La Mettrie's materialism had gone one step too far, especially since his treatise was written in the vernacular and directed at a wide

[79] 'Praelectiones publicae chemicae', vol. 2, p. 273.

[80] Julien Offray de La Mettrie, *L'homme machine* (Leiden, 1748), 93, 12. The body as clock and as machine has been a much studied metaphor. See for example Jessica Riskin, *The Restless Clock: A History of the Centuries-Long Argument over What Makes Living Things Tick* (Chicago, 2015).

[81] Charles T. Wolfe, *Materialism: A Historico-Philosophical Introduction* (Dordrecht, 2016), 10, 52–3; Thomson, *Bodies of Thought*, 180–9.

audience. La Mettrie was duly accused of atheism, exiled, and his book banned from the market.

Gaubius struggled with this legacy since, indeed, his own views on the interrelation between blood and soul appeared eerily similar to La Mettrie's. Although Gaubius firmly stressed the existence of the eternal soul, he had lectured on the disturbances that occurred simultaneously in soul and body, pointing out that the 'life of the human body indisputably consists of some determined motion of liquids through the vessels'.[82] Gaubius nonetheless managed to avoid controversy, by writing within a strictly medical context and in Latin, which ensured that his ideas were only available to an educated audience; by admitting the limits of his knowledge; and by keeping La Mettrie at a distance. In 1763, for example, in his lecture on the healthy influences of the soul on the body, Gaubius publicly condemned La Mettrie: 'I do indeed regret bitterly that a little Frenchman—a Mimus or Momus?—brought forth a repulsive offspring, to wit, his mechanical man'.[83] In the same address Gaubius expressed his views on the physiological mechanism of the interaction between body and soul, this time stressing that, while physical conditions had their effects on the mind, the soul could equally have beneficial and harmful effects on the body. However, when his colleague Paulus de Wind (1714–1771), a physician from Middelburg, asked him in a letter to elaborate further on the connections between the soul and the body, Gaubius struggled to provide an adequate answer: regarding the 'phenomena of the human oeconomy, as a physician or medical man I cannot approve, for I do not know what kind of idea to connect to the soul that is in the blood and distinct from spirit and body'.[84]

* * *

This chapter has examined the changing medical perceptions of blood and the development of rivalling ways to study this bodily fluid in the early eighteenth century in the Dutch Republic, especially the remarkable developments in blood chemistry and haematology not previously

[82] Hieronymus David Gaubius, *Oratio de vana vitae longae, a chemicis promissae, exspectatione* (Leiden, 1734), 12.

[83] Hieronymus David Gaubius, *Sermo academicus de regimine mentis quod medicorum est* (Leiden, 1763), 2; Rather, *Mind and Body*, 115.

[84] Gaubius to Paulus de Wind, 9 August 1758 in Hamers-van Duynen, *Hieronymus David Gaubius*, 216.

considered in historiography. Prompted by the works of Herman Boerhaave, a school of Dutch physicians conducted investigations into the living fluids to answer questions in physiology and pathology.

Understanding the nature of blood was no straightforward enterprise, as new research possibilities complicated the answer as to what constituted valid knowledge about blood.[85] To understand blood and its properties, physicians combined new practical skills of observation, measurement, and chemical experimentation with theoretical knowledge about the animal oeconomy. Some were convinced that medicine necessitated blood chemistry because hydraulic and hydrostatic approaches to the living body could not explain the properties and changes in blood and other bodily fluids. While these mechanist views were, therefore, limited, blood chemistry held out the promise of truly understanding the nature of blood.

Ultimately we can trace an Enlightenment debate on the nature of blood. Physicians had first perceived blood to be an ordinary fluid, focusing on its broad range of sensible and physical properties, for example, their visible transformations, like coagulation. At the turn of the eighteenth century, however, we can see that the materiality of blood was understood on different levels of inquiry.[86] Some characteristics of blood, such as its volume, weight, and temperature, were observed thanks to microscopes, thermometers, and hydrometers. Yet, at another level of inquiry, blood was considered to be composed of constituent parts, and its properties could be explained in reference to four principles. Thanks to chemistry various bodily fluids and their variations became an object of conceptual inquiry, which was employed in order to improve existing theories of physiology and pathology. Whereas Thomas Schwencke argued that blood *in vitro* was fundamentally different from blood *in vivo*, Hieronymus Gaubius merely saw differences in state and composition while also recognising blood as a vital matter.

The different approaches advanced here are crucial to our understanding of how and why medical researchers began to explore the material and physical properties of the human body. Historiography on the Enlightenment commonly equates materialism with atheism, with Julien Offray de La Mettrie as the most commonly mentioned proponent. But as

[85] Lorraine Daston and Elizabeth Lunbeck, eds., *Histories of Scientific Observation* (Chicago, 2011).

[86] Cf. Ursula Klein, 'Shifting Ontologies, Changing Classifications: Plant Materials from 1700 to 1830', *Studies in History and Philosophy of Science*, 36 (2005), 261–329, here 266–7.

Charles Wolfe has recently emphasised, there existed many different forms of materialism.[87] And as this chapter has shown, Dutch physicians also located vital properties in the fluid of blood without adhering to a radical form of materialism. The false binary of materialism and vitalism thus proves rather unhelpful for historiographical attempts to understand blood chemistry in the eighteenth century. Dutch physicians were, in fact, part of a long medical tradition that considered blood and its perambulations in the body the source of life, and a dynamic substance that could be sensed and studied, both in bodies and laboratories.

Moreover, the frequently cited conceptual dichotomy between mechanism and vitalism proves to be false. Focusing on a material such as blood has de-stabilised the conceptual comfort of opposing mechanist and anti-mechanist models of the body. This chapter has shown how physicians presented increasingly complex and hybrid forms of mechanism, which included vital properties and active matter, as demonstrated by chemical investigations of blood. Early eighteenth-century chemistry, then, effectively acted as a go-between by blending mechanist, materialist, and vitalist perceptions of the living body within the field of academic medicine.

[87] Wolfe, *Materialism*.

CHAPTER 4

Piss Prophets and Urine Matters

On Tuesday 2 October 1736 Thomas Humphrey (b. *c.* 1707) defended his doctoral thesis before the board of medical professors at Leiden University and argued that all elements constituting bladder stones originated from ordinary, healthy urine. Calculi frequently tormented early modern individuals, and many physicians pondered their formation and growth. Whereas large bladder stones could cause excruciating pain and obstruct the secretion of urine, Humphrey argued that they consisted of the smallest elements usually found in urine. He justified these statements on the basis of chemistry: he observed how morning urine from a healthy person, poured in a cylindrical glass phial, would change from a homogeneous and transparent liquid into a turbid mixture with white sediment. When collected and dried, this deposit became hard as a rock. He therefore concluded that 'the elements of stones are in the healthiest men'.[1]

Humphrey and his study of bladder stones provide one example for the topic examined in this chapter, namely the thriving of urine chemistry in the eighteenth-century Dutch Republic. The unfortunate occurrence of bladder stones in patients necessitated pathological research. I attribute

[1] Thomas Humphrey, *Dissertatio medica inauguralis de calculi urinosi generatione et incremento* (Leiden, 1736), [Cv]. On 5 December 1735 a 28-year-old Humphrey had enrolled at Leiden University to study medicine. Although Herman Boerhaave had retired from the chair of chemistry already in 1729, he still acted as Humphrey's doctoral advisor.

R. E. Verwaal, *Bodily Fluids, Chemistry and Medicine in the Eighteenth-Century Boerhaave School*, Palgrave Studies in Medieval and Early Modern Medicine,
https://doi.org/10.1007/978-3-030-51541-6_4

the widespread use of chemical methods to analyse calculi and urine to the Boerhaave school. Boerhaave attracted hundreds of students from across Europe to Leiden, of which many delved into chemistry, took lecture notes, and wrote dissertations on urine and urological disorders—a development which continued to thrive when Gaubius succeeded Boerhaave in the 1730s. Indeed, Humphrey was far from the only one writing a dissertation on lithiasis. In the course of Boerhaave's professorship, another 14 doctoral students researched urine in relation to urological disorders, and this interest outlived Boerhaave's death. With no fewer than 36 other students writing their dissertations on urine and urinary diseases while Gaubius held his professorship, the chemical laboratory, where urine was experimented upon by professors and students alike, was a busy place. These investigators were convinced that the discipline of chemistry and its methods could provide them with physiological and pathological insights, and therefore, urine chemistry can be considered a productive field of research in eighteenth-century medicine.

Remarkably, historians of science and medicine have largely bypassed this flourishing of urine chemistry in the early eighteenth-century Dutch Republic. Instead, most scholars have emphasised the development and successes of clinical chemistry around the turn of the nineteenth century.[2] They have argued that figures like Carl Wilhelm Scheele (1742–1786), Tobern Olaf Bergman (1735–1784), Antoine François de Fourcroy (1755–1809), and Nicolas Louis Vauquelin (1763–1829) were the first to isolate different substances in urine and calculi by chemical analysis. These chemists detected uric acid in 1776, believing that all urinary concretions originated from the fluid of urine. At the turn of the nineteenth century, it has been argued, urine chemistry offered valuable support to medicine, ultimately leading to clinical chemistry in the nineteenth century. These studies perfectly fit the narrative of the chemical revolution and the invention of the new chemistry by Antoine Lavoisier and his contemporaries in the last quarter of the eighteenth century. But as most scholarship has endorsed a narrative that characterises urine chemistry as part of

[2] Noel G. Coley, 'Animal Chemists and Urinary Stone', *Ambix*, 18 (1971), 69–93; idem, 'Medical Chemists and the Origins of Clinical Chemistry in Britain (circa 1750–1850)', *Clinical Chemistry*, 50 (2004), 961–72; Louis Rosenfeld, *Four Centuries of Clinical Chemistry* (Amsterdam, 1999), 37–9; E.J. Westbury, 'A Chemist's View of the History of Urinary Stone Analysis', *British Journal of Urology*, 64 (1989), 445–50; the most recent example is chapter 12 'Founding Fathers of Stone Chemistry' in Michael E. Moran, *Urolithiasis: A Comprehensive History* (New York, NY, 2014), 101–16.

nineteenth-century science, it has not paid sufficient attention to the early modern origins of this methodology. In fact, it has been argued that the chemistry of urine in the preceding decades needed no substantial attention, because 'before Scheele, work on urinary stones was more alchemy than science'.[3] How then do we reconcile the flowering of Boerhaave's urine chemistry with some historians' dismissal of this period as 'stagnation' of developments, and as a mere 'interlude' between Robert Boyle and Lavoisier?[4]

I here argue that scholars have not adequately addressed the eighteenth-century proliferation of urine chemistry and its uses for medicine to date. Much can be learned from the much-neglected yet prolific urine chemistry of the early eighteenth century. This chapter follows the story of the fluid of urine in the laboratory, through an examination of the works and ideas of Boerhaavian physicians and chemists. By analysing how and why urine chemistry was taught and studied at universities, to the extent and as early as it was, I argue that the turn of the eighteenth century saw a shift from chymical uroscopy to urine chemistry. While urine chemistry profited from methods developed in the context of chymical uroscopy, Boerhaave and his students applied them in the service of very different aims and applications. The way in which urine chemistry thrived in the course of the eighteenth century was characterised by theoretical and procedural innovations in physiology and pathology. Furthermore, Boerhaavian physicians and chemists provided fertile ground for the revolutionary clinical chemistry arising around the turn of the nineteenth century.

Chymical Uroscopy

This section will demonstrate that, while the sensory and chymical inspection of urine was increasingly discredited, chemical experiments on urine provided a strong basis for researching physiology and pathology. This development from chymical uroscopy to urine chemistry, which took place around the year 1700, has largely gone unnoticed in modern scholarship. In his important study on uroscopy—that is, the practice of making

[3] Gabriel Richet, 'The Chemistry of Urinary Stones around 1800: A First in Clinical Chemistry', *Kidney International*, 45 (1995), 876–86, here 877.

[4] 'Stagnation of Chemical Theory: 1675–1750' in Robert Siegfried, *From Elements to Atoms: A History of Chemical Composition* (Philadelphia, 2002), 56–73; 'Interlude: The Crisis of Inter-Revolutionary Chemistry' in Victor D. Boantza, *Matter and Method in the Long Chemical Revolution: Laws of Another Order* (Farnham, 2013), 115–41.

diagnostic examinations of urine by sensory inspection—the medical historian Michael Stolberg questions the anachronistic perspectives adopted by some scholars who consider uroscopy a direct predecessor of modern urinalysis. Henry Wellcome, for example, had traced an 'evolution' from ancient and medieval uroscopy through to 'the advent of scientific urine analysis' in the nineteenth century.[5] But Stolberg rightly points out certain problems with this approach: uroscopy was the diagnostic examination of urine by simple inspection, offering unrivalled access to the internal processes of the body, and covering the entire spectrum of diseases, and was understood within a Galenic model of the body. Modern urinalysis, by contrast, analyses urine by physical, chemical, and microscopic or photometric means, in order to diagnose particular metabolic disorders, or to test for urological and nephrological diseases. To see uroscopy as the predecessor to urinalysis, Stolberg concludes, 'does justice neither to the very different theoretical foundations of traditional uroscopy nor to its much more wide-ranging diagnostic significance'.[6] I agree with Stolberg with regard to the discontinuity in the way physicians examined urine throughout the ages. But the question about the origin and development of urinalysis remains unresolved. Building on Stolberg's work I suggest that a rigid demarcation between uroscopy and urinalysis necessarily passes over the chronological overlap between chymical uroscopy and urine chemistry at the turn of the eighteenth century. The works and teachings on urine by Dutch physicians of the period in particular point towards an overlap between chemical practices at this time.

The work of Johan van Dueren (*c.* 1664–1726) is a case in point.[7] Like many other Dutch physicians Van Dueren took considerable satisfaction in pointing out the multiple deficiencies of uroscopy.[8] The frontispiece of Van Dueren's voluminous work *De ontdekking der bedriegeryen van de*

[5] Henry S. Wellcome, *The Evolution of Urine Analysis: An Historical Sketch of the Clinical Examination of Urine* (London, 1911), 75–92; see also Leonard Paul Wershub, *Urology: From Antiquity to the 20th Century* (St. Louis, MO, 1970).

[6] Michael Stolberg, *Uroscopy in Early Modern Europe*, trans. Logan Kennedy and Leonhard Unglaub (Farnham, 2015), 2.

[7] Johan van Dueren, *De ontdekking der bedriegeryen van de gemeene pis-bezienders* (Amsterdam, 1688). In addition to the dedication, preface, and poems by friends Stephen Blankaart and Franciscus Sylvius, this work was 426 pages long.

[8] Uroscopy had most famously been criticised by Pieter van Foreest in *De incerto, fallaci, urinarum iudicio* (Leiden, 1589). This work was translated to German, English, and Dutch vernaculars in the early seventeenth century.

gemeene pis-bezienders (*'The Discovery of the Deceptions by Public Uroscopists*', 1688) perfectly illustrated the popularity and widespread practice of uroscopy in the early modern period (see Fig. 4.1). A female messenger (*pis-brenger*), accompanied by a little boy, delivers a patient's urine to the uroscopist in a wicker basket. The 'piss-watcher' (*pis-besiender*) takes out the round-bottomed urinal and holds it up to the sun, in order to scrutinise the urine in a clear light. From its colour, smell, taste, and its other characteristics the uroscopist drew deductions about the state of the body and the stage of the patient's disease. At first sight, this engraving appears to conform to a popular type in Dutch genre painting, depicting uroscopists and alchemists as men who unlock the secrets of nature.[9] The furnace with chemical vessels, retort, and utensils, and the library, support this interpretation. Nevertheless, Van Dueren's frontispiece turns this pictorial tradition on its head by adding unsavoury figures at the top. A carefree satyr appears to drop divinations one after the other, while a backscratching monkey defecates onto the entire scene. Joined by a central head with unmistakable goat's ears, all these figures express the underhand support between messengers and uroscopists, and their mischievous actions at the expense of unsuspecting patients—a topical criticism numerously repeated by academic physicians.[10] Yet if the inspection of urine was not fit for application by learned physicians, the chemical analysis of urine was not entirely dismissed. On the contrary, even Van Dueren's polemical essay against uroscopists talked in favour of urine examination by chemistry: 'One can find this unfailing use of uroscopy (based on Reason itself, and true Experience of Chemistry) in the writings of the Gentlemen Bontekoe, Blanckaart, Overkamp, de Heyde, Muys, Daalamns, &c'.[11] As this section demonstrates, therefore, while the traditional sensory and chymical inspection of urine was increasingly discredited, chemical experiments on urine provided new ground for researching and teaching physiology and pathology in the early eighteenth century.

The history of uroscopy extends back to the middle ages. The late sixteenth century saw a reinvention of the practice of uroscopy at the hands of the Swiss physician Paracelsus. While Paracelsus upheld the general

[9] See 'The Uroscopist as a Truth-Seeker' in Stolberg, *Uroscopy*, 114–7.

[10] More on the academic criticisms against uroscopy, see Michael Stolberg, 'The Decline of Uroscopy in Early Modern Learned Medicine (1500–1650)', *Early Science and Medicine*, 12 (2007), 313–36.

[11] Van Dueren, *Pis-bezienders*, 329.

Fig. 4.1 Frontispiece in Johan van Dueren, *De ontdekking der bedriegeryen van de gemeene pis-bezienders* (Amsterdam, 1688). KB National Library of the Netherlands, The Hague

premise that urine reflected the condition of the body, his method of examination was rather different from that of those that had gone before him. Paracelsus distinguished between an internal urine, resulting from Nature's excretory efforts and mirroring the state of the body, and an external urine, which resulted from the digestion in the organs. The urine flowing from a person's urethra was, therefore, a mixture of two fluids. The fact that Paracelsus aimed at separating these fluids explained his preferred method of examination, distillation: this process enabled him to extract and expose those morbid parts in the urine which were otherwise too subtle to be recognised with the naked eye.[12] Many medical men, including the influential German physician Leonhard Thurneysser (or Thurneisser zum Thurn, 1530–1596) adopted Paracelsus's teachings and distilled their patients' urine. In one treatise Thurneysser depicted the correspondence between the distillate and the part of the body where the disease in question was believed to be located (see Fig. 4.2). The chymical uroscopist gently heated a sample of urine in an alembic, in which the steam would rise slowly, and then condensate in different places of the receiving flask. With the help of anatomical drawings of both female and male bodies, distillation charts mapped out the individual parts of the body, as well as the ways in which a physician would be able to diagnose certain diseases based on the composition of the urine. Thurneysser made quite a reputation for himself as a gifted uroscopist, and well-to-do patients across the Holy Roman Empire sent him their urine with an accompanying letter, asking him to analyse the sample in his laboratory.[13]

The works of the Thüringian physician Christoph von Hellwig (1663–1721) and the Leiden lecturer of chemistry, Jacob le Mort, demonstrate that the chymical approach to uroscopy continued to be popular well into the early eighteenth century. Von Hellwig studied medicine at the universities of Jena and Erfurt, and he enlarged Thurneysser's

[12] Paracelsus, *De urinarum ac pulsuum judiciis* (Cologne, 1568); Stolberg, *Uroscopy*, 66.

[13] Leonhart Thurneisser zum Thurn, *Protokatalēpsis oder Praeoccupatio durch zwölff verscheidenlicher Tractaten gemachter Harm Proben* (Frankfurt (Oder), 1571); idem, *Bebaiōsis agōnismou: Das ist Confirmatio concertationis, oder ein Bestettigung desz jenigen so streittig, häderig, oder zenckisch ist, wie dann ausz Unverstandt die neuwe und vor unerhörte Erfindung der aller nützlichesten und menschlichem Geschlecht der notturftigesten Kunst desz Harnnprobirens ein Zeitlang gewest ist* (Berlin, 1576). See Amy Eisen Cislo, *Paracelsus's Theory of Embodiment: Conception and Gestation in Early Modern Europe* (London, 2010), 80–5; Stolberg, *Uroscopy*, 66–70.

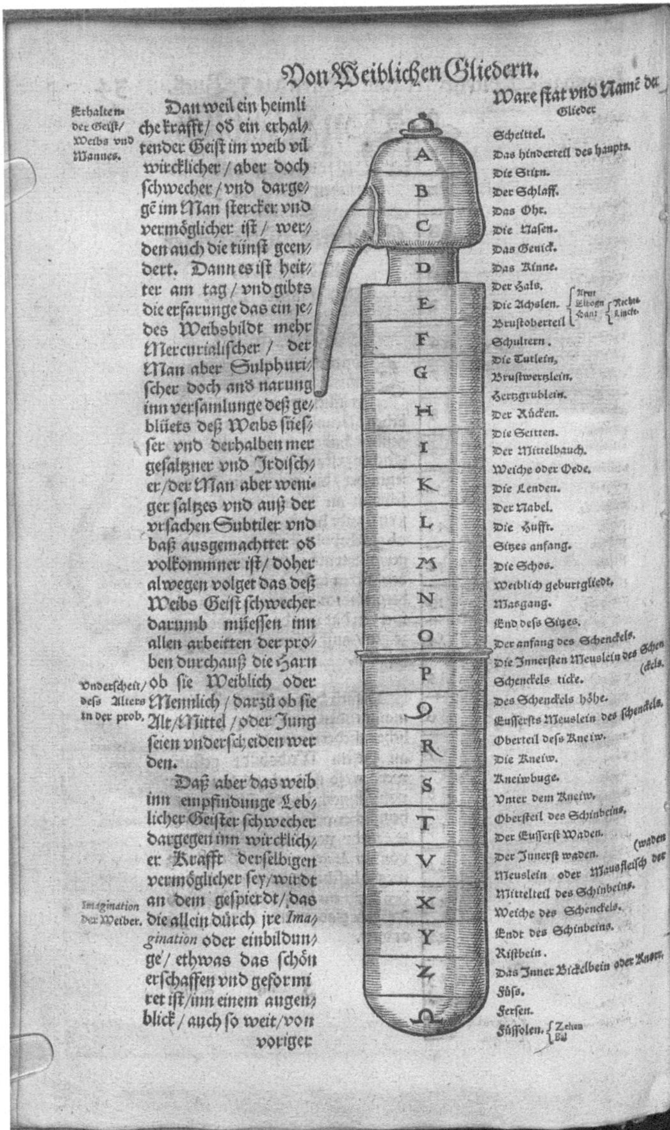

Fig. 4.2 Chymical uroscopy chart by unknown engraver in Leonard Thurneysser, *Bebaiōsis agōnismou* (Berlin, 1576). Bayerische Staatsbibliothek, Munich.

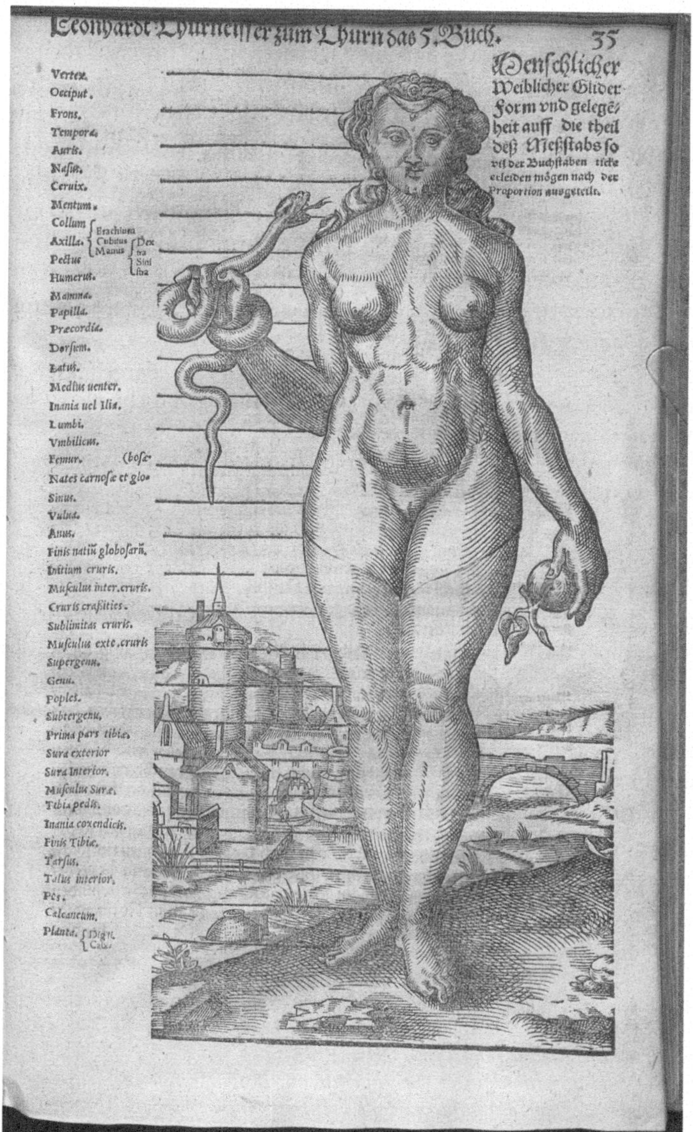

Fig. 4.2 (continued)

gender-specific uroscopy chart with the further details of colour and quality.[14] In Leiden, Le Mort delivered pharmacy lectures in his private laboratory from 1672 onwards. Le Mort was appointed to the post of chemistry lecturer at the university, and of *praefectus* of its laboratory in 1690. Mainly teaching the preparation of medicines, he was appointed professor in medicine and chemistry in 1702 and would keep that chair until his death in 1718.[15] Le Mort defended the practice of uroscopy, particularly uroscopy based on chymical methods. Concerned about the contemporary criticism of uroscopy and uromancy, Le Mort was at pains to distinguish himself from frauds and quacks, and attempted to save the practice from falling into disfavour. He stated that uroscopy:

> Truly requires an accurate knowledge of the art of colouring and chymistry, before a wise doctor can make reasonable judgments, and is not required, like quacks, to first ask the state of the sufferers, from which story he pronounces his opinion of the urine, as an oracle.[16]

It appears that Le Mort continued to perform uroscopy by distillation until his death in 1718. Chymical uroscopy, in short, was similar to ordinary uroscopy in that it presupposed urine to reflect the state of the entire body. But it was believed that the diagnostic practice was improved by the method of distillation, which was supposed to provide a more thorough analysis of the urine's qualities and contents, and hence led to a more accurate diagnosis.

Despite the fact that Von Hellwig and Le Mort were university-trained physicians, many other physicians did not adopt their improvements in chymical uroscopy with enthusiasm. Physicians in general were concerned with their authority and status, and they were particularly keen not to be

[14] Ludwig von Hellwig, *Valentin Kräutermanns Curieuser und vernünfftiger Urin-Artzt* (Arnstadt, 1735); Cislo, *Paracelsus's Theory of Embodiment*, 93.

[15] More on Le Mort, see Dalila Wallé, *Leiden Medical Professors, 1575–1940* (Leiden, 2007), 89–90; H.A.M. Snelders, *De geschiedenis van de scheikunde in Nederland: Van alchemie tot chemie en chemische industrie rond 1900* (Delft, 1993), 37.

[16] Jacob le Mort, *Chymia, rationibus et experimentis auctioribus, iisque demonstrativis superstructa, in qua malevolorum calumniae modestè simul diluuntur* (Leiden, 1688), 161. See also the dissertation of Le Mort's student Henricus Snellen, *Disputatio chymico-medica inauguralis de urinarum inspectione* (Leiden, 1701). On Le Mort and the institutional context of chymistry in Leiden in this period, see Gerhard Wiesenfeldt, *Leerer Raum in Minervas Haus: Experimentelle Naturlehre an der Universität Leiden, 1675–1715* (Amsterdam, 2002), 189–241.

associated with Paracelsus's stained reputation.[17] At the turn of the eighteenth century beliefs and opinions about alchemical practices and ideas were already quite sceptical, and the name and fame of Paracelsus in particular had caused a chasm among chemists and physicians.[18] In his chemistry course Boerhaave explained this to his students: 'To hear the generality of chemists talk, he [Paracelsus] was nothing less than a God [...] yet others represent him as one of the most vile, flagitious, and worthless of the race of men'.[19] Chymistry—defined as comprising all chemical and alchemical practices and theories as they were understood in the seventeenth century—had a bad reputation in general, mainly involving claims of universal medicines.[20] Accusations of quackery and heresy were not uncommon. Writing about the spicy escapades of European diplomats at the Utrecht peace negotiations in 1713, for example, author Augustinus Freschot even described how ladies of pleasure gathered 'like chemists who recognise each other without ever having met, because the similarity of their occupation apparently has something contagious'.[21] In the timespan of a few years, Boerhaave, Gaubius and Abraham Kaau presented orations and public lectures in which they strongly distanced themselves from certain aspects of chymistry, specifically the belief in eternal life, panacea, and the practice of chrysopoeia (the transmutation of base metals

[17] On uroscopy as threat to physicians' authority, see Stolberg, *Uroscopy*, 128–34.

[18] On Paracelsus's bad reputation see, for example, Godfried Arnold, *Historie der kerken en ketteren van den beginne des Nieuwen Testaments tot aan het jaar onses Heeren 1688*, 3 vols (Amsterdam, 1701–1729), vol. 2, 585. Here, Paracelsus was presented as 'the principle founder of heretics'. On Paracelsus's multiple and often contradictory reputations throughout time, see Ole Peter Grell, ed. *Paracelsus: The Man and His Reputation, His Ideas and Their Transformation* (Leiden, 1998).

[19] Herman Boerhaave, *A New Method of Chemistry: Including the Theory and Practice of that Art: Laid down on Mechanical Principles, and Accommodated to the Uses of Life*, trans. Peter Shaw and Ephraim Chambers, 2 vols (London, 1727), vol. 1, 22.

[20] On chymistry, see William R. Newman and Lawrence M. Principe, 'Alchemy vs. Chemistry: The Etymological Origins of a Historiographic Mistake', *Early Science and Medicine*, 3 (1998), 32–65; William R. Newman, 'From Alchemy to 'Chymistry'', in *The Cambridge History of Science*, ed. Katharine Park and Lorraine Daston (Cambridge, 2006); Lawrence M. Principe, ed. *Chymists and Chymistry: Studies in the History of Alchemy and Early Modern Chemistry* (Sagamore Beach, 2007); Lawrence M. Principe, 'Alchemy Restored', *Isis*, 102 (2011), 305–12, here 312.

[21] Augustinus Freschot, *Amoureuze en pikante geschiedenis van het congres en de stad Utrecht: Augustinus Freschots verhaal achter de Vrede van Utrecht*, ed. Erik Tigelaar, trans. Roland Fagel (Hilversum, 2013), 79.

into noble metals, particularly gold).[22] Le Mort was considered part of this chymical tradition. Boerhaave's students, therefore, learned to distance themselves from Le Mort completely, because Le Mort allegedly 'insists on [an] abundance of operations, which have long since been disused'.[23]

Besides being concerned with the possible tarnish of a bad reputation, physicians were sceptical towards uroscopy of other reasons too. First, they held reservations about the method of distillation for this particular use. The physician Konrad Barthold Behrens (1660–1736), for example, expressed his doubts about Thurneysser's procedure. While Behrens agreed that distillation allowed for the sulphurous and the salty parts of urine to be separated successfully, the procedure was nevertheless 'very uncertain', since volatile matter might accumulate in different places of the flask, simply depending on the intensity of fire.[24] Second, many physicians of the Boerhaave school voiced their reservations concerning uroscopy in general.[25] Certain diseases did not reveal any difference in urine at all, and some simple foods could make the urine appear as if produced by a serious disease. Albrecht von Haller, for example, reported that a physician would diagnose internal gangrene and prognosticate imminent death when presented with black urine—only to find out later that the person in question had simply eaten asparagus or rhubarb which, indeed, could render urine very dark.[26] The validity of solely basing one's diagnosis of urine was, therefore, seriously questioned. Boerhaave in particular warned his students of uromancy or the practice of making predictions exclusively on the basis of urine: 'In the passing judgement of urine, we are to have nice [proper] regard of other signs, appearing in diseases; or otherwise this art

[22] See Herman Boerhaave, *Sermo academicus de chemia suos errores expurgante* (Leiden, 1718); Hieronymus David Gaubius, *Oratio de vana vitae longae, a chemicis promissae, exspectatione* (Leiden, 1734); Abraham Kaau, *Declamatio academia de gaudiis alchemistarum* (Leiden, 1737). Marieke M.A. Hendriksen, 'Criticizing Chrysopoeia? Alchemy, Chemistry, Academics and Satire in the Northern Netherlands, 1650–1750', *Isis*, 109 (2018), 235–53.

[23] Boerhaave, *A New Method*, vol. 1, 40.

[24] Konrad Barthold Behrens, *Ob das Wasserbeschawen in Krankheiten etwas nuze, und wie weit demselben zu trauen* (Hildesheim, 1688), 24–6; Stolberg, *Uroscopy*, 139.

[25] The arguments of the so-called urosceptics are discussed in Stolberg, 'The Decline of Uroscopy', 123–34.

[26] Albrecht von Haller, ed. *Praelectiones academicae in proprias institutiones rei medicae*, 6 vols (Göttingen, 1739–1744), vol. 6, 319; idem, *Dr. Boerhaave's Academical Lectures on the Theory of Physic: Being a Genuine Translation of his Institutes and Explanatory Comment*, 6 vols (London, 1742–1746), vol. 6, 216.

is very full of fallacy and deceit'.[27] These warnings were not unwarranted, seeing that three so-called piss prophets were known to be dwelling in Leiden at the time. According to Von Haller:

> One of these prophets is so ignorant and stupid that he hardly knows his own name, but has three or four formulas or notes containing symptoms common to almost all diseases. These he dictates in every disease, not knowing how to write. But the other prophet is a little wiser than the first, having nine set forms of prognosticating, which are always sure to contain something of truth.[28]

Gaubius taught a nuanced view of uroscopy, acknowledging that some symptoms could be seen in the urine, but that these had to be considered always in conjunction with other symptoms. Making a diagnosis solely on the basis of urine was possible, but carried a high risk of misdiagnosis. And indeed, many of his doctoral students at Leiden University repeated this verdict in their dissertations. Frederik de la Fosse (*c.* 1721–*c.* 1791) from The Hague, for example, argued that 'there is nothing more deceptive in the present remedy, than to judge from the sole inspection of urine'.[29] Another of Gaubius's doctoral students, Christiaan Swaving (1717–1795), was of the opinion that though uroscopy was useful for certain diseases, 'it suits reason nor experience to predict everything from the inspection of urine'.[30]

In sum, a clear shift from chymical uroscopy to urine chemistry can be observed when we take a closer look at the works of Von Hellwig and Le Mort on the one hand, and of Boerhaave and his students on the other. Previously, chymical uroscopists took a sample of urine and distilled it in order to diagnose the patient. Urine was a reflection of the humours and could be inspected by means of the senses and distillation. But in the eighteenth century learned physicians studied the urine of healthy and diseased persons as well as patients, with chemical methods, to gain a better

[27] Herman Boerhaave, *Institutiones medicae in usus annuae exercitationis domesticos digestae*, alt. ed. (Leiden, 1713); idem, *Boerhaave's Institutions in Physick*, trans. Joseph Browne, 2nd ed. (London, 1715), 277.

[28] Von Haller, *Praelectiones*, vol. 6, 319; idem, *Academical Lectures*, vol. 6, 216–7.

[29] Frederik de la Fosse, *Specimen medicum inaugurale de aëre vitae et morborum causa* (Leiden, 1743), [D4r].

[30] Christianus Swaving, *Dissertatio inauguralis physiologica-medica de excrementis secundae coctionis* (Leiden, 1744), [F4r].

understanding of the physiology of secretion and the urinary organs. Urine chemistry was more than a diagnosis; it became a means to understanding the pathology of diseases such as lithiasis, diabetes, ischury or strangury, and cystitis (inflammation of the bladder), which all were resulting from pathological changes in the body. Urine chemistry, in other words, provided new insights into the normal and morbid appearances of urine, reflecting only specific parts and systems of the body. Many of the chemical techniques that were developed within chymistry continued to be practised in the eighteenth century since chymical uroscopy had advanced the study of bodily products with a considerable degree of detail. But urine chemistry was not solely aimed at diagnostics. Rather, it provided a new theoretical foundation for a urological physiology and pathology.

Urine Chemistry

In 1753 the dissertation written by Irishman Arthur Magenis (b. *c.* 1728) opened with the statement that 'among the various fluids that are secreted from the blood stream, not one deserves more attention by physicians, than urine'.[31] Magenis gave a short anatomical description of the urinary organs, explaining that urine was isolated from blood by the kidneys, sent via the ureter to the bladder, and then cast away through the urethra. But to really understand the nature of urine, Magenis argued, 'it will be necessary to subject the blood as the source of this liquid to chemical examination'.[32] He performed six experiments on blood and the serum of blood before turning to the chemical examination of urine. He observed that various processes changed the colour of urine and that distillation left a salty residue. His dissertation went on to identify the so-called spirit of urine and described its colour and smell. Above all, Magenis's activities demonstrate how students performed their own experiments, sensing the constituents and properties of bodily fluids and building upon Boerhaave's and Gaubius's work in chemistry.

Magenis was one of a number of doctoral students specialising in urine chemistry at the time. When Boerhaave and Gaubius held their professorships, that is, between 1709 and 1775, on average 15 medical

[31] Arthurus Magenis, *Dissertatio medica inauguralis de urina* (Leiden, 1753), 5. Magenis enrolled at Leiden University on 4 November 1748 and graduated on 2 March 1753.

[32] Ibid., 6.

dissertations were devoted to a particular disorder or disease, for example, smallpox (22 dissertations), pleurisy (27), dropsy (14), epilepsy (13), dysentery (19), and gout (8). Together, the theses on urine physiology and pathology, including lithiasis, ischury, diabetes, and cystitis, make an impressive 47 dissertations which incorporated chemical knowledge and experiments. Reading these dissertations simply as a reiteration of textbook knowledge would miss an important point. Whereas the length, depth, and quality of medical dissertations differed significantly, many contemporaries recognised the fact that they contributed significantly to the advancements of the fields of chemistry and medicine. Dissertations could even become topics of scientific debates and controversy. Gaubius, for example, often send out copies of some of his students' dissertations to his friends and fellow physicians in appreciation.[33] Others, however, could cause him to feel personally attacked. When in June 1763 Samuel Musgrave (1732–1780) finished a provocative thesis in which he argued that Boerhaave's and Stahl's systems should be replaced by empirical methods, Gaubius tried to control the controversy. He decided not to oppose the student too harshly during the defence, and he completely rewrote the preface to his pathology textbook to include a more thorough treatment of medical methodology.[34] Clearly, these medical dissertations were as formative an element in students' education as they were important to the work of university professors.

Thus, the works of Boerhaave, Gaubius, and their students demonstrate that urine chemistry was not a dogmatic system dictated by university professors, but rather a productive method for studying physiology and urinary disorders. Medical students had many questions to face regarding the pathology of profoundly unsettling disorders which demanded further research. Why, for example, were some patients

[33] See for example Gaubius to António Sanches, Leiden, 6 September 1743, 21 August 1744, and 1 December 1746 in Sophia W. Hamers-van Duynen, *Hieronymus David Gaubius (1705–1780): Zijn correspondentie met Antonio Nunes Ribeiro Sanches en andere tijdgenoten* (Assen and Amsterdam, 1978), 76–9, 84–7. Sanches's correspondence can be found as 'Epistolae ad Antonium Ribeiro Sanches', 1735–1783, 2 vols, Vienna, ONB, Cod. 12713–12714 Han. On dissertations as educational practice, see Jaap Harskamp, 'Dissertatio medica inauguralis…Leyden medical dissertations in the British Library, 1593–1746' (London, 1997).

[34] Samuel Musgrave, *Dissertatio medica inauguralis sive apologia pro medicina empirica* (Leiden, 1763); A.M. Luyendijk-Elshout, 'Samuel Musgrave's Attack upon Stahl's and Boerhaave's Doctrines in 1763', *Janus*, 67 (1980), 141–56.

suffering from a urine obstruction, and yet feeling the strong urge to urinate? How did bladder stones come about? Which treatment might best resolve an infection of the bladder and pain during urination? It was precisely these sorts of disorders that students researched with the help of urine chemistry. By combining the text-based study of a specific disease with chemical laboratory experiments on urine, students' appropriation of chemical skills and knowledge played a central role in the development of urine chemistry in the eighteenth century.

An education in physiology was a central facet of urine chemistry. Many medical students at Leiden University witnessed chemical demonstrations and learned about the physiology and pathology of urine before contributing to urine chemistry with their own experiments. The active role of students in this context provides evidence for the fact that this line of research was at times progressive. Students took notes during chemistry classes while sitting on tiered benches all around the chemical laboratory, from whence they could hear the professor and observe his urine experiments.[35] In the preparation of their doctoral theses, many students conducted their own experiments with the help of their professor. As a young chemistry lecturer, Gaubius was particularly eager to help students master the skills necessary for chemical experimentation. During his inaugural lecture in 1731, he argued that medicine and natural philosophy were inextricably bound to chemistry, and he therefore invited all students to join him in the laboratory and 'work up a sweat'. He promised them to be not only their teacher, but also their guide: 'I offer you myself as *supervisor* and, if you wish, as counsellor. If I possess any strength, service or understanding, then use them as you please. Because to all of you I say: to promote your studies, that is especially the high point of my wishes, that is the only purpose of my work'.[36] As this promise clearly shows, Gaubius wished that professors and students worked on one project together, in order to further knowledge of the healthy human body and its diseases.

Indeed, numerous students visited the laboratory, and today many libraries still house copies of students' lecture notes taken for the

[35] A short description of the laboratory's interior can be found in Zacharias Conrad von Uffenbach, *Merkwürdige Reisen durch Niedersachsen, Holland und Engelland*, 3 vols (Frankfurt & Leipzig, 1753–1754), vol. 3, 454–5.

[36] Hieronymus David Gaubius, *Oratio inauguralis qua ostenditur chemiam artibus academicis jure esse inserendam* (Leiden, 1731), 48. Emphasis added.

chemistry courses of Boerhaave and Gaubius.[37] A comparative analysis of their notes reveals fascinating variations in content and use. Some very rare examples were written down hastily while the professor was delivering his lecture.[38] Yet most copies extant today are very neatly written, suggesting that they were composed once the seminar was finished.[39] Georg Michael Wepfer (1692–1774) from Schaffhausen, Switzerland, studied at Leiden University from 1708 to 1710 and finished writing down his notes in a tidy form four years later, in 1714.[40] One leaf from the list of contents, which was written on paper of lighter quality and probably added later, provides an outline of the chemical processes on urine, indicated by the corresponding symbol (⊡). The processes on urine taught students like Wepfer that fresh morning urine that was discharged 12 h after a meal, and had hence gone through an entire circuit through the body, was neither acid nor alkaline. Even if this 'human water' was distilled on a gentle fire, the watery distillate (▽) was neither acid nor alkaline, neither saline nor inflammable. Only as a result of the process referred to as digestion (*i.e.* the keeping of a fluid at body temperature) the urine turned alkaline and changed colour and smell, turning red and foetid.[41] The role of urine chemistry was to support Boerhaave's explanations of physiological processes of secretion as they occurred in the living body. Wepfer noted these insights and copied them, attesting to his aspiration to hold on to this knowledge for as long as possible. He returned to Switzerland with his

[37]'Hermanni Boerhaave Chemia', Oxford, Bodleian Library, Rawlinson Collection, MS. Rawl. C 535; 'Hieron. David. Gaubii Med. Doct. Med. & Chemio Professoris in Academia Leijdensi Dictata in Chemiam', Middelburg, Zeeland Library, MS 6270–6271; George Paterson, 'Gaubii Chemia', Leiden, c. 1755. London, Wellcome Library, MS 2480–2485; and 'Annotationes ex praelectionibus chemicis', Leiden, 1772–1773. Yale, Beinecke Rare Book and Manuscript Library, Mellon MS 117.

[38]For example, 'Collegii Chemici ex D.D. Boerhaaven'. Leiden, 1702. Royal College of Surgeons of England, London, MS0114/1/1; and Matthias van Geuns, 'Ex collegio chemico H.D. Gaubii annotata', Leiden, 1758–1759. Amsterdam, University Library, MSS 1343 I E 9–10.

[39]One manuscript recording Gaubius's lectures was sold by Leiden bookseller Philip Bonk, and titled 'Dictata in Chemiam'. Leiden, University Library, BPL 1477.

[40]Georg Michael Wepfer, 'Collegium Chymicum experimentale', Schaffhausen, 1714. Uppsala, University Library, Waller MS cod-00068. Wepfer appears to have started to produce fair copies of his lecture notes upon his return to Schaffhausen. Another example of neatly written out lecture notes is 'Descripta quaedam ab Ore Hermanni Boerhaavii inter pronunciandum lectiones ejus de Chemia in 2 tomis. Lugd. Bat. Anno 1719', 2 vols, Leiden, c. 1719–1730. London, British Library, Sloane MS 3183–3184.

[41]Wepfer, 'Collegium Chymicum experimentale', ff. 241–243.

notebook, and probably shared it with fellow physicians in Schaffhausen, among them Bernhard Peyer. When Peyer graduated in medicine at Leiden in 1750, he dedicated his dissertation to his 'brothers' Bernhard and Georgius Michael Wepfer.[42]

In addition to demonstrating the physiology of bodily secretions, urine chemistry was particularly important to the study of urinary diseases. After attending Boerhaave's or Gaubius's chemistry course, some medical students studied further towards a medical doctorate, specialising on a specific disorder, such as the difficulty of producing urine, pain in urination, the involuntary drop-by-drop emission of urine, and incontinence. The most popular topic was bladder stones or lithiasis. Cornelius Stalpart van der Wiel (*c.* 1687–1742) from The Hague, for example, tackled lithiasis by means of textual study of the medical treatises by Jan Baptista van Helmont and Boerhaave, and by performing his own chemical examinations.[43] Bladder stones or calculi had been one of the most common illnesses plaguing humankind for centuries.[44] The condition often started with the secretion of cloudy urine with gravel- or sand-like particles and a thick, white, mucous discharge. The Amsterdam anatomist Frederik Ruysch (1638–1731) famously recorded case examples of, among others, the deadly obstruction of urine due to a stone in the urethra, and an extraordinary large bladder stone in a three-year-old.[45] Physicians' unsuccessful efforts to prevent bladder stones dated back to antiquity, so that the surgical removal of the stone remained the only effective treatment. The municipal physicians' and surgeons' guild in Amsterdam recorded all patients who had been subjected to a lithotomy, including an exact description of the shape and colour of the stones.[46] The surgeon and

[42] Bernhardus Peyer, *Dissertatio medica inauguralis de tussi* (Leiden, 1750), [A2r].

[43] Cornelius Stalpart van der Wiel, *Dissertatio medica inauguralis de urinae calculo* (Leiden, 1712).

[44] On the popular subject of lithiasis and lithotomy, see for example Harry W. Herr, '"Cutting for the Stone": The Ancient Art of Lithotomy', *British Journal of Urology*, 101 (2008), 1214–6; Ahmet Tefekli and Fatin Cezayirli, 'The History of Urinary Stones: In Parallel with Civilization', *Scientific World Journal* (2013); Moran, *Urolithiasis*. The best introduction to the history of bladder stones is 'Learning to Cut for the Stone' in Harold J. Cook, *Trials of an Ordinary Doctor: Joannes Groenevelt in Seventeenth-Century London* (Baltimore, 1994), 76–105.

[45] Cases 15 and 57 in Frederik Ruysch, *Alle de ontleed- genees- en heelkundige werken*, trans. Ysbrand Gysbert Arlebout, 3 vols (Amsterdam, 1744), vol. 1, 60–1, 98–9.

[46] 'Aenteekeninge, uijt ordre en volgens instructie van Haare Edele Grootagtbaare de Heeren Burgermeesteren, van die geene, die van den steen syn gesneden, 't zeedert den

physician Johannes Rau perfected this surgical method, making the severe and dangerous procedure more safe and quick. When Rau delivered his inaugural lecture in Leiden in 1713, he claimed that he had performed the operation on as many as 1547 patients.[47] Nevertheless, as one can imagine, lithotomy remained to be an extremely painful procedure. And even if the stone cutter's operation was successful, patients were at risk of further complications, including infections, haemorrhages, fistulae, incontinence, sterility, and even death. Further, lithotomy was not effective in preventing the formation of new stones.

Medical chemists therefore sought ways to better understand the reasons for the development of bladder stones and effective means for their prevention. Eighteenth-century doctoral students like Stalpart van der Wiel searched for clues in the works of earlier physicians and chemists, of which the most notable was Van Helmont's *De lithiasis*, and in Boerhaave's chemical lectures.[48] Van Helmont had proposed that gravel might be generated from the elementary principles in healthy urine: 'The stone formed in our bodies, and the substance which encrusts the bottom and sides of chamber pots, are in every respect the same, and both have the same essence, cause, and properties'.[49] Boerhaave had shared Van Helmont's hypothesis that the raw elements of bladder stones could be discovered in the urine, but he performed his own chemical experiments. In a cylindrical glass roughly a middle finger's width in diameter and six inches of height, Boerhaave collected the fresh morning urine of a healthy man whom he knew never to have suffered from calculi. Even in a close examination by a microscope, no heterogeneity could be detected in the yellow and clear liquid. Once the urine had been kept at 72 °F for about seven minutes, microscopic inspection revealed corpuscles resembling 'little flocks of wool' swimming in the flask. As time passed, these flocks slowly grew into small clouds and descended towards the bottom of the vessel. When parts

23en April 1700 tot den eersten Januarij 1801. Gehouden door de stads doctoren en overluijden van het Chirurgijns gilde', Amsterdam, 1700–1801. Amsterdam, University Library, MS 1403.

[47] Johannes Jacobus Rau, *Oratio inauguralis de methodo anatomen docendi et discendi* (Leiden, 1713), 37. Rau had worked as ship's surgeon and private lecturer in anatomy and surgery for many years before he started teaching anatomy at Leiden University.

[48] Stalpart van der Wiel, *De urinae calculo*, 8.

[49] 'De lithiasi' in Jan Baptist van Helmont, *Ortus medicinae: id est, initia physicae inaudita. Progressus medicinae novus, in morborum ultionem ad vitam longam*, ed. Franciscus Mercurius van Helmont (Amsterdam, 1652), 666.

of these clouds were inspected, the microscope revealed 'extremely minute shining concreted little crumbs'. These gradually stuck to the bottom and sides of the glass, darkened in colour, and 24 hours later were as large as 'grains of mustard seed'.[50] On the basis of these observations Boerhaave concluded that the granulation or crystallisation of urine formed the stone, not any other element. Besides, the stone could potentially form in any healthy person, because the raw elements of the stone were present in regular urine. In Boerhaave's words, the physician 'understands that a true stone may be generated from the urine of the most healthy person, only by its standing quiet'.[51]

Boerhaave repeated the experiment several times to determine the effect of different temperatures on the results. He performed the same sequence of procedures, but varied the temperature to be 33 °F (just above freezing point) and a hot summer's heat (probably around 90 °F). In both cases the vessel was left uncovered in the open air, or merely covered with a sheet of paper to prevent dust to fall in. But the varieties in temperature did not result in a different outcome, except that the duration of the experiment necessarily varied: 'And in proportion, as the heat of the air increases, the more considerable and speedy is this alteration, so that in the hottest summer weather, these *phenomena* appear most remarkable'.[52]

Medical students built upon Boerhaave's work, not simply to confirm previous findings and summarising the results, but to improve their understanding of the pathology of bladder stones. Stalpart van der Wiel performed his own experiments on urine, as we can gather from his dissertation. He used Boerhaave's method and was able to make urine granulate much faster than his master. He compressed the chemical operation from 12 days into only 1.

Overall, eighteenth-century dissertations grew into works through which the medical doctor-to-be presented his skills as a physician and researcher. Whereas in the seventeenth century Franciscus Sylvius's collaborations with his students were largely presented as Sylvius's work, Albrecht von Haller granted his students the sole authorship for their

[50] Gerard van Swieten, ed. *Commentaria in Hermanni Boerhaave Aphorismos de cognoscendis et curandis morbis*, 5 vols (Leiden, 1742–1772); idem, *The Commentaries upon the Aphorisms of dr. Herman Boerhaave*, 18 vols (London, 1744–1773), vol. 16, 105–7.

[51] Herman Boerhaave, *Elementa chemiae, quae anniversario labore docuit in publicis, privatisque scholis*, 2 vols (Leiden, 1732), vol. 2, 323; idem, *Elements of Chemistry: Being the Annual Lectures*, trans. Timothy Dallowe, 2 vols (London, 1735), vol. 2, 224.

[52] Boerhaave, *Elementa chemiae*, vol. 2, 321; idem, *Elements of Chemistry*, vol. 2, 223.

experimental work.[53] Doctoral students including Stalpart van der Wiel and Anthonius Ludovicus Luiscius began to define the pathology of bladder stones in chemical terms. Luiscius stated that nobody doubted that the bladder stone was generated within the body. However, to learn more about its constitution, he pulverised a bladder stone and provided, by means of distillation, more exact measurements of the quantities of salt, earth, and other constituent parts within it.[54] What appeared to be a simple series of chemical processes in the laboratory thus became the central argument in the pathological understanding of lithiasis.[55] Other students performed similar studies and experiments in the cases of ischury (the retention of urine) and strangury (the constant urge to pass water paired with the inability to empty the bladder completely), and in these cases the chemical analysis revealed a certain acridity in the urine, causing the feeling of sharp and hot pain. And the large amounts of urine discharged by diabetes patients were found to be either limpid or aqueous, or a thick, turbid, whitish, and sweet. In the research field of urine chemistry, then, dissertations were no longer straightforward rites of passage but contributed to the project of a better understanding of urinary diseases.

[53] On Sylvius, see Evan R. Ragland, 'Experimenting with Chemical Bodies: Science, Medicine, and Philosophy in the Long History of Reinier de Graaf's Experiments on Digestion, from Harvey and Descartes to Claude Bernard' (Doctoral thesis, University of Alabama, 2012), 14. On Von Haller, see Ku-ming Chang, 'Collaborative Production and Experimental Labor: Two Models of Dissertation Authorship in the Eighteenth Century', *Studies in History and Philosophy of Biological and Biomedical Sciences*, 41 (2010), 347–55.

[54] Anthonius Ludovicus Luiscius, *Disputatio medica inauguralis anatomico chemico practica de calculo renum et vesicae* (Leiden, 1720), 18.

[55] Other Leiden dissertations on bladder stones include Daniel Dolegius de Kedts, *Disputatio medica inauguralis de calculo in genere, sed praesertim renum et vesicae* (Leiden, 1717); Matthaeus de Roode, *Disputatio medica inauguralis de lithiasi* (Leiden, 1720); Christianus Stephanus Scheffelius, *Dissertatio medico-practica inauguralis de lithiasi fellea sive calculo vesiculae biliariae, cujus occasione traditur simul brevis historia lapidis porcini Malacensis et fellis bovis* (Leiden, 1721); Joannes Petrus Kind, *Dissertatio medica physico-pathologica inauguralis de remediis calculum in renibus diffringentibus* (Leiden, 1724); Christopher Farwell, *Dissertatio medica inauguralis de calculo renum et vesciae urinariae* (Leiden, 1729); Humphrey, *De calculi urinosi*; Pieter Bourgeois, *Dissertatio pathalogico-medica inauguralis de calculo et remediis cum solventibus* (Leiden, 1744); Cornelius Canisius, *Dissertatio medica inauguralis de calculo in genere et preacipue renum et vesicae* (Leiden, 1757); Joannes Gulielmus Everwijn, *Specimen medicum inaugurale de lithogenesia in vesica urinaria* (Leiden, 1758); Johannes Christophorus à Bergen, *Dissertatio medica inauguralis de lithiasi* (Leiden, 1765); Leonardus Henricus de Thoméze, *Dissertatio medica inauguralis de calculo biliari* (Leiden, 1773).

Urine chemistry also contributed to research on the efficacy of therapies. One doctoral student working on bladder stones was the Englishman Samuel Horsman (*c.* 1698–1751) from Middlesex. Joining many bright young men from the British Isles seeking a good education he enrolled at Leiden University in 1719.[56] Horsman, whose dissertation at Leiden concerned kidney and bladder stones, would go on to have an impressive career as a physician and Fellow of the Royal College of Physicians.[57] In his dissertation, he sought to find effective treatments for the stone and considered the merits of various kinds of diuretics or drugs which would increase the volume and frequency of passing urine in patients suffering from small stones. Furthermore, he made efforts to visualise his laboratory work (see Fig. 4.3). His title page followed the fashion of the day, showing the black cap of the doctoral student at the top and depicting the allegorical figure of Athena in the middle. Additionally, Horsman's frontispiece included a view into the central place of urine chemistry: the chemical laboratory with chemical vessels, furnaces with alembics, and at a professor and student discussing the experiment they are observing.

The productive research programme of urine chemistry was not only applied to physiology and pathology. Some doctoral students invented new chemical procedures. One illustrative example dates from 1753 when the medical student Johannes Albertus Schlosser (1733–1769) from Utrecht defended his dissertation on 'the native salt of human urine'.[58] This essential salt was of particular interest because Boerhaave had proposed that this salt was produced in the healthy human body through the process of digestion (*i.e.*, at body temperature), and hypothesised that it contributed to the development of lithiasis: 'It should be considered what share this salt has in producing the stone of the bladder or kidneys'. Boerhaave had distilled fresh urine from a healthy man, stored it to let it thicken and congeal. After one year, a 'saline, solid, hard, brown, and somewhat transparent mass will concrete to the bottom' while a 'thick, black, unctuous liquor [will] float above it'. Having decanted the mass and

[56] E. Ashworth Underwood, *Boerhaave's Men at Leyden and After* (Edinburgh, 1977).

[57] Samuel Horsman, *Dissertatio medica inauguralis de calculo renum et vesicae* (Leiden, 1721). Boerhaave acted as the doctoral advisor for this dissertation.

[58] Johannes Albertus Schlosser, *Specimen chemico medicum inaugurale de sale urinae humanae nativo* (Leiden, 1753). For more biographical details and his work in natural history, see Mario Engelmann and Bert Sliggers, 'Johannes Albertus Schlosser, the First Author Describing *Artemia Salina* (L.) (Branchiopoda: Anostraca): A Biographical Sketch', *Journal of Crustacean Biology*, 35 (2015), 571–5.

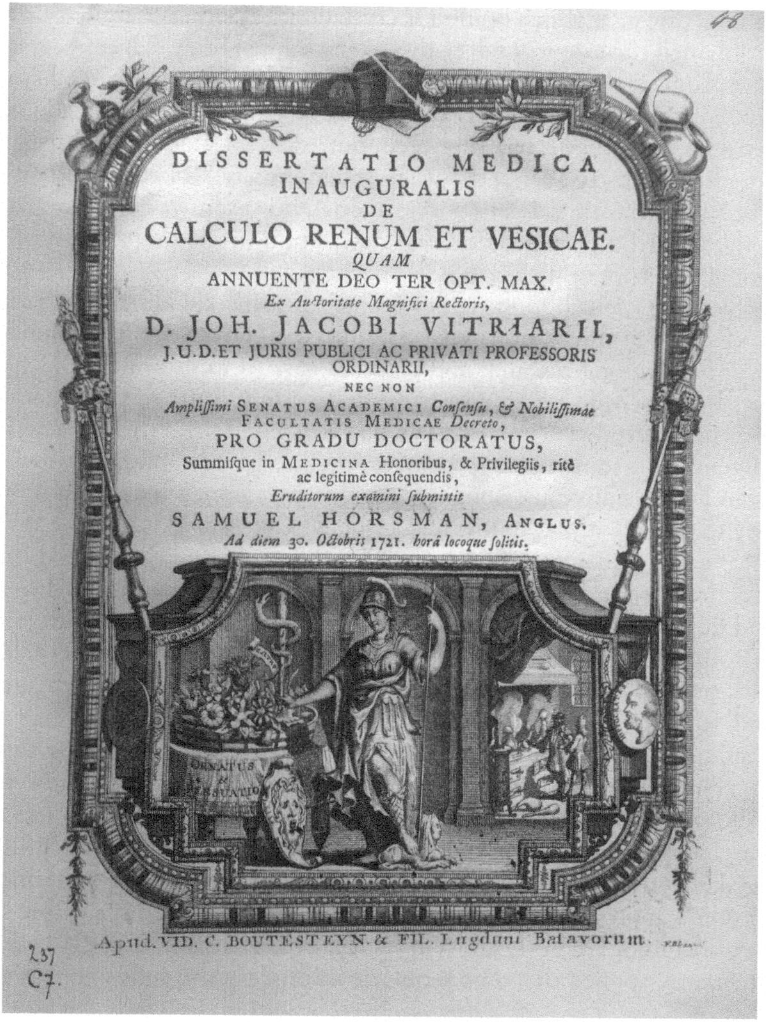

Fig. 4.3 Frontispiece to Samuel Horsman, *Dissertatio medica inauguralis de calculo renum et vesicae* (Leiden, 1721). Leiden University Library, 237 C 7:48

let it cool down, at which point it is crystallised, Boerhaave argued that he had obtained the essential salt native to the human body.[59] But he was not the first to have discovered the native salt of urine. In fact, when Schlosser delved into the subject, he recalled that the essential salt had already been called 'sal urinarium' by Van Helmont, 'sal fausible' by Christianus Marggraf (1626–1687), and 'sal crystallinum' by Robert Boyle. What marked Boerhaave's 'sal nativum' as special, however, was that Boerhaave was the first who provided an exact description of the preparation of the essential salt from urine.[60]

Schlosser's contribution to the project of urine chemistry was largely one of research innovation. He reproduced Boerhaave's experiments as part of his doctoral research and placed a flask of fresh urine in his basement. But when, just 1 month later, he transferred the urine to another vessel, he noticed that salt crystals had formed. Furthermore, when he negligently left some urine to distil on the fire for too long, he observed crystals forming after just one night. Schlosser reasoned that it was unnecessary to wait for a year to produce essential salt. And indeed, he was able to heat a large quantity of fresh urine from healthy individuals and cause the entire surface to be covered with scum or dirt. After clarifying the liquid he placed it into a clean vessel and let it rest. After 24 h, at the sides and bottom of the vessel, hard and slightly transparent crystals formed in a dark red and thick liquor. He repeated this process with the remaining liquid to form more salt crystals.[61] Schlosser's method, then, had the advantage of being quick as well as providing the possibility for repeating the crystallisation process multiple times. Schlosser's dissertation was, therefore, clearly an example of ongoing research and innovation. In researching the nature of urine, Schlosser displayed a critical attitude towards his teachers and he innovated chemical processes. Furthermore, Schlosser's approach proved particularly productive when his fellow students, including Marc-Louis Vullyamoz (1727–1810) from Lausanne, Switzerland, applied the same sequence of processes to milk and was able

[59] Boerhaave, *Elementa chemiae*, vol. 2, 317–8; idem, *A New Method of Chemistry: Including the History, Theory, and Practice of the Art*, trans. Peter Shaw, 2nd ed., 2 vols (London, 1741), vol. 2, 198–9.

[60] Schlosser, *De sale urinae*, 5–6. Boerhaave's 'native salt of urine' was later called urea. For H.J. Backer's attempt to reproduce Boerhaave's process of making the native salt of urine, see 'Boerhaave's ontdekking van het ureum', *Nederlands Tijdschrift voor Geneeskunde*, 87 (1943), 1274–8.

[61] Schlosser, *De sale urinae*, 8–9.

to produce native salt from it.[62] Likewise, the student Petrus Nicolaus Lotichius (*c.* 1736–1792) from Nijmegen continued Schlosser's work on the elements of urine and managed to isolate phosphor.[63]

In short, urine chemistry in the early eighteenth century had many advantages for the advancement of medical research, particularly in the case of urinary disorders such as the stone. The most significant differences between chymical uroscopy and urine chemistry were their respective aims and applications. Uroscopists had applied distillation to diagnosis, for example, to identify bladder stones in a patient. By contrast, students of the Boerhaave school used chemistry to understand *how* bladder stones came about in the first place. Not the diagnosis but the pathology was the issue at hand. Furthermore, the method of urine chemistry proved to be a productive research programme for many Dutch physicians. They performed retrials of Boerhaave's experiments to gain insights into the nature of urine, developed new laboratory methods to perform certain chemical processes, and tested pharmaceuticals and their use for the treatment of patients suffering from calculi. Eighteenth-century physicians affirmed the basic premise that the chemical method could reveal the constituents of bodily fluids and provide clues for the internal processes and substances within the living body.

Urinalysis

Many historians have described the chemical revolution at the end of the eighteenth century as the confluence of pneumatic chemistry and quantitative analytical techniques, especially in the works of the English theologian and natural philosopher Joseph Priestley (1733–1804), the French chemist Antoine Lavoisier, and others.[64] In opposition to the rather

[62] Marc-Louis Vullyamoz, *Dissertatio chemico-medica inauguralis de sale lactis essentiali* (Leiden, 1756). See also Georges Terrier, 'Médecine et chimie dans le contexte des Lumières: la Thèse de doctorat du médecin lausannois Marc-Louis Vullyamoz', *Gesnerus*, 60 (2003), 149–69.

[63] Petrus Nicolaus Lotichius, *Dissertationem physico-medicam inauguralem de phosphoris et phosphoro urinae* (Leiden, 1757).

[64] Notable books that stressed the enlightened chemistry of Lavoisier include Douglas McKie, *Antoine Lavoisier: The Father of Modern Chemistry* (London, 1936); Henry M. Leicester, *The Historical Background of Chemistry* (New York, 1956); J.R. Partington, *A History of Chemistry*, 3 vols (London, 1961–1970). For a more complete historiographical overview, see John G. McEvoy, *The Historiography of the Chemical Revolution: Patters of Interpretation in the History of Science* (London, 2010). More recent literature on the

abstract and internalistic approach to the development of modern chemistry, Jonathan Simon has recently shown how the development of Lavoisian, 'philosophical chemistry' cannot be properly understood without taking into account the central role of pharmacy and the community of French pharmacists at the time.[65] Others have similarly argued that many nineteenth-century changes in the discipline were rooted in eighteenth-century natural philosophy as much as in practices and trades of everyday life, such as mining and agriculture.[66] Along similar lines, I here argue that the revolutionary investigation of urine in the late eighteenth century had already started with the Boerhaave school. Scholarship has often presented the Swedish chemist and apothecary Carl Wilhelm Scheele as the man who discovered uric acid, based on the publication of his chemical analysis of urine in 1776. Scholars have shown that, in the 1790s, the French chemists Antoine de Fourcroy and Nicolas Vauquelin first isolated urea and demonstrated its decomposition with the formation of ammonia. But this section will demonstrate that their achievements ought not to be seen as a dramatic revolution, but rather as a logical continuation of and an improvement upon eighteenth-century urine chemistry. The techniques and premises on which their works were based, I argue, had been a shared methodology among the physicians and chemists of the Boerhaave school. A comparison of the experiments of the Boerhaave school with those of the Swedish and French chemists makes clear that the experiments, as well as the conclusions drawn from them, were remarkably similar, proving the point that the Swedish and French chemists were, perhaps, not quite as revolutionary as has often been claimed: their work had improved in accuracy and analytical skills, but not in aim nor application.

First it is necessary to note that urine chemistry in the tradition of Boerhaave continued until the late eighteenth century. In the 1760s the

Chemical Revolution include the works of Hasok Chang, Jan Golinski, Bernadette Bensaude-Vincent, Mi Gyung Kim, and Ursula Klein.

[65] Jonathan Simon, *Chemistry, Pharmacy and Revolution in France, 1777–1809* (Aldershot, 2005). See also Frederic L. Holmes, *Eighteenth-Century Chemistry as an Investigative Enterprise* (Berkeley, 1989). Focusing on France, Holmes similarly stressed the important role of early eighteenth-century chemists, thereby 'de-centring' the Chemical Revolution from Lavoisier and Priestley.

[66] For example, Hjalmar Fors, *The Limits of Matter: Chemistry, Mining, and Enlightenment* (Chicago, 2015); Joppe van Driel, 'The Filthy and the Fat: Oeconomy, Chemistry and Resource Management in the Age of Revolutions, 1700–1850' (Doctoral Thesis, University of Twente, 2016).

physician and medical author Gerard van Swieten finally finished writing the fifth and final volume of his *Commentaries* (1742–1772) on Boerhaave's *Aphorisms* (first edition 1709) on the topic of the stone.[67] The first volume of his *Commentaries*, which had been published in 1742, had awarded Van Swieten the reputation of a true Boerhaave disciple, and he was offered a position at the imperial court of Maria Theresia (1717–1780) in Vienna. There he achieved a complete reform of the medical curriculum and gathered physicians around him who had studied with Boerhaave or his successor, Gaubius, including Antonius de Haen, Nikolaus Joseph von Jacquin (1727–1817), and Jan Ingenhousz (aka Ingen-Housz, 1730–1799). Van Swieten's *Commentaries* had started as notes on Boerhaave's lectures from 1718 onwards, and even after Van Swieten was promoted to a medical doctor in 1725 he continued to attend and record Boerhaave's lectures, including his 1729 lectures on the stone.[68] Many of the remarks and anecdotes the professor shared with his students found a way into Van Swieten's *Commentaries*. But as the years after Boerhaave's death passed Van Swieten felt compelled to include new research results and his own observations and experiments in his writing, to keep the text up to date. This also applied to Boerhaave's urine chemistry. In his section on the stone, Van Swieten described the chemical experiments Boerhaave had performed to gain insights into the formation of bladder stones, and mentioned that he himself often repeated the experiments for his personal satisfaction of the facts: 'I have frequently examined my own urine, and with pleasure observed these first rudiments of the stone very slowly separated, and sometimes twenty-four hours and above, before they could coalesce into

[67] Gerard van Swieten (ed.), *Commentaria in Hermanni Boerhaave Aphorismos de cognoscendis et curandis morbis* (5 vols, Leiden, 1742–1772). The fifth volume also discussed venereal disease, rickets, and rheumatism. For biographical details on Gerard van Swieten, see Frank T. Brechka, *Gerard van Swieten and His World 1700–1772* (The Hague, 1970); J.K. van der Korst, *Een dokter van formaat: Gerard van Swieten, lijfarts van keizerin Maria Theresia* (Amsterdam, 2003); Luuc Kooijmans, *De geest van Boerhaave: Onderzoek in een kil klimaat* (Amsterdam, 2014).

[68] Gerard van Swieten, 'Praelectiones de lue venerea', Leiden, 1729. Österreichische Nationalbibliothek (ONB), MS 11189. These notes were on Boerhaave's lectures on the stone from 16 June to 8 July 1729, and on venereal disease from 14 March to 30 June 1730. Van Swieten's many other lecture notes can still be consulted at the Austrian National Library too, such as on solvents and blood. Gerard van Swieten, 'Praelectio publica de menstruis', Leiden, 1725. ONB, MS 11170; idem, 'Praelectiones nosocomicae et de sanguine', Leiden, 1737–1738. ONB, MS 11187.

gravel of a larger size'.[69] Indeed, from laboratory logs, we know that Van Swieten at the very least spent the entire summer of 1758 experimenting and observing urine.[70] His elaborate observations on bladder stones finally appeared as the fifth and last instalment of his *Commentaries* in 1772. Van Swieten's volume on bladder stones was soon translated into English (1773, 1776) and German (1775), proving that urine chemistry was still of interest.

Given that the publication dates of Van Swieten's and Scheele's works were so close to each other, it is not surprising that they were similar in certain important respects, especially the assumptions that, first, the chemistry was an academic discipline and essential to medicine; and second, that chemistry could isolate substances in urine and bladder stones. Van Swieten had noticed that some patients showed signs of gravel in their urine but no large stones. How exactly did bladder stones grow? During a chemical experiment, Van Swieten added a quill to the urine, causing the same to get a crust. He added more urine, and yet another layer formed on top of it, from which Van Swieten concluded that any solid body present in the bladder could function as the nucleus of a future stone. Van Swieten named a kidney stone or nephritic gravel as the most probable 'occasional cause'.[71] In addition to examining urine, Van Swieten analysed actual bladder stones to reveal their elementary principles. Distillation in the chemical laboratory revealed that the stones mainly consisted of pure water and a dry, fixed earth, salts, and an oleaginous glue. Kidney stones, by contrast, contained no fixed alkaline salt, nor any calcareous earth, but rather a volatile salt.[72] Chemistry, in other words, enabled Van Swieten to analyse urine, bladder stones, and renal stones down to their elementary particles.

In a similar vein, Carl Scheele gathered bladder stones at the request of his friend Peter Jonas Bergius (1730–1790), a Swedish medical doctor and botanist who had studied under Carolus Linnaeus (1707–1778) at Uppsala. Linnaeus had proficient knowledge of Boerhaave's works, because he had studied medicine with Boerhaave's disciple Johannes de

[69] Van Swieten, *Commentaria*, vol. 5; idem, *Commentaries*, vol. 16, 109–10.

[70] Gerard van Swieten, 'Adversaria experimentorum chemicorum', Vienna, 1732–1758. ONB, MS 11193, ff. 109–111. This manuscript logged Van Swieten's chemical experiments and observations with metal, mineral, vegetal and animal materials, such as antimony, mercury, lead, hepatic calculi, opium, and urine.

[71] Van Swieten, *Commentaria*, vol. 5; idem, *Commentaries*, vol. 16, 112–3, 21.

[72] Van Swieten, *Commentaria*, vol. 5; idem, *Commentaries*, vol. 16, 160–1, 4, 71, 78, 83.

Gorter at the University of Harderwijk, and had become acquainted and corresponded with Boerhaave and many of his students, for example, Johann Bartsch (1712–1738).[73] Scheele, however, had not studied medicine, but mainly worked as an apothecary. He conducted much chemical research and befriended chemists at Uppsala University, including Torbern Bergman (1735–1784), who made use of Scheele's skills in his chemical experiments. To build on the study of the chemical composition of urine Scheele reduced bladder stones to a powder and then combined them with various acids and alkalis to observe the effect. Making a solution of the powder with vitriolic acid and heating the mixture at body temperature—a process he still called digestion—resulted in black coal. Muriatic acid had no effect. What did show some effects, however, was 'aquafortis' ('strong water', a solution of nitric acid obtained by distilling saltpetre). Notably, with the term 'nitrious acid', Scheele began to use the new chemical nomenclature but also continued to use the older terminology. The acid caused an effervescence, dissolved the stone, and gave off 'red vapours' and a 'yellow solution'. Finally, Scheele dissolved bladder stones in 'lime-water', ultimately isolating an acid which later came to be known as uric acid.[74] Bergman elaborated on Scheele's findings and noted some deviations and improvements based on more extensive investigations.[75]

The French chemists Hilaire Marie Rouelle (1718–1779), De Fourcroy, and Vauquelin can further be considered the direct descendants of early eighteenth-century urine chemistry.[76] In 1773 Hilaire Rouelle, the younger brother of Guillaume François Rouelle (1703–1770), published his research comparing human urine with that of cows and horses. According to Rouelle, urine contained a 'saponaceous extract' (*matière*

[73] Lisbet Koerner, *Linnaeus: Nature and Nation* (Cambridge, MA, 1999). Linnaeus was close friends with Boerhaave's doctoral student Bartsch. Linnaeus congratulated Bartsch on his promotion in 1737. See Linnaeus to Bartsch, 10 July 1737, in Johannes Bartsch, *Dissertatio medica inauguralis sistens calorem corporis humani effectum hygraulicum* (Leiden, 1737), [E2r–v]. Linnaeus and Bartsch corresponded until Bartsch untimely death in 1738.

[74] Carl Wilhelm Scheele, 'Undersökning om Blåse-stenen', *Kongliga Vetenskaps Academiens Handlingar*, 37 (1776), 327–32; 'Analysis of the Calculus Vesicae', in *The Chemical Essays* (London, 1786), 145–56. Anders Lennartson, *The Chemical Works of Carl Wilhelm Scheele* (Dordrecht, 2017), 33–4.

[75] Tobern Bergman, 'Tilläggning Om Blåse-Stenen', *Kongliga Vetenskaps Academiens Handlingar*, 37 (1776), 333–8.

[76] On Antoine François de Fourcroy and his central role in the discipline formation of pharmacy in France, see Simon, *Chemistry, Pharmacy and Revolution*. See also W.A. Sweaton, *Fourcroy, Chemist and Revolutionary, 1755–1809* (Cambridge, 1962).

savonneuse). He described this substance as being dark brown in colour and ointment-like, exuding a particularly unpleasant smell, and dissolvable in alcohol.[77] Modern chemists agree that this substance can be identified as urea, albeit in a rather impure form. However, the chemical procedure followed by Rouelle agreed significantly with Boerhaave's experiment on the 'native salt of urine', which I have described in the previous section. Boerhaave and his student Schlosser had already prepared one form of urea in the early eighteenth century. Modern chemists think that Boerhaave's urea was, in fact, purer than the one described by Rouelle, and have awarded Boerhaave the primacy in the discovery of urea.[78]

The point I want to make here is that Rouelle was part of a long tradition of urine chemistry, which was first set out by Boerhaave and his students. Common to all was the notion that elementary substances may be identified by chemistry, whether for the identification of native salt in the case of bladder stones, of acridity in the case of ischury and strangury or of limpidity and sweetness in the case of diabetes. Rouelle's most significant contribution may have been his comparative study of humans and animals. The Boerhaavian research programme also continued in the works of the French chemists De Fourcroy and Vauquelin at the end of the century. These gathered an even larger sample of urine and bladder stones for further analyses of urine. They first coined the term urea (*urée*) and described it as that 'urinary matter' which gave urine its properties colour, smell, and taste. De Fourcroy and Vauquelin had heated a young man's urine and had it reduced to a thick syrup by means of condensation which crystallised when cooled—which essentially was the same process as Boerhaave's and Schlosser's isolation of the 'native salt of urine'. Once again, we may conclude that urine chemistry was a productive research programme that was active throughout the entire eighteenth century. Thanks to the works of many of his students and disciples, Boerhaave's new method of chemistry had a lasting impact. It is no surprise, then, that

[77] Hiliare Marie Rouelle, 'Observations sur l'urine humaine, & sur celles de vache & de cheval, comparés ensemble', *Journal de médecine, chirurgie, pharmacie, &c.*, 40 (1773), 451–68.

[78] Backer, 'Ureum'; Frederick Kurzer and Phyllis M. Sanderson, 'Urea in the History of Organic Chemistry: Isolation from Natural Sources', *Journal of Chemical Education*, 33 (1956), 452.

De Fourcroy and Vauquelin paid tribute to him as the 'illustrious professor of Leiden'.[79]

Larger samples of urine and bladder stones would ultimately enable nineteenth-century chemists to achieve a greater accuracy and more refined measurements—compared with Boerhaave, Schlosser, and Van Swieten—in isolating urine's constituent parts. But the perception that chemistry was the best discipline for isolating individual substances in the urine, bladder stones, and other animal matter remained the same. The works of Scheele and Bergman, Rouelle, De Fourcroy, and Vauquelin eventually led to the development of modern urology and nephrology by Richard Bright (1789–1858). Thanks to the improved equipment and the large number of patients treated in Guy's Hospital, Bright was able to isolate more substances in urine than ever before.[80] But the basic premise on which these studies rested had been introduced much earlier. Eighteenth-century urine chemistry and nineteenth-century urinalysis, then, may be said to have been different in size, but not in kind. New constituent parts continued to be discovered as time progressed, yet the shift in chemical analyses of urine around 1800 was less drastic than previously presented.

* * *

Historians of medicine and chemistry have commonly argued that only in the nineteenth-century physicians started to rely on chemical analysis. But as this chapter has shown, physicians in the early eighteenth century already agreed on the basic premise that chemistry provided insights essential to physiology and pathology. Whereas chymical uroscopists took a urine sample to test it chemically and to diagnose a patient's disorder,

[79] Antoine François de Fourcroy and Louis Nicolas Vauquelin, 'Premier mémoire pour servir à l'histoire naturelle, chimique et médicale, de l'urine humaine', *Annales de Chimie*, 31 (1798), 48–71; idem, 'Second mémoire pour servir à l'histoire naturelle, chimique et naturelle de l'urine humaine dans lequel on s'occupe spécialement des propriété de la matière particulière qui la caractérise', *Annales de Chimie*, 32 (1799), 80–162.

[80] Richard Bright, 'Tabular View of the Morbid Appearances in 100 Cases Connected with Albuminous Urine', *Guy's Hospital Reports*, 1 (1836), 380–402; G.H. Barlow and George Owen Rees, 'An Account of Observations Made under the Superintendance of Dr Bright on Patients Whose Urine Was Albuminous', *Guy's Hospital Reports 2nd Series*, 1 (1843), 189–316. See Steven J. Peitzman, 'Bright's Disease and Bright's Generation: Toward Exact Medicine at Guy's Hospital', *Bulletin of the History of Medicine*, 55 (1981), 307–21.

eighteenth-century physicians studied urine by chemical means aimed at gaining a more specialised and general pathological understanding urinary disorders. Especially the works of Boerhaave and his students reveal this shift from simple inspection to physiological and pathological research; from urine as a reflection of the humours towards urine as a means to understand disease resulting from pathological changes in the body.

CHAPTER 5

Crying Over Spilt Milk

Myn Heer, my dunkt, gy laght, en denkt, wat beuzelingen,
Wat valt 'er over de MELK *te zeggen of te zingen,*
Dat algemeene vogt? wat kan er meer bij elk
In aart en eigenschap bekend zyn, dan de MELK*?*

Sir, it seems to me that you laugh and think, what trifles,
What is there to say or to sing about *milk*?
That common fluid? What could be to the public
More known in nature and property, than *milk*?[1]

These rhyming rhetorical questions open a poem on the merits of milk, published in 1768, by the Dutch physician Abraham Tersier (*c.* 1713–1805) from the city of Dordrecht. What indeed could possibly be interesting and important about that mundane and ordinary fluid, milk? But in this poem Tersier went beyond merely paying tribute to the white fluid flowing from breasts and udders. He endeavoured to moralise those who were, according to him, blinded by prejudice, neither consuming milk themselves nor feeding it to their babies. In 1739, Tersier had written his doctoral thesis

[1] Abraham Tersier, *Brief aan den wel edelen heere N.N. behelzende een vertoog over de melk* (Dordrecht, 1768), 1. Tersier wrote many poems in his life, some of which have been preserved in the appendices to the dissertations of fellow students, such as Nicolaus Stumphius and Petrus Imchoor.

R. E. Verwaal, *Bodily Fluids, Chemistry and Medicine in the Eighteenth-Century Boerhaave School*, Palgrave Studies in Medieval and Early Modern Medicine,
https://doi.org/10.1007/978-3-030-51541-6_5

in Latin under Gaubius's supervision.[2] Yet for his 'moral discourse' on milk, he chose to write in the Dutch vernacular, to reach those ordinary citizens who did not acknowledge milk as the natural nourishment for infants. This was a smart move, because poems were a very popular art form in eighteenth-century society. Popular health advice often appeared in rhyme. Friends and families wrote numerous poems for special occasions, such as graduations, weddings, births, and burials. These cordial songs and verses were expressions of praise and affection, enforcing camaraderie and promoting social cohesion.[3] In this poem Tersier argued that milk was not only 'given to us by God's gentle grace', but that

Dus Reden, en Natuur, en Scheikonst klaar bewyzen,
Dat wy de melk, met regt, als 't beste voedzel prysen,
Als de eerste en eêlste stof tot lighaams onderhoud,
Naar redens licht en kragt in haaren aart beschouwd.

Reason, and Nature, and *Chemistry* clearly prove,
That we rightly commend milk as the best food,
In her essence considered by reason's light and might,
As the first and foremost foodstuff for body's health.[4]

Tersier's curious piece on milk raises many questions: why was Tersier, a physician, not simply treating the medical conditions of his patients but also trying to inform healthy people's eating and breastfeeding habits? What role did chemistry play in his persuading the reader of the nutritious qualities of milk?

Breast milk commonly represented and constituted the early mother-infant relationship, both materially and spiritually. From the thirteenth century onwards, the *Madonna Lactans* was the most prominent image illustrating the sacredness of the nursing mother providing her helpless infant with the necessary and virtuous white liquor to sustain his life (see

[2] Abraham Tersier, *Dissertatio physico-medica inauguralis de vita et morte, quam annuente deo ter optimo maximo* (Leiden, 1739). Tersier built a career as physician in Dordrecht, where he married Quirina Bosbaan. Their son Bartholomeus (c. 1744–1824) followed in his father's footsteps, studied medicine in Leiden, and promoted on the topic of misuse of remedies: Bartholomeus Tersier, *Dissertatio inauguralis medica de intempestivo remediorum usu* (Leiden, 1769).

[3] J.A. Gruys and Adèle Nieuweboer, 'Dutch Occasional Poetry, 16th through 18th Centuries: A Genre Rediscovered', (Leiden, 2004).

[4] Tersier, *Melk*, 2–3, 6. Emphasis added.

Fig. 5.1).[5] Yet the practice of breastfeeding was never self-evident or static. Apart from differences between individual mothers in their ability to give suck or not, medical and cultural norms about mother's milk were constantly reformulated. While early modern medical practitioners were generally in favour of feeding mother's milk, many young mothers did not produce enough milk, suffered from sore nipples, or wanted to be liberated from this time-consuming activity. Popular practice adopted the assistance of wet nurses and the hand-feeding of animal milk or various kinds of porridge and pap. Ideas about the benefits and drawbacks of breastfeeding to the new-borns' and their mothers' health were, therefore, also constantly changing.

As this chapter demonstrates, Dutch physicians consciously attempted to inform breastfeeding practices. Their chemical analyses of milk supplied new insights into the physiology of nutrition, showing the salubrious and nourishing qualities of milk. This provided Dutch physicians with arguments for emphasising the exceptional nutritious qualities of milk, in general, and the importance of mother's milk to the health of the infant, in particular. They promoted maternal breastfeeding by developing innovative breast pumps and new galactagogues, that is, pharmaceutical preparations supposed to promote and increase lactation. This chapter therefore reveals that physicians, surgeons, and apothecaries found themselves increasingly on an equal footing in discussions of societal issues.[6] And indeed, breastfeeding was of much concern. On the basis of the chemistry of milk, this chapter demonstrates that Dutch medical men constructed innovative instruments and employed various other strategies to emphasise the importance of milk and breastfeeding.

The practice of breastfeeding has been given ample attention in social historiography, yet some important questions remain unanswered, predominantly why maternal breastfeeding fell in and out of favour in early modern society. Studies have shown how European cultures challenged the practice of maternal breastfeeding. Instead, the practice of wet-nursing

[5] On the ubiquity of lactation imagery and literary references to breastfeeding in ancient early modern visual culture see Jutta Gisela Sperling, ed. *Medieval and Renaissance Lactations: Images, Rhetorics, Practices* (Farnham, 2013); Sabrina Higgins, 'Divine Mothers: The Influence of Isis on the Virgin Mary in Egyptian Lactans-Iconography', *Journal of the Canadian Society for Coptic Studies*, 3/4 (2012), 71–90.

[6] Harold J. Cook, *Trials of an Ordinary Doctor: Joannes Groenevelt in Seventeenth-Century London* (Baltimore, 1994). Cook demonstrates that Joannes Groenevelt was both a physician and lithotomist, prescribing both internal medicines and treating surgical wounds.

Fig. 5.1 Madonna Lactans, 'and He, by whose providence not even a bird suffers hunger, is fed with a little milk'. Engraving by Adriaen Lommelin, after Peter-Paul Rubens (Antwerp, c. 1650). (Credit: Rijksmuseum, Amsterdam. CC0 1.0)

proliferated, not just among the nobility, but among bourgeois families as well.[7] After all, feeding could be easily handed over to wet nurses, for whom it was a source of income. This also liberated wealthy women from breastfeeding, who were then able to fulfil social obligations and the sexual needs of their husbands—it was generally thought that sexual activity would halt the production of milk. Further, historians have demonstrated that breast milk, either from the mother or a wet nurse, was increasingly replaced by extracts of meat and other chemically produced substances for the artificial feeding of infants by hand. This was especially the case at the turn of the nineteenth century.[8] For the seventeenth to the nineteenth centuries, then, scholars have identified a long-term antipathy against mother's milk. How, then, should we understand eighteenth-century promotions of maternal breastfeeding? The historian Valerie Fildes has, indeed, marked the eighteenth century as a curious exception, and noted that the practice of wet-nursing was seriously questioned.[9] She cites English male midwives, among them Henry Bracken (1697–1764), a Lancaster surgeon, who warned that the nurse's diseases and mental peculiarities might be sucked into the infant via the wet nurse's milk; and William Cadogan (1711–1797), a Bristol physician whose *Essay upon Nursing* advised mothers to suckle their own child. This unprecedented campaign against wet-nursing also spread in France, albeit apparently with only a limited effect.[10] Be that as it may, the question of *why* maternal breastfeeding was so much appreciated in the eighteenth century remains unanswered.

[7] George David Sussman, *Selling Mother's Milk: The Wet Nursing Business in France, 1715–1914* (Urbana, 1982); Valerie A. Fildes, *Breasts, Bottles and Babies: A History of Infant Feeding* (Edinburgh, 1986); ead. *Wet Nursing: A History from Antiquity to the Present* (Oxford, 1988); Deborah Valenze, *Milk: A Local and Global History* (New Haven, 2011), 153–62; Els Kloek, Giesela van Oostveen, and Nicole Teeuwen, eds., *Moederschap en de min* (Utrecht, 1991).

[8] Jacqueline H. Wolf, *Don't Kill Your Baby: Public Health and the Decline of Breastfeeding in the Nineteenth and Twentieth Centuries* (Columbus, 2001); Valenze, *Milk*, 162–77; Barbara Orland, 'Motherhood and Scientific Innovation: The Story of Natural Versus Artificial Baby Food in the 19th Century', in *Gender in Science and Technology*, ed. W. Ernst and I. Horwath (Bielefeld, 2014), 129–46.

[9] Fildes, *Wet Nursing*, 111–4; William Cadogan, *An Essay upon Nursing, and the Management of Children, from their Birth to Three Years of Age* (London, 1750).

[10] Valérie Lastinger, 'Re-Defining Motherhood: Breast-Feeding and the French Enlightenment', *Women's Studies*, 25 (1996), 603–17.

Also historians of medicine and chemistry have discussed the historical antipathy against mother's milk. Looking at the works of French chemists and pharmacists at the turn of the nineteenth century, Barbara Orland argues that the societal debate on this topic motivated chemists to compare the milk of women from those of animals. In 1785 and 1787 the Royal Medical Society of Edinburgh and the Royal Society of Medicine in Paris commissioned chemical analyses of milk, which were expected to support the so-called maternalisation of breast-feeding and strengthen the polemic against wet-nursing.[11] Orland concludes, however, that chemistry failed to deliver these results. Among Orland's insightful contributions to the development of animal chemistry in the late eighteenth century, the question of milk's rise in stature in the eighteenth century remains unresolved.

This chapter contributes to the current historiography by focusing on eighteenth-century debates on the importance of milk. I argue that in the early eighteenth century, chemistry proved and emphasised the merits of milk, which explains the widespread campaign for maternal breastfeeding. In the first section I demonstrate that experiments on milk in the chemical laboratory contributed to physicians' emphasis on the nourishing qualities of the white substance, and in particular on the importance of milk for the health and growth of infants. It was no coincidence that the abovementioned fervent supporters of breast milk, Bracken and Cadogan, were both Leiden alumni and had both studied under Boerhaave.[12] Medical professors performed numerous chemical demonstrations with the use of blood, urine, and, of course, milk. Boerhaave considered milk to be special; he prescribed milk as a treatment for various disorders, and he taught that breast milk contained all the nutritious parts vital for the development of an infant's body. Boerhaave even advised many of his patients suffering from gout to walk the milky way (*via lactea*).[13] With Boerhaave's death in

[11] Barbara Orland, 'Enlightenend Milk: Reshaping a Bodily Substance into a Chemical Object', in *Materials and Expertise*, ed. Ursula Klein and E.C. Spary (Chicago, 2010), 163–97.

[12] Bracken, after studying medicine in London and Paris, matriculated at Leiden in 1730 and graduated MD. Cadogan matriculated at Leiden in 1732 and again in 1737, graduating MD in June. E. Ashworth Underwood, *Boerhaave's Men at Leyden and After* (Edinburgh, 1977), 27, 130–1, 87.

[13] Rina Knoeff, *Herman Boerhaave (1668–1738): Calvinist Chemist and Physician* (Amsterdam, 2002), 203. On early modern medicinal uses of breast milk see Marylynn Salmon, 'The Cultural Significance of Breastfeeding and Infant Care in Early Modern England and America', *Journal of Social History*, 28 (1994), 247–69.

1738, academic attention to milk far from subsided. In fact, from the 1730s onwards, Dutch physicians profusely published on the subject of milk and related issues, such as milk's essential salt and milk fever.[14]

In the second section, I demonstrate how the importance of milk and breastfeeding was communicated in eighteenth-century society. The diverse strategies employed by Dutch medical men in order to increase the practice of breastfeeding by mothers. Prize essay competitions by the Holland Society of Sciences in the 1760s, for example, solicited knowledge and expertise for practical applications. Although the publication of the winning essays enabled dissemination mainly among the educated—physicians, scholars, and philosophers of all disciplines, but also the members of the Holland Society of Sciences—this readership was, notably, simultaneously the group in society most likely to hire wet nurses.

The final section of this chapter continues my exploration of the dissemination of the maternal breastfeeding campaign through popular how-to books and poems. Such treatises provided detailed instructions in the Dutch vernacular, on subjects such as child-rearing and the curing of children's diseases. But these how-to books taken together also showed great discrepancies: while male physicians passionately argued in favour of maternal breastfeeding, female authors, who had the general happiness and well-being of both mothers and their children at heart, stated that breastfeeding was encouraged, but not essential. In other words, this gendered discussion took place between men blaming women for neglecting their so-called maternal duty, and women resisting the notion that good motherhood could be reduced to maternal breastfeeding.

The Merits of Milk

A central place where generations of physicians were trained to study and research milk was the university. At the medical faculty in Leiden at the turn of the eighteenth century, chemical knowledge was incorporated into the curriculum, and students were encouraged to perform their own

[14] Consider the works of, for example, Daniel Sluim, *Dissertatio medica inauguralis de lacte* (Leiden, 1716); Henricus Doorschodt, *Dissertatio medica inauguralis de lacte* (Leiden, 1737); Joannes Stook, *Dissertatio practico-medica inauguralis de febri lactea* (Leiden, 1746); Herman Boerhaave de Gorter, *Disputatio medica inauguralis de lacte et lactatione* (Harderwijk, 1751); Marc-Louis Vullyamoz, *Dissertatio chemico-medica inauguralis de sale lactis essentiali* (Leiden, 1756); Jan Egeling, *Dissertatio chemico-medica inauguralis de lacte* (Utrecht, 1759).

experiments in the laboratory.[15] So when the 20-year-old Abraham Tersier matriculated as a medical student in July 1733, he attended Hieronymus Gaubius's chemical demonstrations on milk in the chemical laboratory, which was located on the north-west side of the botanical garden (see Fig. 5.2). In the main lecture hall, adjacent to the garden and in front of the Rapenburg canal, Tersier listened to Boerhaave's medical teachings. In these classes, both professors declared milk special by arguing that the white fluid had certain nutritious qualities. Chemistry, this section argues, was used to prove and emphasise the merits of milk. And in the context of the university—an exclusively male institution[16]—teachers were able to convince their pupils that maternal breastfeeding was proper conduct surrounding infants.

As far as Boerhaave was concerned, milk was one of the most important fluids of the body, for it played a central role in growth and nutrition. Boerhaave was a fervent chemist, experimented on milk in various ways, and taught his students how they could conduct chemical experiments of their own. Furthermore, Boerhaave's lectures taught Tersier and other medical students how chemical inquiry into milk could yield insights into the physiology of the body, particularly in cases of maturation and nutrition. In one experiment, for example, the professor added an acid, such as the spirit of nitre, to a quantity of fresh cow's milk. He then started heating the mixture in his little furnace, and although no ebullition appeared, the mixture divided into separate thin and thick parts. This process illustrated the formation of curd from milk, from which then cheese could be made. Boerhaave used this demonstration as a basis for his explanation of physiological functions. He taught his students that the extraction of moisture from the coagulum by means of pressure and evaporation would turn it as hard as stone: 'when applied to the fire it grows tough, scorches, fries, and smells perfectly like horn. This is a strange change of so fluid a

[15] Rina Knoeff has argued that Boerhaave stimulated his students to think, observe and experiment themselves, rather than to repeat the professor's words and apply ready-made courses of action to a problem: Rina Knoeff, 'Herman Boerhaave at Leiden: Communis Europae praeceptor', in *Centres of Medical Excellence?*, ed. Ole Peter Grell, Andrew Cunningham, and Jon Arrizabalaga (Farnham, 2010), 269–86.

[16] Women were generally not allowed to matriculate, nor did hold any official teaching positions. Women solely worked at universities in supportive functions, such as housemaster and washerwoman. See 'schaftmeester' and 'wasvrouw' in Ronald Sluijter, *'Tot ciraet, vermeerderinge ende heerlyckmaeckinge der universiteyt': Bestuur, instellingen, personeel en financiën van de Leidse universiteit, 1575–1812* (Hilversum, 2004), 110–1, 4, 6, 293–4.

Fig. 5.2 The Leiden botanic gardens. Engraving in Herman Boerhaave, *Index plantarum* (Leiden, 1710). (Credit: Wellcome Collection. CC BY)

matter as milk! But is it not *the origin of all the solids in the body?*'[17] Milk, in other words, could form into any body part, including solid bones. Indeed, medical men knew from experience that when bones were broken, they would knit together again over time. The bodies of animals might, therefore, be slowly wasting, withering, and breaking, but thanks to the nutrition they might also constantly repair and rebuild. 'Animals', Boerhaave concluded, 'must necessarily be composed of what they take in as aliment; which, by their vital powers, is converted into their own substance'.[18]

Boerhaave reasoned that milk was an apparently mundane but actually very special fluid as it was a major confluence of a series of physiological processes: it represented the result of digestion and secretion in women; it was the initiator of digestion and nutrition in infants. Boerhaave reasoned that all animals fed on vegetable matter, either directly if herbivores, or indirectly as carnivores eating herbivores. Humans, in turn, ate those animals; and hence 'all our nourishment is therefore vegetable juices, prepared by the action of one or more stomachs, according as they are drawn either immediately from the vegetables themselves, or from the broken fibres and vessels of animals'.[19] The next step involved analysing that fluid, which was in the process of losing its vegetable nature and maturing into having an animal nature. 'This part must be such a fluid', Boerhaave explained, 'as proceeding originally from a vegetable, has felt the vital forces of the body, mixed with the blood, passed thro' the arteries and the veins, and been soon separated again. And this can be no other than chyle from vegetables, turned to milk, and separated in the breasts'.[20] Most new-borns took in solely mother's milk for several months after their birth, from which Boerhaave deduced that the bodies of infants—having grown in the meantime—were nothing other than a composition of milk.

[17] Herman Boerhaave, *Elementa chemiae, quae anniversario labore docuit in publicis, privatisque scholis*, 2 vols (Leiden, 1732), vol. 2, 301; Herman Boerhaave, *A New Method of Chemistry: Including the History, Theory, and Practice of the Art*, trans. Peter Shaw, 2nd ed., 2 vols (London, 1741), vol. 2, 188. Emphasis added.

[18] Herman Boerhaave, *A New Method of Chemistry: Including the Theory and Practice of that Art: Laid down on Mechanical Principles, and Accommodated to the Uses of Life*, trans. Peter Shaw and Ephraim Chambers, 2 vols (London, 1727), vol. 2, 179.

[19] Albrecht von Haller, ed. *Praelectiones academicae in proprias institutiones rei medicae*, 6 vols (Göttingen, 1739–1744), vol. 1, 272; idem, *Dr. Boerhaave's Academical Lectures on the Theory of Physic: Being a Genuine Translation of his Institutes and Explanatory Comment*, 6 vols (London, 1742–1746), vol. 1, 224–5.

[20] Boerhaave, *A New Method*, vol. 2, 180.

'Thus, by the vital actions, the milk is made to pass thro' all the changes of blood, serum, lympha, nervous fluid, or animal spirits, which seem to be the immediate matter of nutrition to all the solids'.[21]

Within the context of the university, the importance of milk was eagerly absorbed by medical students. The chemical lectures at Leiden taught medical students that the human body was made from milk. They learned that chyle, originating from vegetables, was the basis of milk. Once vegetable matter was chewed and mixed with saliva, the chyle would combine and circulate with the bodily fluids as an emulsion, to be separated eventually in the breasts as a white, oily substance called milk.[22] To see how Boerhaave's physiology of milk was appropriated by students, we may look at the doctoral thesis of Henricus Doorschodt (b. *c.* 1712), a student from The Hague. Doorschodt performed seven experiments on milk which were very similar to the ones executed by Boerhaave and Gaubius. He may have been assisted by Jacobus van der Spijk (or Van der Speijck, b. *c.* 1706), who was an amanuensis to the Leiden chemical laboratory from 1735 to 1745.[23] Amanuenses were appointed to restore broken vessels, to replenish supplies, and to assist professors and students alike in conducting their experiments.[24] Doorschodt put fresh milk in a glass alembic and slowly distilled it, gathering in the receiver a simple water that smelled and tasted like milk (see Figs IV and V in Fig. 2.2). By applying a more 'violent fire' to this distillate, Doorschodt was able to gain insight into the 'chemical principles of milk'.[25] One of these properties or active principles of milk was its watery nature. Doorschodt boiled the milk and used a mercury-filled thermometer, for precision, on the Fahrenheit scale to keep track of the temperature.[26] Although milk twice as thick as water, it reached a boil-

[21] Ibid., vol. 2, 181.

[22] Ibid., vol. 2, 181–2. On Boerhaave's business of chylification see Barbara Orland, 'The Fluid Mechanics of Nutrition: Herman Boerhaave's Synthesis of Seventeenth-Century Circulation Physiology', *Studies in History and Philosophy of Biological and Biomedical Sciences*, 43 (2012), 357–69, here 357–69.

[23] Van der Spijk succeeded Abraham Tijken as the new 'famulus Laboratorii Chemici' on 21 May 1735. See Willem Nicolaas Du Rieu, ed. *Album studiosorum Academiae Lugduno-Batavae MDLXXV–MDCCCLXXV* (The Hague, 1875), 955.

[24] In 1718 Boerhaave composed a formal job description to state the responsibilities of amanuenses in the chemical laboratory. See Sluijter, *Tot ciraet*, 176–7.

[25] Doorschodt, *De lacte*, 16.

[26] Boerhaave had integrated Fahrenheit's thermometer in his chemical lectures, and other chemistry instructors followed suit. See John C. Powers, 'Measuring Fire: Herman Boerhaave and the Introduction of Thermometry into Chemistry', *Osiris*, 29 (2014), 158–77.

ing point at around 212 °F, that is, at the same point as water. From this *in vitro* experiment, Doorschodt concluded that milk had a watery nature and that *in vivo* 'milk in blood will never rise to this degree of heat, nor, indeed, in the highest degree of fever'. In this extrapolation from an experiment to the living body, we can clearly recognise the resemblance of Doorschodt's to Boerhaave's and Gaubius's chemical analyses of blood, which have been discussed above in Chap. 2. Another central feature that Doorschodt analysed by chemical means concerned the vegetable versus animal nature of milk. He observed that, through distillation, milk produced the same principles as vegetables. He therefore concluded that 'milk is really a vegetable juice'.[27] In other words, chemistry revealed and proved that milk was, literally and figuratively, the cream of the crop.

Medical students acquired specialist chemical knowledge on the nature of milk, and this knowledge was soon related to breastfeeding and lactation disorders. Making the connection between chemistry and therapy was important, because many women throughout the early modern period, and beyond, would suffer from lactation disorders, such as the so-called milk fever. In 1787, for example, when Johanna Helena Halfman (1759–1829) gave birth to her first son Justus de Vrij, she suffered from milk fever. Milk fevers were generally believed to be caused by a shortage or surplus of breast milk ascending inside the body but not properly secreting. As Johanna confided to her friend Christina van Steensel (1758–1801) in The Hague, 'but with regards to the milk, it did not come up until the 5th day and it was not much, but I have had a rise or 3 milk fevers and no milk came anymore. So I have had to feed the little Jusje with porridge, for I'm very against a wet nurse'.[28]

We do not know whether Halfman successfully recovered from her milk fever and lactation problems. But we do know that Boerhaave's students researched the disease and particular kinds of food as its potential causes. Doorschodt's experiment on the changing colours of milk is a case in point. He knew that wealthy parents could afford to eat meat. As meat was predominantly alkaline, Doorschodt performed an experiment which involved adding an alkaline salt to fresh milk. He boiled the mixture,

[27] Doorschodt, *De lacte*, 16–7.

[28] Johanna Helena Halfman to Christina van Steensel, Nijmegen, January 1788 in Anje Dik and Dini Helmers, eds., *'Het is of ik met mijn lieve sprak': De briefwisseling tussen Jean Malherbe en Christina van Steensel, 1782–1800* (Hilversum, 1994), 94. Halfman had married Johan Eliza de Vrij in 1786.

which resulted in a kind of coagulum. He closely examined the changes in the colour of this coagulated mass as 'first white, turning yellow; secondly, from yellow to green; thirdly, obscuring more and more, until finally, going across to the colour red'.[29] This proved that milk, when it circulated through the body with an excessive amount of alkaline and not a certain amount of acid, could turn sour and putrefy, and hence cause obstructions and fevers. With this experiment Doorschodt successfully verified one of Boerhaave's demonstrations, in which the latter had boiled cow's milk with the salt of tartar, which also had turned from yellow to an intense red. Boerhaave had used this investigation to explain the contagious disease which had caused the milk of Dutch cows to thicken and turn yellow in 1714. When animal fluids circulated through animal bodies they did not contain any acid, but rather inclined towards an 'alkaline nature'. When cows suffered from a fever, their abnormally high body temperature would coagulate the milk and turn it 'from its genuine whiteness to a yellow'. Boerhaave had taught his students that this *in vitro* experiment could be observed *in vivo* as well: when a nurse would eat meat only, it would be almost impossible for her milk to keep from putrefying, causing the infant to suffer from fevers. This 'is too often the case of the infants of wealthy parents'. By contrast, 'children of the poor, whose mothers are principally sustain'd by food of an acid nature, are much less subject to fevers; but oftener afflicted with those diseases which owe their rise to acids'. A change in diet in both cases was regarded the best course of action: the consumption of more vegetables in case of a meat-heavy diet, and of more meat in case of a predominantly vegetable diet, and this diet was to be followed until the colour of the milk of mother or nurse had returned to a perfect white.[30] Doorschodt devised additional medical treatments from his chemical investigations on milk. He thought that water could homogeneously combine with milk, without causing any sourness, and so he advised that when 'the thickness of milk is wrong in the nurse, a watery diet is prescribed'.[31] Within the context of the university, then, chemical experiments on milk explained its physiology and pathologies.

Doorschodt's case history shows that chemical knowledge and laboratory expertise was successfully communicated to the next generation of physicians in the Low Countries. But Doorschodt was far from the only

[29] Doorschodt, *De lacte*, 19.
[30] Boerhaave, *A New Method*, vol. 2, 184–5.
[31] Doorschodt, *De lacte*, 20.

one to communicate this knowledge. Gaubius's nephew Johannes David Hahn, the professor of medicine, botany, and chemistry at Utrecht University, supervised many doctoral students in their chemical research. When Hahn was appointed to Leiden University to succeed Gaubius in 1775, his student Floris Jacobus Voltelen (1754–1795) also moved to Leiden and conducted research on milk. Voltelen performed elaborate chemical experiments, examined the milk of mothers, and compared it with the milk of asses and ewes by applying different salts, solutions, metals, and plants to it, in addition to distilling the liquids slowly on his Boerhaavian furnace. Following this series of experiments, Voltelen concluded that, of course, mother's milk was healthier for human infants than the milk of animals.[32] The bodily fluid expressed by mothers and suckled by their babies as nourishment was, therefore, a rich research subject for Boerhaave students throughout the eighteenth century. They used chemistry to gain physiological knowledge of women's bodies as well as of children's health.

Promoting Lactation

Chemical knowledge of milk was not confined to the chemical laboratory. In fact, once they had graduated and were working as practitioners in their community, many of Boerhaave's students actively tried to translate their knowledge to normative medical prescriptions and dietary advice. Six years prior to the publication of his laudatory poem on milk Abraham Tersier published two medical essays, one on a disease that befell cows and the other on the production of breast milk.[33] The essays had been entered into prize essay competitions launched by the Holland Society of Sciences in Haarlem, requesting the best method for increasing the flow of mother's milk. Tersier's essay and those of many other medical men who

[32] Floris Jacobus Voltelen, *Observationes chemico-medicae de lacte humano, ejusque cum asinino et ovillo comparatione* (Utrecht, 1775). When Voltelen succeeded Hahn as professor of chemistry in 1784, he mentioned in his inaugural address that Hahn had also performed experiments with milk in the summer of 1779. Floris Jacobus Voltelen, *Oratio aditialis de chemiae hodiernae pretio rite constituendo* (Leiden, 1784). On Hahn and Voltelen, see H.A.M. Snelders, *De geschiedenis van de scheikunde in Nederland: Van alchemie tot chemie en chemische industrie rond 1900* (Delft, 1993), 62–3; Dalila Wallé, *Leiden Medical Professors, 1575–1940* (Leiden, 2007), 120–1, 4–5.

[33] Abraham Tersier, *Responsa ad rogata bina medica, unus de morbo bubulo, &c., alternus de lactis conformatione, &c., ab Collegis scientiarum Harlemensi proposita* (Dordrecht, 1762).

participated in the competition were, therefore, clearly part of the societal utilisation of knowledge that was encouraged by Dutch academies. As this section shows, while the essays unanimously agreed that mother's milk was the best nutriment for a new-born, they differed in opinion on the kinds of therapies best used to treat mothers with lactation problems or related disorders. Some suggested herbal medicines, but many others argued for a therapeutic diet of dairy products—an argument based on a chemical understanding of milk. Yet still others proposed a mechanical solution.

Founded in 1752, the Holland Society of Sciences was the oldest society in the Dutch Republic formed to promote the sciences (Dutch: *wetenschappen*) and the production of useful knowledge (see Fig. 5.3).[34] The Amsterdam engraver Simon Fokke (1712–1784) depicted the Society's mission in its emblematic logo by surrounding the Greek goddess Athena with scientific and artistic objects and instruments, such as globes, books, and a pair of compasses, while at the same time placing both a shield portraying Stadtholder William V (1748–1806) and a sheet of paper or vellum with the words 'for one's country' (*pro patria*) into her hands. Knowledge was not meant to be produced for its own sake, but for solving societal problems. To achieve active participation in knowledge production as well as the social application of knowledge, the Society's main instrument was its annual prize essay competition, which often spurred much discussion among experts. And in 1760, the Society launched a competition on the question 'how can lactation best be increased and decreased?'[35] This resulted in the submission of more than a dozen essays suggesting all kinds of milk-inducing medicines.

The topic had not been chosen by chance, for much more was at stake here than scientific curiosity. Many young mothers experienced complications when breastfeeding. Women from all strata of society and across Europe struggled with a shortage of breast milk, thought to be caused by

[34] On the historical context of Dutch learned societies, see Wijnand Mijnhardt, *Tot heil van 't menschdom: culturele genootschappen in Nederland, 1750–1815* (Amsterdam, 1988); Joost Kloek and Wijnand Mijnhardt, *1800: Blueprints for a National Community,* trans. Beverley R. Jackson (Basingstoke, 2004).

[35] J.G. de Bruijn, ed. *Inventaris van de prijsvragen uitgeschreven door de Hollandsche Maatschappij der Wetenschappen 1753–1917* (Groningen, 1977). The question of lactation fitted in with other relevant research questions, such as 'What are the causes of the general illnesses of our seafarers?' (1758/1759) and 'How [can we] keep the body of children to live long and healthy?' (1761).

Fig. 5.3. Allegory of the foundation of the Holland Society of Sciences Haarlem. Engraving by Simon Fokke (Amsterdam, 1752). (Credit: Rijksmuseum, Amsterdam. CC0 1.0)

malnutrition, disease, mental imbalance, or a combination of some or all of these. Three examples may illustrate the gravity of the issue at the time. In 1736 in Utrecht, the patrician couple Maria Jacoba Meinertzhagen (1712–1787) and Cornelis Jan van Royen (1711–1774) became the parents of a daughter, Anna Catharina (1736–1783).[36] Anna was not breastfed by her mother, but by a wet nurse. Maria Meinertzhagen wrote in her diary: 'as I was not able to suckle myself, mother Van Royen has lavishly treated our child's wet nurse with clothes and given us much love in many ways'.[37] It appears that Maria may have had the intention to breastfeed, but was simply unable to do so. Effective therapies were in high demand, and perhaps unsurprisingly, the main question of the essay competition was how lactation might be managed and controlled. Another contemporary mother, now one who tried to be treated for her lactation disorder, was Anna Charlotta Bäck (1737–1767). She gave birth in February 1759 but experienced troubles breastfeeding. Her husband, Abraham Bäck (1713–1795), physician and president of the Collegium Medicum in Stockholm, corresponded with Linnaeus to inquire about possible treatments. Linnaeus's preferred method was to let the mother sweat out the obstruction, and consequently, he prescribed a sudorific, that is, a sweat-inducing medicine. He reasoned that it was important to remove the obstacles that would make the milk stay in the breasts, and sweating was believed to dissolve the sourness or obstruction. 'If the swelling in the breasts does not recede, mix 3 ounce *Album Graecum* (Greek white, *i.e.* the dog dung that had turned white) with *Robes Sambuci* (Red of Sambucus, *i.e.* concentrated syrup of elderberries) to induce sweating; it has always been a sacred anchor to me, for it has never easily failed, unless the evil is old'.[38] The fact that lactation disorders could be even fatal was reported by the surgeon and obstetrician Jacobus de Puyt Jz. (1740–1812), who practised medicine in the city of Middelburg and was an active contributor to the journals of the Amsterdam Society for the Advancement of

[36] On Maria Meinertzhagen see J.N. van der Meulen, 'Maria Jacoba Meinertzhagen (1712–1787), een patriciërsvrouw', in *Utrechtse biografieën* (Utrecht, 1998), 107–12; Els Kloek, ed. *1001 vrouwen uit de Nederlandse geschiedenis* (Nijmegen, 2013), 661–2. Meinertzhagen and Gaubius were related through Gaubius's mother-in-law, Maria Louise Meinertzhagen.

[37] J.H. Scheffer, 'Het dagboek van eene merkwaardige vrouw', *Algemeen Nederlandsch Familieblad*, (1884), here 3.

[38] Carolus Linnaeus to Abraham Bäck, February 1759, *The Linnaean Correspondence*, linnaeus.c18.net/Letter/L2490, Letter L2490 (consulted 23 November 2016).

Surgery and the Zeeland Society of Sciences in Flushing. De Puyt supervised the pregnancy of one Mrs Pagter, who was expecting triplets. After the delivery, although she at first appeared to be fine for a few days, she was soon badly affected by a milk disorder, which eventually caused her death.[39]

The three women discussed above were just a few of many mothers suffering from lactation problems. The Holland Society of Sciences therefore issued the prize essay competition in 1760 and then reissued it in 1761. Among the contributors were Hendrik van Someren, a surgeon in Haarlem, and J.M.V. Berkman, a physician in Amsterdam, whose impressive Latin essay of 91 pages only arrived at the Holland Society on 14 March 1762, and was therefore too late to be considered. Ultimately the jury awarded the first prize, the gold medal, to the Paris physician Jean-Pierre David (fl. 1763), and the silver medal to Lambertus Bicker (1732–1801), a Dutch physician in Rotterdam (see Fig. 5.3).[40] One jury member, Joannes Grashuis (1699–1772), had been very complimentary about Bicker's essay, saying that 'this [essay] is very detailed, has great accuracy and clarity, and also a very appropriate division and arrangement; and it complies, in my understanding, not only with the question, but even gives here something more'.[41] Another jury member, however, added a more critical note: 'Is that treatise not more a fruit of erudition, than of judgement and experience?'[42]

The essay prize winners, David and Bicker, soon shared their essays with the general public, as both went on to publish their work on their own accord. The Society only accepted unpublished essays, which, when awarded, were being published as part of the Society's journal. But David had already published his essay in France, apparently unaware of the Society's procedures.[43] Bicker also entered into negotiations with the

[39] Jacobus de Puyt, 'Observatie van een verlossinge van drie kinders', Middelburg, 1777. Middelburg, Zeeland Library, Handschrift 4940.

[40] The jury consisted of J. Engelman, Hieronymus Gaubius, Joannes Grashuis, Jacobus Hovius, P. Sannié, and Thomas Schwencke. See de Bruijn, *Inventaris*, 35–6. The golden medal awarded to monsieur David was designed and made by Amsterdam goldsmith Marchand. Louis Marchand to C.C.H. van der Aa, Amsterdam, 18 June 1762 in Noord-Hollands Archief (NHA), 444, Hollandsche Maatschappij der Wetenschappen te Haarlem, 1752–1975, no. 373.

[41] Joannes Grashuis to C.C.H. van der Aa, Hoorn, 22 May 1761. NHA 444, 373.

[42] [anonymous], NHA 444, 373, n.d.

[43] As suggested by Johan Abraham Bierens de Haan, *De Hollandsche Maatschappij der Wetenschappen, 1752–1952* (Haarlem, 1952), 185. David's book appeared as Jean-Pierre

academy about the publication of his essay. He was not aware of the Society's intentions to publish the silver medallist's in addition to the winner's essay and had, therefore, already approached the academic publishers Samuel and Johannes Luchtmans.[44] While asserting control over the publication strategy, Bicker held on to the society's validation of his work by mentioning it was awarded the silver medal. In addition, he dedicated his work to professor Gaubius, his former teacher at Leiden, because 'nobody than Your Noble Highly learned is more able in the subjects discussed here, and everything I bring forth is built on the weighty lessons, which I have received from Your Noble Highly learned mouth'.[45] Once the authors had made their essays available as separate, stand-alone publications, the reading public beyond the Society learned about lactation and milk-promoting medicines. Bicker's *Zog der vrouwen* ('The Milk of Women') was advertised in newspaper *Opregte Haerlemsche Courant* on 25 October, 1 and 15 November 1763, and it received an overall laudatory review in the journal *Vaderlandsche letter-oefeningen* ('Fatherland's Literary Exercises').[46] Measuring the impact of Bicker's and David's work is difficult to ascertain to any certain degree, but both were often cited in subsequent studies on milk.[47]

To answer the question of how a sufficient supply of good-quality breast milk could be achieved, David and Bicker evaluated the contemporary use and efficacy of galactagogues, that is, substances promoting and increasing the flow of mother's milk. While the use of various herbs and minerals proved popular throughout the early modern period, eighteenth-century physicians tended to promote milk-based products as more effective lactation stimulants. Whereas herbal galactagogues remained in use,

David, *Dissertation sur ce qu'il convient faire pour diminuer ou supprimer le lait des femmes* (Paris, 1763).

[44] On the history of academic publishers Luchtmans see Sytze van der Veen, *Brill: 325 Years of Scholarly Publishing* (Leiden, 2008).

[45] Lambertus Bicker, *Verhandeling van het zog der vrouwen, ter beantwoordinge op de vraage by de Hollandsche Maatschappye der Weetenschappen opgegeeven in de jaaren 1760 en 1761* (Leiden, 1763), [*3].

[46] *Vaderlandsche letter-oefeningen*, 4 (1764), 201–212. This journal was one of the most prominent Dutch literary journals.

[47] For example, David and Bicker were both referenced in Joannes Guilielmus Thorvarth, *Dissertatio practico-medica inauguralis de lactis defectu* (Leiden, 1764), 19, 30–1, 3. Readers were directed to Bicker for all children's diseases relating to milk in the Dutch edition of Nils Rosén von Rosenstein, *Handleiding tot de kennis en geneezing van de ziekten der kinderen*, trans. and ed. by Eduard Sandifort (The Hague, 1768), 52.

David and Bicker nevertheless encouraged mothers to use lacteal products to augment their breast milk production.

In the seventeenth century, galactagogues were almost exclusively derived from herbal and mineral sources. A Dutch series of household manuals titled *Het vermakelijck landt-leven* ('The Pleasurable Country Life', 1683) included all kinds of advice on gardening and husbandry. The third part is particularly interesting, for it contains instructions for distilling waters and oils from flowers and herbs, as well as making one's own medicines for the treatment of various disorders 'of men and beasts'. Among the disorders and diseases mentioned are also breast infections and painful nipples. The very first medicine recorded here for mammiform disorders is devoted to the increase of lactation: 'To increase the Milk in the breasts. Take a half loot [1/4 oz.] Aniseed or Fennel seed/ drink it with Cabbage juice when going to bed/ Or mix green Fennel and Dill with barley water/ make it sweet with sugar and drink it'.[48]

The ingredients for these remedies were all herbal in nature: seeds of anise and fennel, dill, coriander, and mint. Although these galactagogues could be made at home, from materials from one's own garden, the list of ingredients does not differ greatly from recipes recorded in other pharmacy books and pharmacopoeia. In 1681, for example, the famous *Medicina pharmaceutica* by the Flemish physician Robert de Farvacques (fl. 1600–1637) was published in Brussels.[49] In this treatise of 1144 pages an overview was given of the basic raw materials necessary to make remedies, together with instructions on their preparation, and details of the instruments required for the task. In chapter 13 De Farvacques stated that milk-making medicines (*lac generantia*) 'are of a nature, which are drawn out of the best nourishment of food and drink, as some do such proficient medicines, which are of thin and fine parts, and naturally hot, such as fennel, dill, chick pea, onion'.[50] Another pharmacopoeia, the *Nieuwe Nederduitsche apotheek* ('New Netherlands Apothecary'), promoted

[48] J. van der Groen, *Den Ervaren huys-houder; zijnde het III. deel van het Vermakelyck landt-leven* (Amsterdam, 1683), 16.

[49] Robert de Farvacques, *Medicina pharmaceutica, oft drôgh-bereydende ghenees-konste: met besondere aenmerckingen op verscheyde misbruycken, die soo wel in de medecyne als chymie zyn voor-vallende* (Brussels, 1681).

[50] Robert de Farvacques, *Medicina pharmaceutica, of Groote algemeene schatkamer der drôgbereidende geneeskonst*, 3 vols (Leiden, 1741), vol. 1, 21.

ingredients such as lettuce, oregano, and sow thistle (*sonchus oleraceus*, Dutch: *zagte zeugdistel*), also known as milk thistle because of its milky sap.[51]

Recipes for herbal galactagogues continued to be circulated in eighteenth-century treatises, but they no longer constituted the primary solution for a lack of breast milk. Jean-Pierre David's submission essay to the Holland Society of Sciences lists a number of treatments to first support the flow of fluids towards the mother's breasts. The main aim was to increase the 'fullness of the veins' through good nourishment, an avoidance of heavy physical activities to prevent excessive perspiration, armbands to keep the blood in the torso as much as possible, bathing in warm water, and many other treatments.[52] David ended his elaboration on treatments with the suggestion that 'one can add the use the seeds of anise, fennel, caraway, and dill, of which some authors say it is beneficial in promoting the milk of women'.[53] In other words, while David was aware of the well-known herbal galactagogues, they were merely mentioned in passing as alternative treatments to the detailed dietary prescriptions and externally applied treatments.

Lambertus Bicker concurred with David that anise, fennel, and other carminative seeds could be consumed for the promotion of lactation, yet he preferred those 'foods rich in chyle', such as milk, whey, and cream. Alternatively, one could prepare a porridge by boiling bread, barley, oats, groats, and rice with milk, beer, or water.[54] Yet as far as Bicker was concerned, mothers had to strictly abstain from galactagogues of a mineral nature, such as the so-called milk of the moon (*lac lunae*). One fervent supporter of moon-milk had been Johann Daniel Major (1634–1693), the professor of medicine at the Christian Albrechts University in Kiel, where he taught medical theory, botany, and chemistry. According to Major the substance of moon-milk, which was found in silver mines, derived its name from its white and shiny appearance: 'Silver is indeed known as a gift of the

[51] *Nieuwe Nederduitsche apotheek: Op eene klaare en verstaanbaare wyze onderwys gevende omtrent de beste dagelyks gebruikt wordende geneeskundige bereidingen* (Leiden, 1753), 26, 9, 34. This work listed Boerhaave's and Gaubius's chemical preparations, or so the preface argues.

[52] Jean-Pierre David, 'Wat 'er behoore gedaan te worden, om het Zog der Vrouwen te vermeerderen, te verminderen, of te doen verdwynen', *Verhandelingen uitgegeeven door de Hollandse Maatschappy der Weetenschappen te Haarlem*, 7 (1763), 2–76, here 24–8.

[53] Ibid., 30.

[54] Bicker, *Zog der vrouwen*, 108.

Moon, just as iron stems from Mars, gold from the Sun and other metals or minerals from planets which lend them their names because sympathies persist between certain heavenly objects and subterranean substances'.[55] Moon-milk was therefore thought to originate from the moist effluvium of the moon, and its sticky and pasty texture was similar to that of soft cheese made from unskimmed milk and cream, hence moon-milk.[56] According to Major, moon-milk worked best as a galactagogue when harvested at the full moon and taken together with the juice of herbs, such as thistle, fennel, parsley, anise, or basil. One century later, however, Bicker radically dismissed the use of moon-milk as a false galactagogue: 'Because everything which is here prescribed as peculiar milk-making medicines, similar to the notorious milk-stone (*melksteen,* the Dutch name for moon-milk) and others, I regard to mere fables and superstitions based on a deceptive imagination'.[57] Although moon-milk appeared to have many of the same outer characteristics and much of the appearance of milk, this example of sympathetic medicine was falsified by experience, and moon-milk was dismissed as a false galactagogue. Based on eighteenth-century chemical understandings of milk, like those of Boerhaave and Doorschodt, physicians and apothecaries promoted chylous galactagogues as aids for a sufficient supply of good-quality breast milk.

Within the context of making scientific knowledge and expertise useful to society, physicians even started experimenting with so-called breast pumps, which mechanically expressed the milk from the breasts and collected it in a glass receiver. In 1773, George Wilhelm Stein (1737–1803), the physician and professor at the Collegium Carolinum in Kassel in northern Hesse, collaborated with the instrument maker Johann Gottlieb Stegmann (1725–1795) to develop the 'breast or milk pump', which mechanically drew milk from the breasts into a glass receiver.[58] Various

[55] Johann Daniel Major, *Dissertatio medica de lacte lunae* (1667); cited in Bruce T. Moran, 'Lac Lunae and Breast Milk', *Pharmacy in History*, 41 (1999), 33–4.

[56] Today, moon-milk is known to be a white deposit consisting of carbonate materials, for example calcite, hydro-magnesite, and gypsum. It is slowly formed in caves of aggregates of very fine crystals of varying composition.

[57] Bicker, *Zog der vrouwen*, 108.

[58] George Wilhelm Stein, *Kurze Beschreibung einer Brust- oder Milchpumpe* (Kassel, 1773); Johann Gottlieb Stegmann, *Kurze Beschreibung einer Saug- und Drukpumpe* (Kassel, 1774). On the issue whether Stein or Stegmann had first invented the pump, and on its commodification in France, see Margaret Carlyle, 'Breastpump Technology and 'Natural' Motherly Milk in Enlightenment France', *Women's Studies International Forum*, 60 (2017), 89–96.

devices to assist the external extraction of milk from the breasts were already in use. Nipple cups, for example, had the shape of thimbles and elongated the nipple, making it easier for the baby to suck.[59] Breast relievers or glass bowls were used to provide suction, but they had very little suction, so they could do not much more than collect surplus milk. This led to the development of a tobacco pipe-shaped instrument made of glass, called the 'milk-suctioner' (*Milchsauger*), by which a woman could apply suction to her own breasts (see Fig. 5.4).[60] Stein, however, was not satisfied with this device, for it required the mother to do physical exercise. But when observing that his colleague Stegmann was experimenting with English vacuum pumps in his workshop, Stein soon had the idea to create a 'machine that sucks' milk mechanically. Stegmann took an ordinary breast glass, made an opening in the top centre, and screwed on a brass cylinder vacuum pump. After some trials, he perfected the design with valves and seals, and the breast pump was born (see Fig. 5.5).[61]

According to Stein, this breast pump exceeded all expectations. It sucked on and shaped the nipple, it drew milk from the breasts more effectively and more forcefully than earlier models, and it was believed to treat and prevent lactation disorders like milk fever since the breasts were emptied fully. News about this innovative technology soon spread to the Dutch Republic, where Stein's treatise was published in translation and, together with the instrument, advertised in newspapers.[62] The pumps were, however, very expensive: a simple, tin breast pump was 12 guilders and a more elaborate and beautiful brass pump even 16 guilders.[63]

[59] A nipple cup or *petit chaperon* looked like a little thimble and was a well-known device among midwives. See François Mauriceau, *Traité des maladies des femmes grosses* (Paris, 1682); Pierre Dionis, *Traite general des accouchemens* (Paris, 1718), 358; Jacques Mesnard, *Le guide des accoucheurs, ou Le maistre dans l'art d'accoucher les femmes* (Paris, 1743), 357. Ordinary nipple shields were often made of tin. More luxurious versions were made of silver, glass, or ivory, such as A641255, A606830, and A606829, c. 1800, Wellcome Library, London.

[60] Johann Christoph Sturm, *Collegium experimentale, sive curiosum* (Nuremberg, 1676), vol. 2, 56. See for example A606855 and A606862, Science Museum, London.

[61] Eighteenth-century breast pumps are still extant in museum collections today. See for example A124874, Science Museum, London, and V05517, Museum Boerhaave, Leiden.

[62] George Wilhelm Stein, *Korte beschryving eener borst- of zog-pomp* (Utrecht, 1775). See also Isaac Henri Gallandat, 'De zogpomp, voorgesteld en aangeprezen', *Verhandelingen van het Zeeuwsch Genootschap der Wetenschappen*, 13 (1786), 538–58.

[63] In comparison, 10 guilders would get you 100 kg of rye bread or 155 l of beer. See Jan Luijten van Zanden, 'The prices of the most important consumer goods, and indices of wages and the cost of living in the western part of the Netherlands, 1450–1800', International Institute of Social History, http://www.iisg.nl/hpw/brenv.php (accessed 1 June 2020).

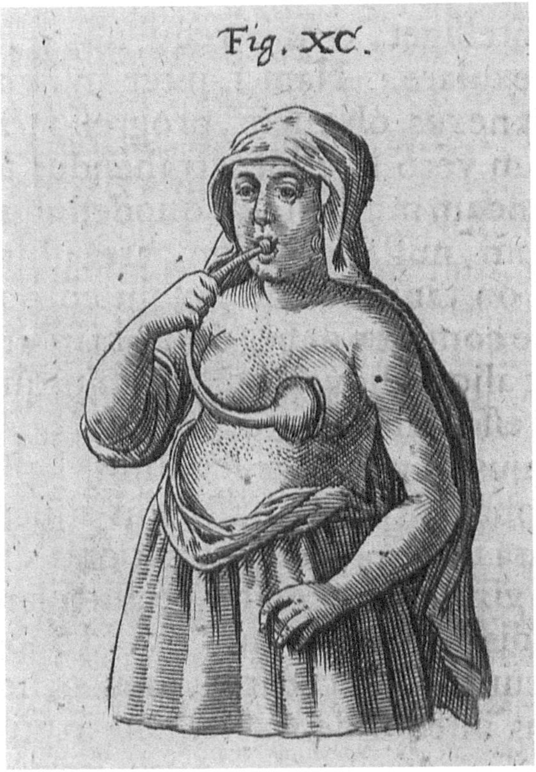

Fig. 5.4 Milk-suctioner in Johann Christoph Sturm, *Collegium experimentale* (Nuremberg, 1676). SLUB / Deutsche Fotothek. CC-BY-SA 4.0

Although it is unlikely that many ordinary women made use of this instrument, its invention nevertheless underscores the contemporary physicians' commitment to finding solutions for lactation problems. Learned societies played an important part in this: they explicitly aimed at generating and encouraging the production of knowledge that was of benefit to society at large. This problem-driven approach to knowledge allowed groups of professionals who ordinarily worked separately from each other to enter a direct conversation. Boerhaavian physicians emphasised the benefit of cheese, butter, beer, and other chylous galactagogues over that of herbal and mineral medicines for addressing milk flow problems—a claim supported by the knowledge of the chemical qualities of milk. But besides encouraging the production of mother's milk through the ingestion of dairy products, its expression from the breasts was also promoted via new technologies, such as milk pumps.

Fig. 5.5 Breast pump, Europe, 1771–1830. (Credit: Science Museum, London. CC BY)

Moralising Mothers

In addition to academic works and prize essays, popular poems and how-to books also fervently debated the importance of milk and maternal breastfeeding. In the course of the eighteenth century physicians and surgeons increasingly joined in the business of writing short pedagogical treatises in the Dutch vernacular, to provide detailed and practical advice about everything from gestation and giving birth, to early complications for mother or infant. While these works differed from the academic publications in approach and style, they show striking similarities in their perception of maternal breastfeeding. The efficacy of these works is difficult to ascertain, and the motivations for their publication would certainly have included financial gain. Nevertheless, the intention to change the population's conduct and perception of breastfeeding and wet-nursing is easily perceptible in these how-to books. But even though most mothers would have been aware of these texts, they were not merely passive prisoners to normative texts of conduct. Some women wrote their own poems and treatises, which might provide conclusions on the issues of right and wrong in the same way that medical tracts did, and yet with very different outcomes. This final section of this chapter analyses the textual strategies employed by male physicians and female authors to moralise mothers on the issue of motherhood and breast milk.

A number of popular publications clearly demonstrate the link between the Boerhaavian chemical discourse and popular literature. Take, for example, Abraham Tersier's poem on milk, which incorporated arguments from Boerhaave's chemical research:

> *De Schei-konst, niet gerust in 't uiterlyk beschouwen,*
> *Wil al, wat zy ontmoet, van deel tot deel ontvouwen,*
> *En toont, door veelerlei proef-neeming duid'lyk aan*
> *Der dingen aart en kragt, naar 't innige Bestaan.*
>
> Chemistry, not satisfied with external observation,
> Wants to uncover all it encounters, from part to part,
> And clearly show by plentiful experimentation
> The nature and force of the intimate existence.[64]

Tersier's poem went on to versify chemical experiments mixing milk with acids and alkalis. To help the reader, Tersier made a direct reference to Boerhaave's *Elementa chemiae* and concluded that 'Reason, and nature, and chemistry clearly prove,/ That we rightly commend milk as the best food'.[65]

Other than the genre of occasional poetry, books of conduct had been the most thriving in the Dutch Republic since the early seventeenth century. These works featured, to a more or lesser extent, physiological and chemical discourse. The most famous publication was *Houwelyck* ('Marriage', 1625) by Jacob Cats (1577–1660), which became the standard wedding gift and was reissued 40 times by the end of the century. The many rules of conduct for husbands and wives, fathers and mothers, naturally included breastfeeding etiquette: 'One who gives birth is a

[64] Tersier, *Melk*, 5.
[65] Ibid., 6.

mother partly; but who suckles her child is a mother completely'.[66] In the late seventeenth century the physician Steven Blankaart (1650–1702) published numerous popular books and treatises, aimed at both the profession and to the general public. In his *Verhandelinge van de opvoedinge en ziekten der kinderen* ('Treatise on the Education and Diseases of Children', 1684), Blankaart's first chapter was on the breastfeeding and raising of children. He emphasised that the mother was the best to give suck because, on the one hand, her nourishment was the best, while on the other, the milk of wet nurses could be too 'heavy' and transmit illnesses caused by the nurses' misbehaviour. But practically minded as he was, Blankaart provided a list of criteria for wet nurses to comply with in case the mother really was unable to breastfeed.[67]

In the eighteenth-century, these books were reissued and new titles produced. In 1771, Johann Hendrik Schutte (1694–1774) who, following his studies at Utrecht, had made his career as the court physician of the Duchy of Cleves in Westphalia, published *De wel onderwezene vroedvrouw* ('The Well-Educated Midwife').[68] Schutte was already celebrated for the discovery and chemical analysis of a mineral spring in Cleves, and for communicating its health-giving properties.[69] Despite the fact that many handbooks for midwives were already available, Schutte thought that these offered too broad a range of topics and hence were not very useful for putting their advice into practice. Consequently, Schutte himself presented a how-to book with short articles. His book was cast as a pedagogical work, directly addressing the reader as the midwife as well as the *pater familias*. Schutte had, therefore, designed the work to be useful for families living in the countryside who did not find themselves in the close vicinity of a physician.

Schutte's *The Well-Educated Midwife* succeeded in being a non-academic publication, and supposedly more accessible to midwives, but it

[66] Jacob Cats, *Houwelyck: Dat is de gansche gelegentheyt des Echten staets*, 7 vols (Middelburg, 1625), vol. 5, 45.

[67] Steven Blankaart, *Verhandelinge van de opvoedinge en ziekten der kinderen* (Amsterdam, 1684), 1–10. A similar example is Johan van Beverwijck, *Wercken der genees-konste, bestaende in den Schat der gesontheyt, Schat der ongesontheyt, Heel-konste* (Amsterdam, 1680).

[68] Johann Hendrik Schutte, *De wel onderwezene vroedvrouw, of grondig en beknopt onderwys, van het geen een vroedvrouw weeten, en by alle voorkomende natuurlyke en tegennatuurlyke zwaare geboortens door vaardige handgreepen, verrichten moet* (The Hague, 1771).

[69] Johann Hendrik Schutte, *Beschreibung des Clevischen Gesund-brunnens* (Cleve, 1742); idem, *Beschryving van de nieuw uitgevondene Cleefse gezond-bron* (Amsterdam, 1742).

adopted an explicitly moralising tone. The style of Schutte's text contrasted sharply with, for example, that of Albrecht von Haller's commentaries on Boerhaave's medical lectures. On the topic of childbirth and breastfeeding, Von Haller was descriptive and informative, framing his opinion for the reader to make up his or her own conclusions:

> Most women of quality imagine they shall preserve their beautiful shape by not suckling their own children. But we ought not to presume ourselves wiser than nature, who never changes her laws with respect to the fabric of the human body. [...] Those who suckle their own children have a less discharge of the lochia, and have all the symptoms of their lying-in milder without any return of their menses so long as they suckle.[70]

The advice that maternal breastfeeding would save the mother from a certain degree of discomfort certainly pointed in a clear direction. Yet the weighing of both the benefits and drawbacks of breastfeeding left room for readers to make up their own mind. This contrasted sharply with Schutte's more proscriptive language and explicitly prescriptive content. Structured in the form of questions and answers, his reply to query 334, 'Is a mother obliged to suckle her child?', obviously was a very normative 'yes', because:

> Nature gives the mother, after deliverance, to that end, the milk in the breasts, so that she would feed her child with it. If she does not, yet able to do so, she acts against the intention of the all-wise Creator, and against the laws of nature, and she robs the child of the nourishment which best matches its nature, is the most appropriate, and given by the Creator. 2. If she presents her child to a wet-nurse, [she leads the child] towards many hazards, of the mind, as well as of the body, and of health.[71]

In other words, mothers who refrained from breastfeeding were committing a sin, breaking the law of nature, and were not having their child's best interest at heart. Subsequent topics in Schutte's discussion all pointed into the same direction: why is mother's milk the best nutriment for the infant? What negative effects does the milk of a wet nurse have on the health and soul of the infant? Only illness would render mothers exempt

[70] Von Haller, *Praelectiones*, vol. 5, pt. 2, 439–40; idem, *Academical Lectures*, vol. 5, 210.
[71] Schutte, *De wel onderwezene vroedvrouw*, 199.

from their 'maternal duty'.[72] This was a lesson that was taught from a very young age. Johan Hendrik Swildens (1745–1809), who became known for his pedagogical works and political pamphlets, published *Vaderlandsch A-B boek voor de Nederlandsche jeugd* ('National A-B Book for the Dutch Youth', 1781), which presented young people with an alphabet of virtuous topics.[73] The letter 'M' stood for 'mother', but might as well have been milk, for it presented breastfeeding as the key feature of any good mother (see Fig. 5.6). Furthermore, the engraving facing the text depicted a father directing his son's gaze towards the desirable sight of his mother breastfeeding. In the bottom left corner a little girl practises breastfeeding on a doll. This may be seen as the example *par excellence* of the paternalistic attitude towards maternal breastfeeding in the eighteenth century.

Granted, infants continued to die from malnutrition, either through the milk of wet nurses or by bottle-feeding. Once again, an example from the Van Royen family illustrates how families were plagued by such tragedies. On 9 January 1747 Maria Meinertzhagen gave birth to a daughter, Justina Maria. Of Maria's offspring, Justina was the fourth daughter and the ninth child. But by the time she was born, her siblings Johanna Cornelia, Cornelis Jan, Bernard Jacob Isaac, and Isaac Pieter had already passed away between the ages of one and three. Justina, too, would not be saved from this fate. In the spring of 1748, Justina became terribly ill after bottle-feeding. And on '15 April at 11.30 in the morning, our beloved little daughter Justina Maria van Royen, born 9 January 1747, died because of a languishing disease'.[74] Four days later, Justina's coffin joined those of her brothers and sisters in the family grave in the Buurkerk in Utrecht.

But some took issue with Schutte's and Swildens's equation of breastfeeding with good motherhood. In the course of the eighteenth century, female writers also partook in the discussion on women's well-being and infant health. The Dutch author and poet Elisabeth Bekker (1738–1804), the widow of the reverend Adriaan Wolff (1707–1777), is a famous example. One of her earliest works, to which her partner Agatha Deken (1741–1804) also contributed, appeared in 1779 as *Proeve over de*

[72] Ibid., 202.

[73] Johan Hendrik Swildens, *Vaderlandsch A-B boek voor de Nederlandsche jeugd* (Amsterdam, 1781), [B2r]. On Swildens, see Barry J. Hake, 'Between Patriotism and Nationalism: Johann Hendrik Swildens and the 'Pedagogy of the Patriotic Virtues' in the United Provinces during the 1780s and 1790s', *History of Education*, 33 (2004), 11–38.

[74] Scheffer, 'Het dagboek', vol. 105, 4.

Moeder heeft u ook zo lief, O zy is zo goed!
Zie hoe teder aan haar hart zy het Kindje voedt.

𝔐𝔪 Mm

Is 'er iets, dat de waare *Moeder* zó eigenaartig kenmerkt, en dat voor haare natuurlyke tederheid zó zoet is, als het zoogen van haar Kroost? Is 'er iets in de geheele Natuur, dat het oog van eenen rechtſchapenen Vader zó aanminnig kan ſtreelen, als zyne lieve Vrouw met haar teder Kindje aan haaren boezem? —

„ Vrouw!... gy zyt geſchapen om uw Kroost te zoogen. Welk eene verantwoording by den SCHEPPER, zo gy, buiten den dringendſten nood, hetzelve aan eenen vreemden boezem overgaaft! — Ook zoudt gy daardoor uw Kind ontſlaan van die tederſte pligten, die elk Kind aan zyne *Moeder*, door welke het gezoogd wierdt, in het byzonder ſchuldig is. — Ook lydt het Welzyn der Maatſchappy onbedenklyk veel nadeel door zulk eene onnatuurlyke gewoonte. Menig Kind van den gemeenen man wordt immers daardoor van zyne natuurlyke Voedſter beroofd; menig huwelyk, menig huisgezin min of meer bedorven, zowel onder den minderen als voornaamen ſtand." —

(In het plaatje ziet men tevens de *toekomende Moeder.*)

B 2

Fig. 5.6 Nursing mother. Etching by Cornelis Bogerts in Johan Hendrik Swildens, *Vaderlandsch A-B boek* (Amsterdam, 1781). (Credit: F.G. Waller Bequest, Amsterdam. CC0 1.0)

opvoeding, aan de Nederlandsche moeders ('Essay on Childrearing, Addressed to Dutch Mothers').[75] Bekker's marriage with Wolff had remained childless, but she nevertheless had much experience with children through looking after the children in church and neighbourhood. In her essay she reflected upon motherhood and the raising of children, arguing that women ought to take a greater part in childrearing than was common at the time. Aristocratic women had many social obligations, while middle- and lower-class women often assisted their husbands in their professions and had no staff themselves. Nevertheless, Bekker encouraged her readers to think differently about the importance of motherhood, both for their own happiness and for the welfare and wellbeing of their children. In that sense, Bekker's essay may be considered as moralising as those of Schutte and Swildens.

Bekker's essay did not feature medical or chemical discourse, and argued differently on the case of breastfeeding than the treatises of physicians. Giving suck was not at all a point of contention. The main aim of motherhood, according to Bekker, was a woman's happiness and that of her children. Women did not have a different soul than men, but their nerves were more delicate, which did not make them weaker but of a more tender character. Mothers were, therefore, ideal for teaching confidence to their boys, and joyfulness to their girls, Bekker argued. Mothers should not doubt or worry, for that would have harmful effects both on them and on the vital spirits of their children. In addition to her maxim that mothers should embrace their natural talent to love, care, and educate their children, Bekker depicted 'the theatre of domestic life as it truly is in a peaceful house of enthusiasts of virtue' (see Fig. 5.7).[76] The engraving shows playing children, a nurse holding a baby, and a mother overseeing the household. Bekker's argument that maternal breastfeeding was far from a necessary condition for a happy mother-child relationship also appears in other publications. Giving suck could be part of a happy and peaceful motherhood, but so could the practice of wet-nursing. In the same year as

[75] Elisabeth Wolff-Bekker, *Proeve over de opvoeding, aan de Nederlandsche moeders* (Amsterdam and The Hague, 1779). A French translation appeared as *Essai sur l'éducation. Dédié aux Mères de Famille Hollandoises* (The Hague, 1785). More on Bekker and Deken, see Peter Altena and Myriam Everard, eds., *Onbreekbare burgerharten: De historie van Betje Wolff en Aagje Deken* (Nijmegen, 2004). On the context of Dutch literary authors, see Inger Leemans and Gert-Jan Johannes, *Worm en donder: Geschiedenis van de Nederlandse literatuur 1700–1800. De Republiek* (Amsterdam, 2013).

[76] Wolff-Bekker, *Proeve over de opvoeding*, 85.

Fig. 5.7 'The domestic stage of happiness' by N. van der Meer, after I.G. Waldrop in E. Wolff-Bekker, *Proeve over de opvoeding* (Amsterdam, 1779). KB National Library of the Netherlands, The Hague

Swildens's ABC of virtues, Bekker and Deken published a collection of poems and songs, which started with an illustration and a verse praising 'the good nurse':

Zie daar, voldoe uw lusjes,
Och, had myn Liefje dorst?
Wat krygt dit handje kusjes,
Dus leggende op myn borst!

Behold, satisfy your appetite,
Oh, was my sweetheart thirsty?
What gives this hand kisses,
So resting on my breast![77]

The far-reaching impact of mental stability and happiness on both mother and child was particularly clear in cases of an imbalance. Aagje Luijtsen (1756–1797), who had married Hermanus Kikkert (1749–1806), an officer in the Dutch East India Company, was a case in point. Kikkert had left on a mission to China and was away for almost two years. His wife Aagje regularly wrote him long letters, telling him about life on the island of Texel, sharing her experiences, and pouring her heart out to him. On 7 March 1777, Luijtsen gave birth to her son Lammert, but her strong sentiments reportedly affected the flow of healthy mother's milk. She missed her husband so much, she wrote, 'that my flesh is consumed and my dear child could not suck no more and had no desire to eat or to drink'. Malnutrition caused little Lammert to misbehave. 'My child is so unruly [...]. But I hope he will be a bit sweeter, because he sucks the dismay from me'.[78]

Perhaps the most explicit dismissal of moralising lessons forcing mothers to breastfeed came in the form of the French *Adèle en Theodoor*, a book on raising children translated by Elisabeth Bekker. The author was Stéphanie Félicité Ducrest de Saint Aubin comtesse de Genlis (1746–1830), better known as Madame de Genlis. She had already written several works on education and was therefore regarded as an expert on the matter. Her book argued that a mother who did not suckle her own child was by no

[77] Elisabeth Wolff-Bekker and Agatha Deken, *Economische liedjes* (The Hague, 1781), 4.

[78] Aagje Luijtsen to Hermanus Kikkert, Den Burg, 20 September 1777 in Perry Moree, ed. *Kikkertje lief: brieven van Aagje Luijtsen, geschreven tussen 1776 en 1780 aan Harmanus Kikkert, stuurman in dienst van de VOC* (Den Burg, 2003), 91–3.

means a bad mother but truly loved her child. 'Nature surely seeks from us that we suckle our own children, unless even greater obligations prevent us from doing so'.[79] A mother would do better not to suckle her child than to do it poorly. Both mother and child were better off when, in such circumstances, the child was left with a caring wet nurse who stayed home all day to take care of the infant. Breastfeeding at the cost of the wellbeing of the mother was an issue, for example, in the case of Christina van Steensel. Grandmother Johanna Maria van Steensel urged her daughter Christina not to suckle her child anymore. 'Dear daughter', she wrote, 'don't suck your child so much, but wean it during the day; it will grow better and you can once again go out and it is now so old that it can eat'.[80] By this time, Christina had been giving suck to Frans for almost six months. She did not listen to her mother, however, as we learn from a letter a month later: 'You better take your convenience and do not suckle your child during daytime, please do yourself some good. I know your character and fear that you do not take care of yourself, so provide for yourself in eating and drinking'.[81] The concern for Christina's health, in other words, was enough reason for her to stop breastfeeding and start accustoming her baby to pap and porridge.

* * *

As this chapter has argued, the eighteenth century witnessed a stream of lacteal knowledge all pointing towards maternal breastfeeding. Dissertations, essays, and popular publications contained knowledge on milk and breastfeeding, increasingly constituting distinctive normative stands on the issue. We have seen that the university provided a safe and untroubled transfer of knowledge and know-how between medical professors and their students. Contrary to the historiography that suggested that the chemistry of milk flourished at the turn of the nineteenth century, this chapter has demonstrated that many of Boerhaave's students developed chemical knowledge on milk throughout the eighteenth century. Chemical analyses proved to be a fertile practice for many Boerhaavian physicians,

[79] Félicité de Genlis, *Adèle en Theodoor, of Brieven over de opvoeding*, trans. Elisabeth Wolff-Bekker, 3 vols (The Hague, 1782–1783), vol. 1, 105–8.

[80] Johanna van Steensel to Christina van Steensel, Leeuwestijn, 28 April 1787 in Dik and Helmers, *De briefwisseling*, 88.

[81] Johanna van Steensel to Christina van Steensel, Leeuwestijn, 12 May 1787. Ibid., 91.

who were able to expand on the importance of milk and develop solutions for lactation disorders. Besides the widespread critique of wet nurses, this chapter has presented milk chemistry as a critical factor in the explanation of the maternalisation of breastfeeding in the eighteenth century.

The foundation of scientific societies, such as the Holland Society of Sciences in Haarlem in 1752, stimulated the utilisation of lactational knowledge in society. Through regular gatherings and prize essay competitions, men of all kinds of professions encountered each other on the topic of milk, each promoting his favourite solution to lactation problems. Eighteenth-century men applied their medical, chemical, and pharmaceutical knowledge to actively support maternal breastfeeding and explicitly warn against the health risks of wet-nursing and hand-feeding other foods. Mother's milk, in other words, became explicitly politicised.

Ideas of milk and breastfeeding moved into the domestic sphere of private homes, specifically the nurseries of ordinary families. In this space, not only physicians but also philosophers and authors mapped out their ideas about maternal duties and natural fashions. Arguments and treatises took on a wide variety of forms: how-to books, essays, and poems. These all contributed to the corpus of cultural ideas about and social norms around breastfeeding. Physicians generally appealed to nature—it is natural to breastfeed your child, and hence one ought to do it—to make their normative and moralising statements. Their arguments struck a chord because the close, sensuous proximity of lactation to the material of milk signified the truthfulness and importance of their claims. Mother's milk mattered.

In this context, as maternal breastfeeding became politicised, the female authors Elizabeth Bekker and Madame de Genlis provided some resistance. They dismissed the notion that fulfilling one's maternal duty necessarily entailed breastfeeding. Instead they gave prominence to motherhood as the task of caring, protecting, and educating a child. But in their argument, when representing women as tender and caring, and assigning them social and cultural roles as mothers, they made just as normative claims as to their male counterparts and similarly constructed gendered identities.

The question that remains is, of course, whether all these pedagogical teachings, moralising essays, popular how-to books, and poems referencing chemical and medical studies, had any effect on attitudes towards the practices of maternal breastfeeding and wet-nursing. Even Christina van Steensel, who herself had been breastfeeding her eight children throughout the 1780s and 1790s, reacted very surprised when she learned that the young and pregnant Princess Wilhelmina of Prussia (1774–1834), wife of Prince William VI (1772–1843), decided to nurse her own child. As Christina confided to her husband: 'The old [Nicolas] Gautier has told mother that our Hereditary Princess will nurse her child herself, which is something miraculous for such greats'. Van Steensel was very familiar with the customs at court, and hence knew that maternal breastfeeding among queens and princesses was highly unusual. Nevertheless, she bore witness to a general, though perhaps small, change in the *zeitgeist*: 'But that natural fashion is widespread, so I guess it is true'.[82] And indeed, the fact that Princess Wilhelmina nursed her own children had such an impact that all biographies which appeared after Wilhelmina's death in 1834 mentioned this political feat as an example of her kind and caring character.[83] Hiring a wet nurse, especially in wealthy and aristocratic families, would continue to remain common practice in the late eighteenth and early nineteenth centuries. Maternal breastfeeding, as well as galactagogues and breast pumps, continued to be in and out of fashion at different intervals.

[82] Christina van Steensel to Jean Malherbe, The Hague, 4 September 1792. Ibid., 122.

[83] See for example J.C. Voorduin, *Proeve eener karakterschets van nu wijlen Hare Majesteit, de Koningin der Nederlanden, Frederika Louisa Wilhelmina* (Utrecht, 1837), 35–6; and C.P.E. Robidé van der Aa, *Frederika Louise Wilhelmina van Pruissen, eerste koningin der Nederlanden: als een voorbeeld ter navolging, der Nederlandsche meisjes aangeprezen* (Amsterdam, 1838), 46.

CHAPTER 6

Sweat It Out

In the early modern period perspiration played a pivotal role in the preservation of health. In the 1730s, when the Dutch Republic was struck by an outbreak of catarrh (a defluxion, from the Greek ***katarrhous***, from ***katarrhein***, 'flow down'), men and women across the country suffered from symptoms including mucus running down their noses and phlegm stuck in their throats, causing trouble breathing and light-headedness. Confronted with this common yet severe cold, Johannes de Gorter (see Fig. 6.1), a physician and professor of medicine at the University of Harderwijk, championed ***sal ammoniac*** as 'the method which draws out the matter of disease by ***insensible perspiration***'.[1] De Gorter's treatment was based on an intricate pathological theory in which the obstructed 'nervous juice' (***liquor nervosus***) could not be perspired, which then caused the disease. Yet the chemical properties of sal ammoniac, De Gorter

[1] Johannes de Gorter, *Morbi epidemii brevis descriptio et curatio per diaphoresin* (Harderwijk, 1733), [*2v]. Emphasis added. De Gorter was a physician in Enkhuizen from 1712 to 1725. Between 1725 and 1754 he was the professor of medicine at the University of Harderwijk. On the life and works of De Gorter, see Aegidius W. Timmerman, 'Johannes de Gorter: Een schets van zijn leven en werk', *Nederlands Tijdschrift voor Geneeskunde*, 112 (1968), 35–41. On De Gorter's surgery books and education, see Daniel de Moulin, *A History of Surgery: With Emphasis on The Netherlands* (Dordrecht, 1988), 157–63, 72–73. On the University of Harderwijk: J.A.H. Bots et al., eds., *Het Gelders Athene: Bijdragen tot de geschiedenis van de Gelderse universiteit in Harderwijk (1648–1811)* (Hilversum, 2000).

R. E. Verwaal, *Bodily Fluids, Chemistry and Medicine in the Eighteenth-Century Boerhaave School*, Palgrave Studies in Medieval and Early Modern Medicine,
https://doi.org/10.1007/978-3-030-51541-6_6

Fig. 6.1 Johannes de Gorter. Line engraving by Jacob Houbraken after Jan Maurits Quinkhard, 1735. (Credit: Wellcome Collection. CC BY)

argued, would dissolve thick and slimy mucus, open pores, and enable the so-called insensible perspiration to normalise, thereby allowing the patient to recover.

While applauded by some, De Gorter's work also met with fierce criticism, most notably from physician Antonius Stochius (1681–1752) of Enkhuizen, a city right across the Zuiderzee. Referencing Hippocrates, who had argued that diseases ought to be treated with remedies whose efficacy lay in the application of substances with qualities opposing that of the disease, Stochius questioned both the efficacy of the pungent sal ammoniac and De Gorter's pathological theory concerning the obstructed nervous juice and suppressed perspiration. Stochius sarcastically observed that 'if the renowned professor had *kept this secret to himself* and within the walls of Harderwijk, the Republic would not have suffered such damage'.[2] He claimed that the condition of one of his patients had actually worsened when he had administered sal ammoniac. De Gorter's physiology of insensible perspiration and the criticism he received illuminate the argument of this chapter, namely that at the turn of the eighteenth-century physicians considered the role of the nerves, developing a neurological theory about the physiological process of perspiration.

For much of the seventeenth century, the study of insensible perspiration had mostly focused on digestion. The Venetian physician Santorio Santori (1561–1636), in particular, meticulously measured everything he ate, drank, all his urine, and faeces, and compared these figures with the changes in his body weight throughout the day and night. He was convinced that health constituted a harmonious equilibrium between ingestion and excretion. Although his practice of quantification may appear modern in hindsight, Santorio's approach to insensible perspiration was firmly grounded in ancient humoral theory, which was the basis of the tradition of regimen known as 'the six things non-natural' (*sex res non-naturales*). Because healthy or unhealthy perspiration correlated with these six categories, Santorio measured the secretion of perspiration

[2] Antonius Stochius, *Dissertatio medico-practico de morbo epidemico hac hyeme grassato, necdum cessante cum animadvers. in cel. J. Gorteri brevem descriptionem et curationem morbi epidemii per diaphoresin* (Enkhuizen, 1733), 18; idem, *Geneeskundige verhandeling over eene algemeene volk-ziekte, die deze winter gewoedt heeft en nog niet ophout*, trans. Antonius Stochius, jnr (Enkhuizen, 1733), 17. Emphasis original.

depending on air, food and drink, sleeping and waking, exercise and rest, sexual intercourse, and the passions of the soul.[3]

But at the turn of the eighteenth century medical perceptions of insensible perspiration experienced a transformation, as physicians began to focus on the role of microscopic nerves and arteries, as well as the nature of bodily fluids. De Gorter continued to make measurements with the help of the weighing chair. De Gorter's measurements, it must be noted, were far less extensive than Santorio's had been. Whereas Santorio minute recorded his weight for over thirty years, De Gorter's data were collected in the course of just one year and which, by his own admission, were incomplete. But contrary to Santorio, De Gorter incorporated neurological descriptions of the internal functioning of perspiration into his medical treatises.[4] Moreover, he developed a chemico-pathological understanding of the catarrh, based on the concept of obstructed perspiration and stagnating particles, which turned sharp. This shift in emphasis from balance and digestion on the one hand, to mechanical and chemical explanations of the motion of the fluids and the nerves on the other, had far-reaching effects as physicians, anatomists, and natural philosophers began to map out the physiology of perspiration on the basis of the anatomy of the body in minute detail. Likewise, doctoral students at medical faculties across the Dutch Republic studied the effects of suppressed perspiration as well as the drugs and treatments to relieve it.

Scholarship on sweat and perspiration has not yet addressed this shift. Historians of medicine and science including Lois Magner, Nancy Siraisi, and Fabrizio Bigotti have focused on the emergence of calculation and quantification in medical research. They present Santorio as part of a

[3] Jerome J. Bylebyl, 'Nutrition, Quantification and Circulation', *Bulletin of the History of Medicine*, 51 (1977), 369–85, here 377–8. On the six non-naturals, see Lelland J. Rather, 'The 'Six Things Non-Natural': A Note on the Origins and Fate of a Doctrine and a Phrase', *Clio Medica*, 3 (1968), 337–47; Peter H. Niebyl, 'The Non-Naturals', *Bulletin of the History of Medicine*, 43 (1971), 486–92; Chester R. Burns, 'The Nonnaturals: A Paradox in the Western Concept of Health', *The Journal of Medicine and Philosophy*, 1 (1976), 202–11; Sandra Cavallo and Tessa Storey, eds., *Conserving Health in Early Modern Culture: Bodies and Environments in Italy and England* (Manchester, 2017); Rina Knoeff, ed. *Gelukkig Gezond! Histories of Healthy Ageing* (Groningen, 2017); James Kennaway and Rina Knoeff, eds., *Lifestyle and Medicine in the Enlightenment: The Six Non-Naturals in the Long Eighteenth Century* (London, 2020).

[4] The knowledge of nerves, or a treatise on and description of the nerves, was called neurography or neurology. See, for example, Willem Séwel, *A Compleat Dictionary English and Dutch*, ed. Egbert Buys, 2 vols (Amsterdam, 1766), vol. 2, 513.

medical revolution that rested on experiments and personal observation rather than abstract theory.[5] Santorio then fits nicely into the narrative of the scientific revolution, which identifies the mechanisation of the world as the primary novelty of the time, and a fundamental departure from the ancients. But as Andrew Cunningham has shown, although early modern physicians and anatomists like Andreas Vesalius (1514–1564) and Hieronymus Fabricius ab Aquapendente (1537–1619) may appear modern from our perspective, they were, in fact, working within an ancient Galenic framework.[6] The same was true for Santorio, who aimed to confirm long-standing theories on perspiration. It was only at the turn of the eighteenth century that these physiological ideas were gradually called into question, making De Gorter's work more truly modern than Santorio's ever was. Similarly, although Lucia Dacome has rightfully drawn attention to the longevity of Santorio's famous weighing chair, she, too, does not discuss the shift in physicians' perception of insensible perspiration, but continues to underline the ancient association between ingestion and excretion.[7]

Thus, by looking at Dutch physicians, and De Gorter in particular, this chapter highlights eighteenth-century Boerhaavian medicine as having formulated an alternative view of insensible perspiration. I first discuss the cultivation of multiple methodologies that supported the development of theories on an internal physiological process of perspiration. These methods included measurements with the help of the weighing chair, but also microscopic observations, anatomical studies, and chemico-botanical

[5] Lois N. Magner, *A History of Medicine*, 2nd ed. (Boca Raton, 2005), 263–6; Nancy G. Siraisi, 'Medicine, 1450–1620, and the History of Science', *Isis*, 103 (2012), 491–514, here 504–5; Fabrizio Bigotti, 'A Previously Unknown Path to Corpuscularism in the Seventeenth Century: Santorio's Marginalia to the *Commentaria in Primam Fen Primi Libri Canonis Avicennae* (1625)', *Ambix*, 64 (2017), 1–14; idem, 'The Weight of the Air: Santorio's Thermometers and the Early History of Medical Quantification Reconsidered', *Journal of Early Modern Studies*, 7 (2018), 73–103; Fabrizio Bigotti and David Taylor, 'The Pulsilogium of Santorio: New Light on Technology and Measurement in Early Modern Medicine', *Societate si Politica*, 11 (2017), 55–114.

[6] Andrew Cunningham, *The Anatomical Renaissance: The Resurrection of the Anatomical Projects of the Ancients* (Aldershot, 1997); Fabrizio Bigotti, *Physiology of the Soul: Mind, Body, and Matter in the Galenic Tradition of the Late Renaissance (1550–1630)* (Turnhout, 2019).

[7] Lucia Dacome, 'Living with the Chair: Private Excreta, Collective Health and Medical Authority in the Eighteenth Century', *History of Science*, 39 (2001), 467–500; ead., 'Balancing Acts: Picturing Perspiration in the Long Eighteenth Century', *Studies in History and Philosophy of Biological and Biomedical Sciences*, 43 (2012), 379–91.

experiments. I then attend to pathology, to demonstrate how, in the case of the cold, a link was made between the chemical properties of sweat and pharmaceutical drugs. For it was only in De Gorter's new framework of perspiration that it became possible to study the chemical properties of nervous juice and sal ammoniac.

Balancing Ingestion and Excretion

Santorio was interested in perspiration and its possible suppression, because he deemed the quantitative imbalance of bodily fluids crucial in foreseeing the onset of a disease. It was a priority, therefore, to balance ingested food and the excretion of waste matter in an attempt to prevent disease. To analyse this balance in more detail Santorio had designed a scale to measure his *ingesta* and *excreta*. The scale consisted of a large steelyard with an adjustable weight hanging on one side, and a chair on the other (see Fig. 6.2). Santorio was able to obtain detailed information on the body's relative mass through systematic measurements made with this chair over a long period of time. As the total weight of food and drink far exceeded the weight of the visible discharges, he concluded that the discrepancy could only be explained by insensible perspiration. Assuming that insensible perspiration could be determined by systematic weighing, Santorio argued that it was heavier than all other forms of excretion combined, and that it was not constant but varied depending on sex, age, internal factors—like sleeping and digestion—and external conditions, such as hot or cold weather.[8]

Santorio's aphorisms on insensible perspiration in relation to food and drink reflected long-standing views on digestion, held by medieval and early modern physicians alike, and inspired by the Greek philosopher Aristotle (384–322 BC). When awake, the stomach was filled with food and drink, but once asleep these nutrients were believed to be heated, broken down, and altered in the body. Thanks to the heat of the body, hot vapours or fumes would arise from the stomach, transporting a warm humidity up to the cold brain. As these vapours condensed, the moisture descended back into the body, nourishing the internal organs, or leaving

[8] Santorio Santori, *Ars de statica medicina, sectionibus aphorismorum septem comprehensa* (Venice, 1614).

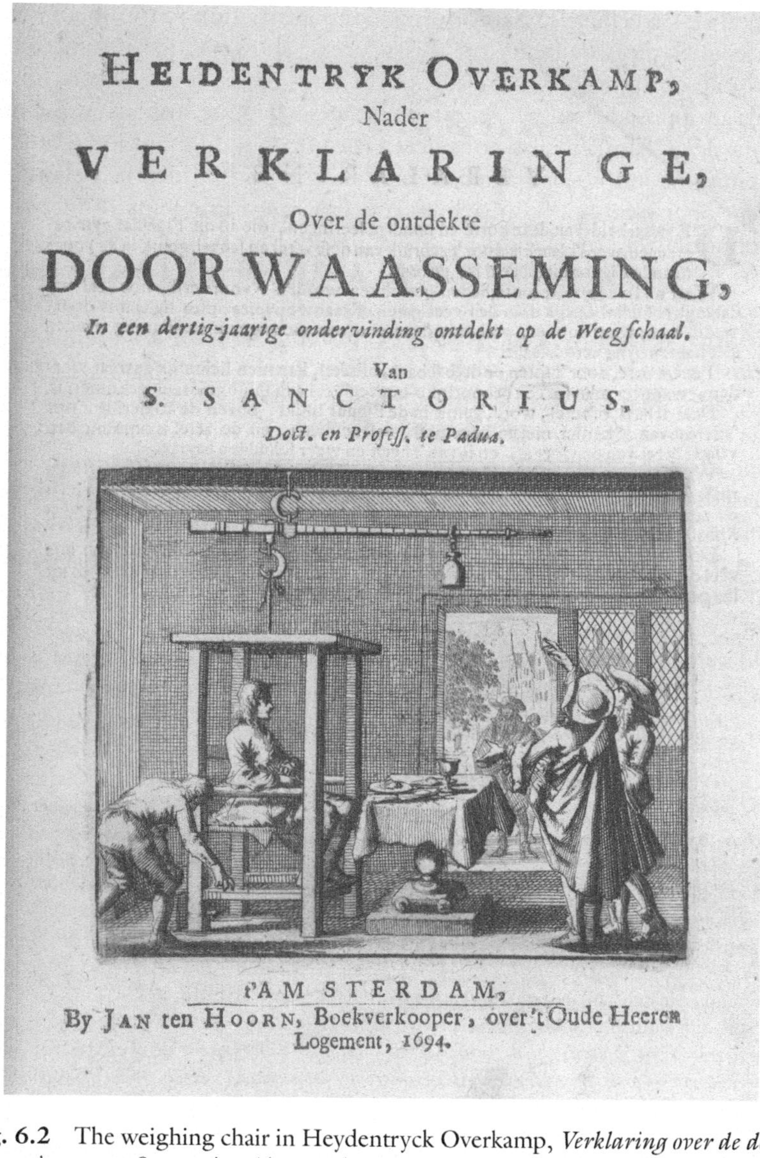
HEIDENTRYK OVERKAMP,
Nader
VERKLARINGE,
Over de ontdekte
DOORWAASSEMING,
In een dertig-jaarige ondervinding ontdekt op de Weegschaal.
Van
S. SANCTORIUS,
Doct. en Profess. te Padua.

t'AMSTERDAM,
By JAN ten HOORN, Boekverkooper, over 't Oude Heeren Logement, 1694.

Fig. 6.2 The weighing chair in Heydentryck Overkamp, *Verklaring over de doorwazeming van Sanctorius* (Amsterdam, 1694). KB National Library of the Netherlands, The Hague

the body if superfluous.[9] Santorio's major contribution to this theory were his detailed measurements on this process, showing, for instance, that insensible perspiration could add up to as much as forty ounces in one night, as opposed to merely eighteen ounces when the stomach was empty during sleep. Santorio also considered the effect of different foodstuffs on perspiration, and warned against foods such as pork, which he believed to obstruct perspiration.[10] But while Santorio's measurements of insensible perspiration were revolutionary, and although they would be used by physicians for generations to come, they were still firmly rooted in ancient humoral physiology. Indeed, as the professor of theoretical medicine Santorio was mainly occupied with explaining and commenting on the aphorisms of Hippocrates, Galen, and Ibn Sina.[11] Like other Renaissance physicians, then, Santorio's intention was not to replace old ideas, but rather to prove the Aristotelian notion of digestion and Galen's concept of perspiration through detailed and exact measurements.

Based on these ancient understandings of digestion and insensible perspiration, Santorio looked for ways of sustaining health, arguing that this rested on a perfect balance between ingestion and excretion. This equilibrium depended less on the quality of ingested food than on its quantity, which could be precisely controlled by using the weighing chair. Physicians could calculate the correct quantity of food for any given person as follows: before a meal, one placed a weight corresponding to the quantity of food on the other end of the beam. The moment the subject had eaten

[9] On premodern notions and medical perceptions of sleeping and digestion, see Sasha Handley, *Sleep in Early Modern England* (New Haven, 2016); Sandra Cavallo and Tessa Storey, *Healthy Living in Late Renaissance Italy* (Oxford, 2013), 113–44; Karl H. Dannenfeldt, 'Sleep: Theory and Practice in the Late Renaissance', *Journal of the History of Medicine and Allied Sciences*, 41 (1986), 415–41.

[10] Santorio Santori, *De statica medicina et de responsione ad staticomasticem* (Leiden, 1642), 55–80; idem, *De ontdekte doorwaasseming des menschen lichaams*, ed. Heydentryck Overkamp (Amsterdam, 1686), 66–100. The ounces refer to weight, not volume.

[11] Santorio Santori, *Commentaria in Artem medicinalem Galeni* (Venice, 1612); idem, *Commentaria in primam fen primi libri Canonis Avicennae* (Venice, 1625); idem, *Commentaria in primam sectionem Aphorismorum Hippocratis* (Venice, 1629). Santorio also sent copies of these books to his friends. See, for example, Santorio Santori to Senatore Settala: 'I send to His Lordship the two books on the Avicenna's text and I pray His Lordship to read them carefully, because He will read new thoughts, yet based on the authorities of Hippocrates and Galen, as far as both practice and experience are concerned'. Santorio Santori to Senatore Settala, 27 December 1625, in Carlo Castellani, Alcune lettere di Santorio Santorio a Senatore Settala, Estratto dalla Rivista Castalia, Milan, 1958.

and drunk sufficiently, the scale would tip over and the chair drop down, signalling the end of the meal (notice the food on the table in Fig. 6.2).[12] As the encyclopaedist Ephraim Chambers (c. 1680–1740) succinctly put it in 1728, the function of Santorio's weighing chair was 'to determine the quantity of food taken at a meal; and to warn the feeder when he had eat his quantum'.[13] This confirmed that Santorio's weighing chair was aimed less at gaining insights into the physiology of perspiration, than at assisting early modern weight-watchers in avoiding overeating or, conversely, under-perspiring.

Santorio's work had a long-lasting impact on early modern medicine, just as the association between perspiration and digestion continued to be relevant.[14] In the Dutch Republic, Santorio's ground-breaking work *Ars de statica medicina* was reprinted several times by booksellers David Lopez de Haro (Leiden, 1642), Adriaen Vlacq (The Hague, 1657, 1664), and Cornelis Boutesteyn (Leiden, 1703, 1711, 1713, 1728). In 1683 a Dutch translation by Philippe La Grue (*b.* 1658) appeared, while later editions included commentaries by Cartesian physicians Steven Blankaart and Heydentryck Overkamp (Amsterdam, 1684, 1686).[15] Further, medical students at Dutch universities, such as Thomas Secker (1693–1768) and Herman Hulshof (*c.* 1715–1742), defended their dissertations on the topic of Santorio's insensible perspiration.[16]

Although these names and publications hardly constitute an exhaustive reception history of Santorio in the Dutch Republic, they clearly reveal the widely shared fascination with Santorio's work among Dutch physicians. They also testify to the continued existence of long-standing ideas regarding perspiration, because the notion of an invisible vapour continuously leaving the body was widely accepted. In a discussion on respiration, Galen had stated that one went via the lungs, but 'the other [respiration], which

[12] See preface in Santori, *De statica medicina*; idem, *De ontdekte doorwaasseming*.

[13] Ephraim Chambers, *Cyclopaedia, or, An Universal Dictionary of the Arts and Sciences*, 2 vols (London, 1728), vol. 2, 359.

[14] On Santori's long-term influence, see Dacome, 'Living with the Chair'; ead. 'Balancing Acts'; and outside the realm of medicine, Lucia Dacome, 'Resurrecting by Numbers in Eighteenth-Century England', *Past and Present*, 193 (2006), 73–110.

[15] Santorio Santori, *De ontdekte doorwaasseming of de leidstar der genees-heeren*, trans. Philippe La Grue (Amsterdam, 1683); idem, *De ontdekte doorwaasseming des menschen lichaams*, ed. Steven Blankaart, trans. Philippe La Grue, 2nd ed. (Amsterdam, 1684).

[16] Thomas Secker, *Disputatio medica inauguralis de medicina statica* (Leiden, 1721); Hermannus Hulshof, *Dissertatio medica inauguralis sistens febrem diariam benignam ex suppressa Sanctoriana perspiratione ortam* (Groningen, 1740).

has no name because it is not commonly known, since it escapes observation on account of its tenuousness, is called insensible perspiration (*adelos diapnoe*)'.[17] These ideas proved remarkably enduring, and were still evoked in the 1740s by Albrecht von Haller, who argued that

> If an intense cold could be suddenly produced in a close chamber full of company, one person would not be capable of seeing another through the fog or vapours which exhale from their own bodies, almost in the same manner as the poets feign the gods to be hid each in their proper cloud. (see Fig. 6.3)[18]

The method used to measure the amount of perspired mass also remained constant throughout the early modern period. Santorio's weighing chair continued to fascinate physicians across Europe. Physicians like the Scotsman James Keill (1673–1719) in England and De Gorter in the Dutch Republic performed similar though less prolific retrials of Santorio's weighing experiments.[19]

In addition to the enduring perception of perspiration as insensible, and the continued success of measuring, De Gorter's approach nevertheless attests to a substantial departure in general conception. What set De Gorter's work on insensible perspiration apart from that of his contemporaries was his aim to shed new light on the internal physiological process of perspiration. As I demonstrate in the next section, De Gorter combined measurements made with the weighing chair and hydrometer with new evidence from other fields of medical knowledge, thereby shifting the focus from the role of digestion to the role of the nerves.

[17] As quoted in Edward Tobias Renbourn, 'The Natural History of Insensible Perspiration: A Forgotten Doctrine of Health and Disease', *Medical History*, 4 (1960), 135–52, here 135–6.

[18] Albrecht von Haller, ed. *Praelectiones academicae in proprias institutiones rei medicae*, 6 vols (Göttingen, 1739–1744), vol. 3, 576; idem, *Dr. Boerhaave's Academical Lectures on the Theory of Physic: Being a Genuine Translation of his Institutes and Explanatory Comment*, 6 vols (London, 1742–1746), vol. 3, 307.

[19] See Dacome, 'Living with the Chair'; ead., 'Balancing Acts'. For biographical details on Santorio see Fabrizio Bigotti, 'Santorio, Sanctorius', in *Encyclopedia of Early Modern Philosophy and the Sciences*, ed. Dana Jalobeanu and Charles T. Wolfe (Cham, 2020); Mirko D. Grmek, 'Santorio, Santorio', in *Complete Dictionary of Scientific Biography*, ed. Charles Coulston Gillispie, Frederic L. Holmes, and Noretta Koertge (Detroit, 2008), 101–4.

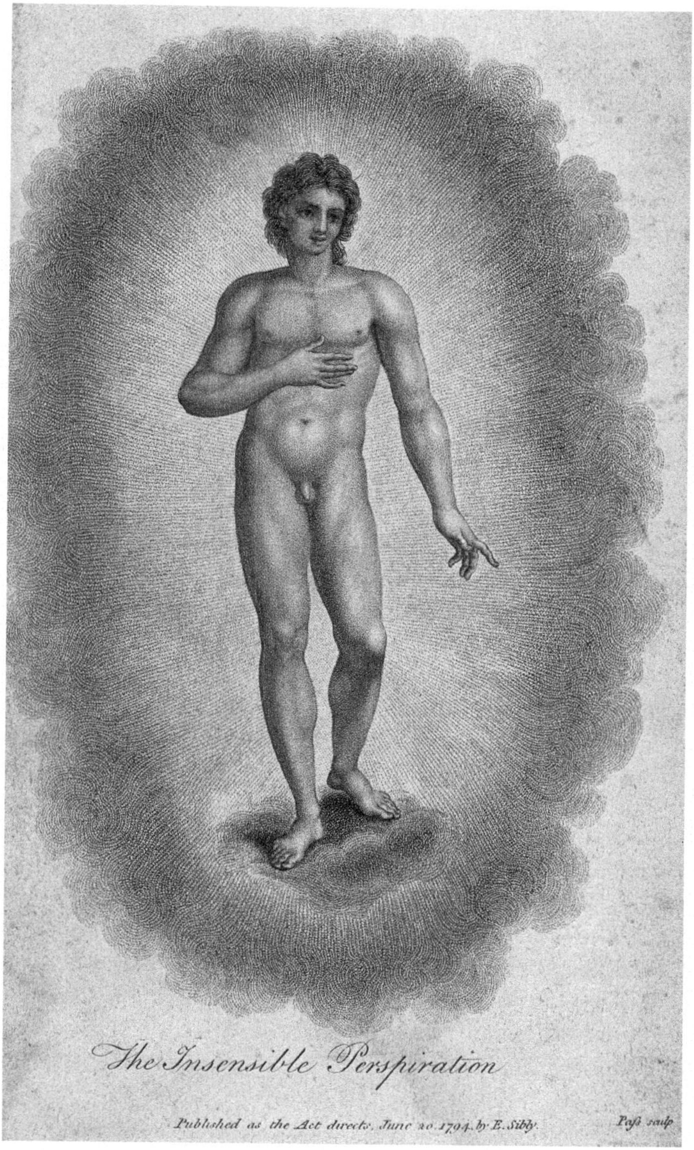

Fig. 6.3 A young man emanating insensible perspiration. Colour stipple engraving by John Pass, in Ebenezer Sibly, *The Medical Mirror* (London, 1794). (Credit: Wellcome Collection. CC BY)

Perspiration and the Nerves

While in the eighteenth century perspiration remained closely linked to ingestion and excretion, the functioning of the nerves took an increasingly prominent role in explaining the internal physiology of perspiration. Besides relying on experimental observations with the weighing chair, medical researchers of the Boerhaave school, in particular Johannes de Gorter and Abraham Kaau (also known as Kaau-Boerhaave), began to link perspiration to the nerves and the anatomy of the skin, which, they argued, was as important as the process of digestion.

One of the reasons why physicians began to doubt the close link between perspiration and digestion as put forward by Santorio was the development of new theories of digestion. In perfect harmony with the Galenic humoral theory, as we have seen above, Santorio had promoted a good night's sleep as beneficial to digestion and perspiration. Physicians in the seventeenth century, however, increasingly perceived digestion as a chemical and mechanical process. As I discussed in Chap. 2, Jan Baptist van Helmont and Franciscus Sylvius moved away from the notion that digestion started with internal body heat. Instead, they argued that digestion was instigated by an active acid in the stomach, causing fermentation that separated nutrients from watery and excremental parts. As such, digestion and nutrition came to be perceived as a series of chemical reactions in the body's internal organs.[20] These chemical theories marked a radical departure from Galenic thermal digestion. Most importantly, the intake of food and drink was no longer related to the origin of vapours inside the body. Consequently, the physiology of perspiration needed to be rethought as well.

This reevaluation of perspiration was also necessary because measuring results could be inconsistent. Physicians adopted Santorio's weighing chair, but the information obtained by De Gorter sometimes contradicted that of Santorio. While still a practicing physician in the port town of Enkhuizen, De Gorter constructed his own weighing chair, known as the *sella statica*, or 'static chair', with the help of friend and colleague Henricus Ris (1687–1727).[21] De Gorter adapted Santorio's experiments by applying

[20] Antonio Clericuzio, 'Chemical and Mechanical Theories of Digestion in Early Modern Medicine', *Studies in History and Philosophy of Biological and Biomedical Sciences*, 43 (2012), 329–37.

[21] See 'Vita auctoris' in Johannes de Gorter, *Praxis medicae systema*, ed. David de Gorter, 2nd ed. (Harderwijk, 1767), [**4r].

them to different regions and he published his findings in 1725 in *De perspiratione insensibili* (see Table 6.1).[22] In accordance with the Galenic notion that digestion—and consequently the creation of vapours and perspiration—occurred during sleep, Santorio had discovered that he perspired considerably more at night than during the day: in the timespan of one day, Santorio had measured a total of 50 ounces of insensible perspiration, of which an average of 29 ounces was perspired at night. Yet as early as in 1718 James Keill published results that deviated from Santorio's. Keill was a Scottish physician and anatomist who practiced in Northampton, and maintained a mathematical and mechanical approach to medicine. Embracing the view that the body was a hydrostatic machine, and with himself as the experimental subject, he performed his own experiments to recreate and question Santorio's ingestion and excretion measurements.[23] Keill perspired just 30 ounces during the day, but contrary to Santorio, he found that he perspired more during the day than during the night (20 ounces versus 10 ounces). Whereas the colder English climate might explain the overall smaller amount of perspiration, the very different ratio between day and night was more problematic. De Gorter, following Keill's experiments, put these contradictory values to the test: his own measurements produced 45 ounces perspiration in total, of which an average of

Table 6.1 Measurements of insensible perspiration

	Time frame	*Santorio Santori*	*James Keill*	*Johannes de Gorter*
I	24 hours	50 oz	30 oz	45 oz
II	At night (average)	29 oz	10 oz	15 oz
III	In the daytime	20 oz	20 oz	30 oz

Source: Johannes de Gorter, *De perspiratione insensibili* (Leiden, 1725)

[22] Johannes de Gorter, *De perspiratione insensibili Sanctoriana-Batava tractatus experimentis propriis in Hollandia* (Leiden, 1725). In good Boerhaave school-fashion, this work was dedicated to Boerhaave, who was praised as the source of all new insights De Gorter may have gained with this study. The book also included a laudatory endorsement by the professor. See De Gorter's 'dedicatio' and Boerhaave's letter in ibid., [*2r–*4v], [**r–**2v].

[23] Anita Guerrini, 'James Keill, George Cheyne, and Newtonian Physiology, 1690–1740', *Journal of the History of Biology*, 18 (1985), 247–66; ead., 'Keill, James (1673–1719),' in *Oxford Dictionary of National Biography* (Oxford: Oxford University Press, 2004); James Keill, *Tentamina medico-physica ad quasdam quaestiones quae oeconomiam animalem spectant, accomodata: Quibus accessit Medicina statica Britannica* (London, 1718).

30 ounces during the day, and 15 ounces at night. Although the total amount was closer to Santorio's than Keill's, De Gorter's proportions confirmed the latter's breakdown into daily and nocturnal perspiration. This corroborated with the slower circulation of blood at night—an important feature which Santorio could not have known as in Santorio's time circulation had not been discovered yet.[24] De Gorter therefore concluded that the human body perspired more during the day than at night, thereby disagreeing with the theory that profuse perspiration was caused by digestion during sleep.[25]

These two criticisms, derived from both theory and experiment, demanded further research on the presumed impact of food and drink on perspiration. De Gorter thus decided to weigh his insensible perspiration before and after lunch. He measured his body weight during the day, and subtracted the amount of perspiration in the morning, to find out the amount of perspiration after midday. Interestingly, De Gorter concluded that the body perspired twice as much in the morning as in the afternoon, that is, more before lunch than after.[26] Moreover, he stated that highest amount of insensible perspiration occurred approximately four hours after eating, from which he deduced that insensible perspiration happened when food and drink had already been digested, thus *independent* of nocturnal sleep. Once again, De Gorter's trials with the weighing chair provided measurements that contradicted Santorio's—and were hence at odds with ancient Galenic physiology.

Once he had disproved the relationship between perspiration and digestion, De Gorter developed a new medical theory on the internal functioning of perspiration. His re-interpretation, presented in *De perspiratione insensibili*, was based on the anatomy and functions of the skin and the nerves. De Gorter based his discussion of the mechanics of insensible perspiration on the ground-breaking work of Dutch anatomists. Santorio had already hypothesised that insensible perspiration evaporated through the pores, but it was only later in the seventeenth century that anatomists, with the help of Van Leeuwenhoek's microscopes, were able to investigate the pores in greater detail. In doing so, they radically transformed medical perceptions of the skin. It was particularly Govard Bidloo (1649–1713),

[24] De Gorter, *De perspiratione insensibili*, 103.

[25] Ibid., 10–1. De Gorter compared his measurements to those found in Santori, *Ars de statica medicina*; Keill, *Medicina statica Britannica*.

[26] De Gorter, *De perspiratione insensibili*, 12–3.

Antoni van Leeuwenhoek, and Frederik Ruysch who minutely investigated, described, and illustrated this porous tissue.[27] In contrast to the macroscopic dissections of larger bodily tissues, which could be observed with the naked eye, De Gorter reported, anatomists found that innumerable minute openings and sweat glands were situated in the dermis of the skin, and that these pores covered the entire surface of the body. Van Leeuwenhoek once calculated as many as 125,000 pores in a surface area the size of a single grain of sand. In 1685 Bidloo's grand anatomical atlas was published, which contained detailed engravings of the skin's pores.[28] In one of these illustrations, Bidloo showed that a layer of skin consisted of sweat glands and excretory ducts; hairs arising near the pores; and papillary glands shaped not unlike pyramids, which were composed of blood veins, lymphatic ducts, and, of course, nerves. The precise structure and function of these glands were keenly discussed by anatomists and physiologists alike.[29] De Gorter used all of these anatomical observations and physiological discussions to construct a new internal physiology of insensible perspiration. Given the large amounts of perspiration that any human would exude in a day, De Gorter reasoned that the secretion of insensible perspiration occurred on all bodily surfaces, both externally through the outer skin and internally via the lungs, throat, mouth, and nose. Perspiration was secreted wherever the finest blood veins and nerves reached the body surface.[30]

[27] Mieneke te Hennepe, 'Of the Fisherman's Net and Skin Pores: Reframing Conceptions of the Skin in Medicine 1572–1714', in *Blood, Sweat and Tears*, ed. Manfred Horstmanshoff, Helen King, and Claus Zittel (Leiden, 2012), 523–48.

[28] Govard Bidloo, *Anatomia humani corporis* (Amsterdam, 1685), Tabula 4. On Bidloo's atlas, see Rina Knoeff, 'Moral Lessons of Perfection: A Comparison of Mennonite and Calvinist Motives in the Anatomical Atlases of Bidloo and Albinus', in *Medicine and Religion*, ed. Ole Peter Grell and Andrew Cunningham (Aldershot, 2007), 121–43; Marian Fournier, 'De microscopische anatomie in Bidloo's Anatomia humani Corporis (1685)', *TGGNWT*, 8 (1985), 197–208.

[29] For example, while Ruysch maintained that the glands were simply the extremities of veins, and functioned to mechanically separate the fluids into smaller particles, Boerhaave perceived the glands as membranaceous follicles in which chemical processes prepared the fluids for secretion. Herman Boerhaave and Frederik Ruysch, *Opusculum anatomicum de fabrica glandularum in corpore humano* (Leiden, 1722); De Gorter, *De perspiratione insensibili*, 20. On the debate between Boerhaave and Ruysch, see Rina Knoeff, 'Chemistry, Mechanics and the Making of Anatomical Knowledge: Boerhaave Vs. Ruysch on the Nature of the Glands', *Ambix*, 53 (2006), 201–19.

[30] De Gorter, *De perspiratione insensibili*, 19–22.

A similar emphasis on the anatomy of the skin and the physiology of perspiration emerged in the works of other physicians of the Boerhaave school. Students at Leiden learned about insensible perspiration in lectures on medical theory. Insensible perspiration was still named after Santorio (*perspiratio Sanctoriana*), but it was explained on the basis of the newly discovered anatomical structure of the skin. According to Boerhaave's textbook, via these smaller vessels transpired a very thin and 'very subtle humour from every point of the body, called from its inventor the Sanctorian perspiration'.[31] Abraham Kaau, who had studied medicine under the tutelage of his uncle Boerhaave, because of a hearing impairment, discussed perspiration at great length. To the new physiology of the skin he added a detailed discussion of the central role of the nerves.[32] In 1738, Kaau's book *Perspiratio dicta Hippocrati per universum corpus* ('Perspiration over the Whole Body, as called by Hippocrates') was published. It presented a series of anatomical observations and experiments to elucidate the insensible perspiration in all parts of the body, both internally and externally, in sickness and in health. Kaau meticulously observed the reticular structure of the nerves in the epidermis, and the nipples to the alveoli—or tiny air sacs in the lungs—through which the transudation was believed to occur.[33] He also confirmed the permeability of the skin and its excreting properties with the help of physiological experiments. He injected the stomach and hepatic artery with water, and when he observed the sample through a microscope, he noticed the liquid oozing out of every small pore in the stomach and liver—it transuded out more beautifully, he noted, when gentle pressure was applied.[34] Von Haller performed the same experiment with fish glue, a substance extracted from fish by heating the skin or bones in water. Von Haller probably used isinglass (from the Dutch *huysenblaas*, sturgeon's bladder), a kind of gelatine

[31] Herman Boerhaave, *Institutiones medicae in usus annuae exercitationis domesticos digestae*, 5th ed. (Leiden and Rotterdam, 1734), 224; Von Haller, *Academical Lectures*, vol. 3, 306.

[32] On Abraham Kaau, see Irina Sjtsjedrova, 'Abraham Kaau-Boerhaave: Bladzijden uit de biografie van een academicus', in *Noord- en Zuid-Nederlanders in Rusland*, ed. E. Waegemans, J.S.A.M. Koningsbrugge, and Nadja Louwerse, Baltic Studies (Groningen, 2004), 293–312; Luuc Kooijmans, *De geest van Boerhaave: Onderzoek in een kil klimaat* (Amsterdam, 2014).

[33] Abraham Kaau, *Perspiratio dicta Hippocrati per universum corpus anatomice illustrata* (Leiden, 1738). This title, 'Perspiration over the Whole Body, as called by Hippocrates', was a double reference: a paraphrase of Galen's discussion of the expiration and inspiration of the body which, in turn, was referencing Hippocrates, *Epidemics*, book VI, section VI.

[34] Ibid., 251–2.

obtained from sturgeons or other species of fish, which was widely used, among other things, as in food production, by artists, and to seal letters.[35] Once the fish glue was injected into the skin, the liquid soon sweated out from all openings. Von Haller therefore concluded that the epidermis was 'perforated by an infinite number of pores, some larger for the sweat, and others smaller for the perspirable vapours'.[36]

Kaau's emphasis on the function of the nerves echoed De Gorter's increased attention to the operation of the nerves and the nervous juice. De Gorter based his ideas on the work of the Oxford physician Thomas Willis (1621–1675), who was the first to publish comprehensive books on the structure, activity, and purpose of the brain and nervous system. Most notably, Willis proposed the existence of a nervous juice, distilled from blood in the arteries of the brain, in order to explain motion and sensation. Movement, for example, was understood as the agitated fermentation of nervous juice in the muscles.[37] De Gorter, in turn, expanded on nervous juice to explain its central role in the internal working of perspiration. He argued that blood vessels discharged a jelly-like sap or 'gelatinous liquid' (*liquorem gelatinosum*) into the nervous system. This thin juice or spirit flowed inside the nerves, lubricating, moistening, and nourishing the nervous fibres as well as the outermost membrane enveloping the brain and spinal cord. De Gorter argued that the nervous juice would find its way into the body and, once it had fulfilled its purpose and was used, had to be discharged. Crucially, in healthy bodies this occurred in the form of insensible perspiration. De Gorter's theory that nervous juice was the source of perspiration was supported by his observation that nerves permeated the skin in much greater numbers than blood veins—more than seemed actually necessary. In addition, the nerves were more concentrated

[35] Floyd L. Darrow, *The Story of an Ancient Art: From the Earliest Adhesives to Vegetable Glue* (Lansdale, PA and South Bend, IN, 1930). Von Haller may have used fish glue to imitate the 'condensed glue' between the dermis and epidermis. A similar interchange of sweat and fish glue can be found in ancient recipes for plasters. John Scarborough has described Andromachus's strigil plaster, which was based on the dirty sweat gathered from the scrapings of a strigil—instrument with curved blade used by athletes—as an alternative to the isinglass plaster, in 'Fish Glue (Gr. ΙΧΘΥΟΚΟΛΛΑ) in Hellenistic and Roman Medicine and Pharmacology', *Classical Philology*, 110 (2015), 54–65, here 64–5.

[36] Albrecht von Haller, *Primae lineae physiologiae*, 2nd ed. (Göttingen, 1751), 265–6; idem, *Physiology: Being a Course of Lectures upon the Visceral Anatomy and Vital Oeconomy of Human Bodies*, trans. Samuel Mihles, 2 vols (London, 1754), vol. 2, 4.

[37] Thomas Willis, *Cerebri anatome: Cui accessit nervorum descriptio et usus* (Amsterdam, 1664), 149–89; J.T. Hughes, *Thomas Willis, 1621–1675: His Life and Work* (London, 1991).

around the sweat glands than elsewhere. De Gorter concluded that nervous juice, once it had been used, was likely to evaporate as insensible perspiration at the body's surface, via the glands or directly through the pores.[38]

With his emphasis on the properties of nervous juice De Gorter made an important contribution to the study of both the nervous system and the physiology of perspiration. Boerhaave, for example, once he had resigned from the chair of botany and chemistry in 1729, decided to continue his investigation of the nerves and nervous juice, incorporating De Gorter's work. Boerhaave's neurology had always been part of his physiology, but from September 1730 to June 1735 Boerhaave delivered a new series of lectures 'on the diseases of the nerves'.[39] In these lectures he paid particular attention to the spirituous particles contained in the nervous juice. He carefully defined the spirit's main properties, and argued that they eventually left the body via perspiration: first, spirituous particles were exceedingly small, invisible even when using a microscope; they could only be detected by smell. Second, spirits in the nervous juice were extremely volatile. And finally, spirituous particles were endowed with a force (*vis*) to enable motion and sensation.[40] In agreement with De Gorter,

[38] De Gorter, *De perspiratione insensibili*, 20–1; idem, *Morbi epidemii*, 14–6. For his anatomical knowledge, De Gorter relied on the work of other anatomists and most likely his own observations. The University of Harderwijk had had a dissection room since the early eighteenth century, and De Gorter himself specialised in teaching anatomy and surgery to surgeons. For his anatomical lectures, De Gorter commented on the *Anatomische Tabellen* (first published in 1725) by Johann Adam Kulmus (1689–1745), the Leiden alumnus and professor of medicine at the Akademische Gymnasium in Danzig. See the lecture notes of Gerhardus Vermeer, 'Comentaria ex ore Clarissimi Viri J. De Gorter excerpta', Harderwijk, *c.* 1740. Leiden, University Library, BPL 1478.

[39] Herman Boerhaave, 'Praelectiones publice habitae de morbis nervorum', Leiden, 1730–1735. S.M. Kirov Military Medical Academy, St Petersburg, MS XIII 11. Microfiche copy stored at University Library, Leiden, F 699. These lecture notes were transcribed, annotated, and translated by B.P.M. Schulte, *Hermanni Boerhaave Praelectiones de morbis nervorum, 1730–1735: Een medisch-historische studie van Boerhaave's manuscript over zenuwziekten* (Leiden, 1959). A contemporary version, based on students' lecture notes, appeared as Herman Boerhaave, *Praelectiones academicae de morbis nervorum*, ed. Jacob van Eems (Leiden, 1761). See also Rina Knoeff, 'Herman Boerhaave's Neurology and the Unchanging Nature of Physiology', in *Blood, Sweat and Tears*, ed. Manfred Horstmanshoff, Helen King, and Claus Zittel (Leiden, 2012), 195–216.

[40] Boerhaave, 'De morbis nervorum', 21 March 1732 in Schulte, *De morbis nervorum*, 152–5; Rina Knoeff, *Herman Boerhaave (1668–1738): Calvinist Chemist and Physician* (Amsterdam, 2002), 191–2.

Boerhaave taught his students that blood exuded this peculiar spirit into the nervous juice, but also into other bodily fluids, such as saliva and semen. Depending on the location of the body, whether it was the head, ears, armpits, between the toes, or the genital area, the spirit eventually left the body as perspiration, sweat, or the oily, waxy secretion called sebum or cutaneous fat.[41]

In sum, De Gorter drastically changed the way in which physicians conceived of insensible perspiration, and shifted the focus from its direct relation to digestion to the involvement of the nerves. Although De Gorter made use of Santorio's innovative weighing chair, his measurements tended to contradict the ancient conception of perspiration as internal vapours directly resulting from digestion. On the basis of anatomical studies of the skin and the nervous system, De Gorter disproved a direct relationship between digestion and perspiration, and developed an internal physiology of perspiration, based on the nerves and nervous juice, instead. This proved highly fertile, because the nerves continued to be an important topic of medical research throughout the eighteenth century. Anatomists like Abraham Kaau, Albrecht von Haller, and Bernard Siegfried Albinus (1697–1770) devoted much time and effort to investigating the structure, use, and function of the nerves in the following decades.[42] Boerhaave, too, continued his research on the nerves, and incorporated the link between the nervous juice and perspiration in his lectures. Neurological concepts of the body circulated widely among the Dutch medical elite, and as such provided a persuasive explanation for the subtle yet significant transformation of the physiology of perspiration.

Transpiring Juices, Distilling Spirits

So far this chapter has outlined the shifting medical understandings of perspiration, from its close alignment with digestion to the association with the flow of juices in the nerves. However, as medical practitioners developed new theories on perspiration, one important question remained: what exactly was this nervous juice? This section will demonstrate that the

[41] Boerhaave, 'De morbis nervorum', 1 April 1732 in Schulte, *De morbis nervorum*, 154–5.

[42] Kaau, *Perspiratio dicta Hippocrati*. On Albinus's and von Haller's work on the nerves and ideas of nervous juice, see Hendrik Punt, *Bernard Siegfried Albinus (1697–1770) on 'Human Nature': Anatomical and Physiological Ideas in Eighteenth Century Leiden* (Amsterdam, 1983); Hubert Steinke, *Irritating Experiments: Haller's Concept and the European Controversy on Irritability and Sensibility, 1750–90* (Amsterdam, 2005), 68, 110.

prioritisation of nervous juice and its spirituous particles in the study of perspiration was only possible thanks to the contemporary importance of chemistry for medicine, which was an essential component of the Boerhaave school. For although the nervous system and the skin were best studied in anatomy and through microscopy, only chemistry could provide insights into the actual properties of the nervous juice and perspiration, both in healthy and in morbid perspiration.

The spirits in the nervous juice, it turned out, were best studied via the chemical analysis of plants, because just like the human body, they also transpired a watery vapour with a unique odour. Naturally, the collection of individual drops of human sweat for chemical investigation was complicated. However, as the nervous juices in plants were considered similar to the nervous juices in human bodies, this opened up the possibility for elaborate chemical investigations. The notion that vegetal transpiration was similar to perspiration in humans and animals is likely to have been based on early modern comparative anatomy, namely the notion that animals could serve as substitutes for human bodies in the attempt to clarify human anatomy.[43] Furthermore, the words perspiration and transpiration were used interchangeably. In historical English dictionaries, the words 'perspire' (from Latin *perspirare*, from *per- 'through'* and *spirare 'breathe'*) and 'transpire' (from *transpirare*, from *trans- 'through'* and *spirare 'breathe'*) were synonymous prior to the nineteenth century, from which point onwards the former was used for human bodies and the latter in botany.[44] Moreover, the property of having a smell, which was intrinsic in the spirits of plants, would lay at the heart of new notions about the scent of healthy and unhealthy human perspiration.

[43] On the continuity between animals and plants in medical studies, see Fabrizio Baldassarri, 'Botany and Medicine', in *Encyclopedia of Early Modern Philosophy and the Sciences*, ed. Dana Jalobeanu and Charles T. Wolfe (Cham, 2020). See also Anita Guerrini, *The Courtiers' Anatomists: Animals and Humans in Louis XIV's Paris* (Chicago, 2015).

[44] Abraham Kaau occasionally used the term 'transpirare' in his *Perspiratio dicta Hippocrati*. Some medical students preferred the term 'transpiratio insensibilis', see e.g. in Henricus Petrus Sigismundus Zehenphenningh, *Dissertatio medico therapeutica inauguralis sistens quaedam therapiae specialis notamina circa abusus remediorum vomitoriorum, laxantium et sudoriferorum* (Leiden, 1750), 10; Arthurus Magenis, *Dissertatio medica inauguralis de urina* (Leiden, 1753), 10. According to a contemporary English-Dutch dictionary, 'perspiration' and 'transpiration' were translated similarly. Compare 'to Perspire, Uitwaassemen, uitdampen door de zweetgaten' with 'Transpiration, De ongevoelige uitwaasseming door de Huid' in Séwel, *A Compleat Dictionary*, vol. 2, 575, 851.

The similarities between vegetable and animal physiology are also obvious in Boerhaave's landmark textbook on chemistry, the *Elementa chemiae* (1732). As historian of science Ursula Klein has suggested, the chemical theory of plants is also referred and extended to animal bodies.[45] Boerhaave explained the anatomy of plants in terms of animal physiology. Once a plant absorbed nourishment from the earth through its porous roots, the juice in the stems was considered crude, heterogeneous; this was called the 'chyle of the plant', because 'the roots, and the body of the plants [...] answer to the stomach, and intestines of an animal'. As this vegetal chyle ascended the stem of the plant, it slowly but surely transformed into the plant's proper nature by 'force of the air' in 'the real lungs of the plant' (*i.e.* the leaves). The analogy between plants and animals continued further: Boerhaave referred to the seed in the fruit as the embryo of the plant; and when, for example, an aloe was bruised, it bled a yellow and bitter sap, the 'blood of plants'.[46] In short, Boerhaave, just as most botanists and chemists of his day, perceived the physiology of plant juices as similar to the movement and properties of animal fluids.

The connection between animal perspiration and vegetal transpiration was quickly made. Boerhaave, who had grown proverbial green fingers following his appointment to the chair of botany in 1709, taught his students that the leaves of plants gave off a scented water vapour: 'There are emunctory [*i.e.* excretory] vessels in rosemary, for instance, which is continually diffusing an odorous atmosphere around it. And thus, in the summer season, when the air is greatly heated, all aromatic and odoriferous plants come to exhale vast quantities of their fine, spirituous and volatile parts'.[47] Similar to the spirit in nervous juice, which could only be detected by smell, the oily and volatile substance from plants had a unique scent, believed to be excited by the force of its 'guiding spirit' or *spiritus rector*. Based on comparative physiology, then, the way in which fluids moved through processes of absorption, nourishment, and transpiration in plants

[45] Ursula Klein, 'Experimental History and Herman Boerhaave's Chemistry of Plants', *Studies in History and Philosophy of Biological and Biomedical Sciences*, 34 (2003), 533–67, here 543.

[46] Herman Boerhaave, *A New Method of Chemistry: Including the Theory and Practice of that Art: Laid down on Mechanical Principles, and Accommodated to the Uses of Life*, trans. Peter Shaw and Ephraim Chambers, 2 vols (London, 1727), vol. 1, 150–62; idem, *Institutiones et experimenta chemiae*, 2 vols ('Paris', 1724), vol. 1, 121–7.

[47] Boerhaave, *A New Method*, vol. 2, 8; idem, *Institutiones et experimenta chemiae*, vol. 2, 14.

ran parallel to that of fluids in animals and humans, going through the processes of ingestion, digestion, and eventually perspiration.

To analyse the phenomenon of perspiration in even more depth, Boerhaave went beyond mere smelling, and subjected plants to numerous chemical examinations to extract their spirit. For example, he gathered fresh samples of rosemary leaves and placed them in a copper cold-still with a cone-shaped head, which he placed on a simple brick furnace (see Fig. VI in Fig. 2.2). The mild and gentle heat created in this furnace matched the heat from the sun, hence this installation was perfectly suited to 'come to know what it is which spontaneously exhales from vegetables by the warmth of the summer's sun'.[48] The well-regulated fire slowly vaporised the fragrant and most volatile parts from the rosemary leaves. These dewy vapours then condensed against the head of the still, and trickled down the sides of a conical neck, being transmitted to and collected in a receiver. By means of this method, chemists had found a way to capture the scent of plants and preserve it.[49] Ordinarily, Boerhaave elaborated, when people walk in scented gardens they breathed the vapours exhaled by rosemary, basil, jasmine, and many more aromatic plants and trees in the form of insensible perspiration. But by means of distillation chemists were able to extract and obtain this juice, which was 'the most volatile, fragrant and aromatic part of the plant, wherein its specific virtue resided'.[50] The unique scent of each plant and animal could be distinguished by this virtue. Boerhaave defined this spirituous substance in terms of its chemical properties: 'By spirit we mean any sulphurous, or oily matter, so attenuated and subtilised, as to become volatile, by the smallest fire, and miscible with water; which characters, where they concur in the same subject, denominate it a spirit'.[51] The *spiritus rector*, in other words, was perceived as an actual particle or substance that was spirituous, subtle yet pervasive, and the cause of an animal or plant's unique scent.[52]

[48] Boerhaave, *A New Method*, vol. 1, 379–80.

[49] Ibid., vol. 2, 13. See also Alain Corbin, *Le miasme et la jonquille: l'odorat et l'imaginaire social, 18e-19e siècles* (Paris, 1982); translated as *The Foul and the Fragrant: Odor and the French Social Imagination* (Leamington Spa, 1986).

[50] Boerhaave, *A New Method*, vol. 2, 12–3, 8.

[51] Ibid., vol. 1, 168.

[52] Cultural historians have studied the human experience of smell and stench, such as Corbin, *The Foul and the Fragrant*. But the case of *spiritus rector* shows how smell was not just a sensation in individuals, but rather considered a real, material thing.

On the basis of the distillation of rosemary and other plants Boerhaave's students learned about the smells particular to living bodies, as well as what constituted the exhalations. What was called *spiritus rector* in vegetal juices was called the *effluvia* in the nervous juices of animals. As Boerhaave explained: 'Exhalations or effluvia; viz. this same spirit, together with the elementary water which contains it, which is continually flying off from them, during the day, in the form of an insensible perspiration'.[53] In other words, animal scents were explained in terms of the spirits or *effluvia*, the most subtle parts that were present in the nervous juices and perspiration. Their existence and unique character was demonstrated with the help of a very visual and evident anecdote. As Boerhaave recalled, 'I saw a dog at Amsterdam having lost his master in a crowd of people, was therefore obliged to run about continually after the steps of his master 'till he came to the house in which he had taken up his residence'.[54] This provided evidence that everyone had their own smell in their perspiration.

Supported by these chemical investigations of the airs and scents exuded by plants, physicians increasingly regarded particular scents and smells of plants and animals as unhealthy. Although many well-to-do ladies used scented spirits to dress their hair and wore perfume from lavender, oranges, cloves and herbs, physicians took a much more sober view. The physician Jan Ingenhousz (or Ingen-Housz, 1730–1799), for example, warned about the negative properties of the transpiration of plants. Born in Breda, Ingenhousz had studied at the universities of Leuven and Leiden, in 1768, and after working as a physician for a few years, he was appointed as a court physician to the Habsburg Empress Maria Theresa (1717–1780). Ingenhousz continued the rich tradition of the Boerhaave school in Vienna, which had begun when Maria Theresa had convinced Gerard van Swieten to come to the Austrian capital in 1745. Together with Antonius de Haen (1704–1776) Van Swieten reformed the curriculum of the medical faculty. They added a chair in botany and chemistry, to which Nikolaus Joseph von Jacquin (1727–1817) was appointed, a Leiden alumnus like De Haen and Van Swieten.[55] While researching the insensible

[53] Boerhaave, *A New Method*, vol. 2, 18.

[54] Von Haller, *Academical Lectures*, vol. 3, 325. Boerhaave shared a variation of the same anecdote during his lectures on the diseases of the nerves, saying that a dog can distinguish a single deer from the herd solely on the basis of the smell of its perspiration. Boerhaave, 'De morbis nervorum', 1 April 1732 in Schulte, *De morbis nervorum*, 154–5.

[55] Frank T. Brechka, *Gerard van Swieten and His World 1700–1772* (The Hague, 1970); J.K. van der Korst, *Een dokter van formaat: Gerard van Swieten, lijfarts van keizerin Maria*

transpiration of plants by means of chemistry, Ingenhousz noted that during the day flowers produced a beneficial air, whereas at night they exuded noxious air. He therefore warned his patients not to keep flowers near the bedside: 'I make no doubt but a great quantity of plants, kept in a close and small place during a night, or by day in the dark, may do some material mischief, and even occasion death, to any person who should be imprudent enough to remain in such a place'.[56]

Ingenhousz' warning may be considered exemplary of the shifting medical perception of the smell of insensible perspiration. In the late medieval and early modern periods, physicians such as Gabriele Zerbi (1445–1505) had recommended scattering rose blossoms or other sweet-scented flowers across beds to assure a good night's sleep.[57] By the eighteenth century, however, the nocturnal transpiration of plants was thought to be noxious, if not deadly. Boerhaave provided a long list of examples for the dangers of the spirits in plants: smelling the blossoms of beans, roses, oleander, and the tuberous hyacinth would cause fainting; and a stable servant who had slept under a layer of saffron was known to have died.[58]

Since the spirits and *effluvia* in the insensible perspiration of plants and animals were essentially the same phenomenon, the heavy smell of human sweat now underwent a shift in perception as to its salubrity as well. According to Galen and Santorio, insensible perspiration released superfluous moist and cleansed the body of potentially harmful and dangerous matter. Santorio's measurements of large amounts of perspiration confirmed the theory that sweating was beneficial to health, and prevented harmful substances from accumulating in the body.[59] In fact, for centuries the scent of sweat was regarded positively. The perhaps best-known examples are the pleasant scent of the skin of Alexander the Great, whose natural bodily odour apparently required no added decoration of ointments,

Theresia (Amsterdam, 2003); Jacob Boersma, *Antonius de Haen, 1704–1776: Leven en werk* (Assen, 1963).

[56] Jan Ingenhousz, *Experiments upon Vegetables* (London, 1779), 47–9; Handley, *Sleep*, 42. Ingenhousz is often heralded as the discoverer of photosynthesis. See for example Geerdt Magiels, *From Sunlight to Insight: Jan IngenHousz, the Discovery of Photosynthesis & Science in the Light of Ecology* (Brussels, 2010).

[57] Catrien Santing, 'Sleeping and Waking', in *Gelukkig Gezond! Histories of Healthy Ageing*, ed. Rina Knoeff (Groningen, 2017), 80–97.

[58] Boerhaave 'De morbis nervorum', 13 May 1732 in Schulte, *De morbis nervorum*, 164–5.

[59] Michael Stolberg, 'Sweat: Learned Concepts and Popular Perceptions, 1500–1800', in *Blood, Sweat and Tears*, ed. Manfred Horstmanshoff, Helen King, and Claus Zittel (Leiden, 2012), 503–22, here 510–1.

and sweat and oil scraped from the skin of Greek and Roman athletes with strigils. This mixture was sometimes bottled and sold as a medical treatment, called *gloios*, to relieve pains and sprains.[60] In ancient times, therefore, sweat and its scent enjoyed both the positive connotation of physical ability related to athleticism, and the thought of containing great medicinal value.

However, supported by chemico-botanical analyses of the nervous juice in the early modern period, the spirits were gradually demystified, and the smell of human sweat came to be a sign of disease. One of the physicians who had specialised in the smell of sweat as a diagnostic tool was Henricus Buisen (*c.* 1678–*c.* 1724), who claimed that the intensity of smell in patients' sweat indicated how sick they were. Born in Coevorden, Buisen studied medicine at the University of Groningen before practicing medicine in Haarlem. In 1706 he published a treatise on human secretions and excretions, which was reprinted several times in the course of the eighteenth century. According to Buisen, a heavy smell in someone's sweat was a sign that it contained 'degenerate and some already corrupted particles'.[61] This was caused either by the retention of perspiration, or by the presence of putrefied blood and other circulating fluids in the body. Although each individual exhibited a unique body odour, those who were ill were believed to exude 'heavy' smells. Buisen was convinced that melancholic individuals in particular suffered from bilious juices and a 'bad and heavy smell' in their sweat.[62]

Other eighteenth-century Dutch physicians also took the foul smell of sweat both as a sign of and as sensory evidence for disease. In their lectures on symptoms and signs of illness Boerhaave and Von Haller taught that the healthy body did not generally stink. When an odour arose from the body, this was caused by extreme heat or excessive physical exercise, which caused the oily and saline parts of the fluids to dilute and evaporate.[63] This was potentially harmful to the nervous system of the affected individual, and even that of others. The spirits of a mother, for example, could negatively affect the nervous system of her infant in a remarkable way, Boerhaave

[60] Boerhaave 'De morbis nervorum', 1 April 1732 in Schulte, *De morbis nervorum*, 156–7; Onno van Nijf, 'Exercise and Rest', in *Gelukkig Gezond! Histories of Healthy Ageing*, ed. Rina Knoeff (Groningen, 2017), 60–79.

[61] Henricus Buisen, *Verhandelinge van de uitwerpingen des menschelyke lighaams, bestaande in pis, afgang, zweet, kwyl; en braaking* (Rotterdam, 1731), 115–6.

[62] Ibid.

[63] Von Haller, *Academical Lectures*, vol. 6, 111.

argued. When her spirits were excessively inflamed by heat and motion, and exuded by panting and sweating while the infant was sleeping under the same sheets, this could incite various nervous diseases in the child, including epilepsy, paralysis, and mental impairment.[64]

Buisen and De Gorter even went as far as to provide a classification of the smell and properties of sweat, so that physicians might diagnose their patients' diseases. The location of sweat on the body, for example, often indicated the seat of the disease. Heat and excessive sweating in the course of a fever indicated a prolonged illness, which in the case of smallpox, for example, could be deadly. Malodorous sweat was clearly visible on someone's shirt, which would become stiff and smell. 'I have seen many times', Buisen wrote, 'that when such a sweat persisted for long, it drowned out the sufferer's strength, causing death'.[65] But not all sweat was unwholesome. According to De Gorter, warm and odourless or slightly fragrant sweat was a sign of health if not continuously flowing. And when a patient was suffering from an inflammation or was salivating, the patient's sweating could indicate the expulsion of superfluous and malignant parts, and the dwindling of the disease.[66]

Eighteenth-century physicians, in short, studied perspiration with some understanding of comparative physiology and the practice of chemistry. Via rosemary and other plants, chemists of the Boerhaave school further developed the concept of *spiritus rector* and *effluvia* in vegetal and animal bodies to explain the unique character of each species as expressed in a unique smell.[67] Early botanists, natural historians, and physicians had always been aware of the smells in nature, but they generally regarded them as safe and healthy. This drastically shifted with Boerhaave's and Ingenhousz' chemical studies on the juices. The nocturnal transpirations of flowers were now considered dangerous to health. Likewise, the heavy smells exuded by humans via insensible perspiration could be harmful and cause nervous diseases.

[64] Boerhaave, 'De morbis nervorum', 12 May 1732 in Schulte, *De morbis nervorum*, 162–3.

[65] Buisen, *Uitwerpingen*, 115–6.

[66] De Gorter, *Praxis medicae systema*, 171–2.

[67] In the late eighteenth century, Joseph Franz von Jacquin, who inherited his father's position as professor of botany and chemistry at the University of Vienna in 1797, still referred to Boerhaave's concept in his discussion of the volatile, ethereal oil transpiring from plants. 'The newer chemists regard it as a simple substance and they call this smell, *aroma*. Boerhaave called it the spirit of plants, *spiritus rector*'. Joseph Franz von Jacquin, *Leerboek der algemeene en artsenijkundige scheikunde*, trans. Gerardus Plaat, 2 vols (Leiden, 1794), vol. 2, 3.

Diagnosis by smelling and inspecting sweat is unlikely to ever have been as common as diagnosis by uroscopy—as has been pointed out by Michael Stolberg.[68] Other than the properties of good and light sweat versus the bad and heavy, physicians did not distinguish between many other qualities of perspiration. Descriptions of urine, by contrast, proved more diverse, with colours, taste, and smells ranging from heavy and sharp to sour and fetid. Moreover, the availability and transportability of urine, and the anonymity of urine collection and inspection, made it far easier to analyse than sweat.[69] Nevertheless, the odour of sweat was part of a physicians' diagnostic toolbox. De Gorter, Buisen, and other Dutch physicians had a nose for smelling disease in their patients' sweat. More importantly, their research on the spirits of plants prompted the clear departure of medical understanding of the smell of perspiration from the Galenic conception, as sweat came largely to be viewed as unhealthy.

Sweating It Out

In the early decades of the eighteenth century, Johannes de Gorter was confronted with an epidemic outbreak of catarrh. Not unlike today's common cold, catarrh caused trouble breathing, coughing, chest pains, lightheadedness, and lapses into sleep. Sometimes these symptoms coincided with an increased body temperature. Patients soon started spitting up phlegm. And as long as the mucilage was stuck in the windpipe and in the larynx under the epiglottis, it continued to irritate these parts, causing patients to cough uncontrollably in an attempt to spit out the mucus.[70] Dutch newspapers even reported on European royalty suffering from catarrh. In the night of 2 February 1706 Frederick I, King in Prussia (1657–1713), was 'struck by a dangerous catarrh'. Although his conditions had improved four days later, he continued to suffer from a 'heavy cold'.[71] For the development of the medical theory of the time, the disease of catarrh is important because it is another example for a shift towards a neurological understanding of perspiration. A professor *and* medical

[68] Stolberg, 'Sweat', 510–1.

[69] On the benefits of uroscopy, see Michael Stolberg, *Uroscopy in Early Modern Europe*, trans. Logan Kennedy and Leonhard Unglaub (Farnham, 2015).

[70] De Gorter, *Morbi epidemii*, 5–7; idem, *Korte beschryving van een algemene doorgaande ziekten, in deze tijd nog woedende, en desselfs genezing door sweetinge*, trans. Amos Lambrechts (Amsterdam, 1733), 10–4.

[71] *Opregte Leydse courant*, 15 February 1706.

practitioner, De Gorter did not limit himself to the development of physiological theories on the internal functioning of the body. On the contrary, he also aimed at applying his knowledge to actual patients. Searching for a treatment to the outbreak of catarrh, De Gorter explained its aetiology with the extreme suppression and retention of perspiration, which brought about a morbid sharpness in the nervous juice. The primary treatment for this severe and unpleasant illness was the administration of sudorific and diaphoretic drugs, in order to, literally, sweat out the disease. As this section will demonstrate, De Gorter based his investigations upon the use of measuring instruments. However, it was only within the context of the chemistry of the nervous juice that he could really develop and define the theoretical underpinnings of the efficacy of pharmaceuticals. At the same time, the application of a neurological understanding of perspiration to actual diseases and drugs also made De Gorter more vulnerable to failure and criticism. Published pamphlets arguing against his proposed therapy and theory demonstrate that new physiological ideas were not immediately or universally adopted at the time.

As De Gorter investigated the catarrh epidemic, he soon realised that patients from various social and geographical backgrounds all suffered from the same illness during the same season, even though they had very different habits and diets. The only common denominator, De Gorter reasoned, was the chilly air blowing on their faces and the cold rains on their bare skin. Indeed, it was in this period that Europe experienced some of its coldest winters in history, also known as the Little Ice Age. The winters in the first half of the eighteenth century were so cold that the Zuiderzee froze over, and Dutchmen were able to sleigh from Enkhuizen to Stavoren, from one side of the sea to the other.[72] In 1707 the botanical garden of Harderwijk, initially at sea, was relocated right next to the academy building where the plants were better protected against the wind.[73] De Gorter hypothesised that the wintry weather conditions were to blame for the epidemic, and used Santorio's weighing chair to measure the fluc-

[72] A news item reported that a group of horse-drawn sleighs crossed the frozen sea ('gisteren tusschen de 30 en 40 Paerden met Sleden van Stavoren tot Enckhuysen zijn overgekomen') in *Oprechte Haerlemsche Courant,* 9 February 1709. See also *Gebeurtenissen, voorgevallen in de maanden January en February, anno 1740 in, en veroorzaakt door de nooit meer gehoorde vehemente en strenge winter* (Enkhuizen, 1740).

[73] J.J. Meinsma, 'Het scheikundig onderwijs aan de Gelderse Hogeschool (1648–1811) en Rijksathenaeum (1815–1818) te Harderwijk', *Scientiarum Historia*, 14 (1972), 201–16, here 203.

tuation in perspiration in relation to variations in temperature and humidity.

With his weighing chair, De Gorter discovered that the body emitted less insensible perspiration in cold and moist than in hot and dry air. He reached this conclusion by adding a thermometer and a hygrometer to his experiments, which indicated temperature and humidity, respectively. Designed by De Gorter's student Petrus Belkmeer (1703–1763) from Medemblik, this particular kind of hygrometer had a cone-shaped body with a spiralling groove to fit a long wire. Belkmeer had matriculated from the University of Harderwijk in 1732, and for his dissertation, which he defended in 1735, he researched motion as the cause and general treatment of all diseases.[74] For the second edition of *De perspiratione insensibili* (1736), De Gorter had Belkmeer's instrument engraved for inclusion in his book (see Fig. 6.4). A sponge and a weight were hung on either end. As the moisture in the air increased, the soft, porous body of the sponge would absorb the moisture, and its weight increase. Because of the conically shaped pulley, the slightest change in weight caused it to turn.[75] As De Gorter measured less perspiration when he increased the humidity, he deduced that cold and moist air caused the smaller vessels in the body to contract, which would congeal the fluids, and the pores to close, thereby suppressing perspiration.[76]

Based on these experiments and his ideas on nervous juices, De Gorter was able to explain the cause of catarrh. When in cold and moist conditions the retention of insensible perspiration persisted, an excess of the 'thin fluid' (*tenuis humor*) of the nervous juice was kept inside the body, and encouraged the growth of internal sores: since it was unable to exit the body, the redundant fluid would flow to other parts within it. It agitated and became lodged in the lungs, trachea, and mucous membranes due to its increasing 'sharpness' (*acer*), ultimately generating the inflammation typical for catarrh, with its characteristic dripping mucus and coughed-up slime. This identification of coldness and dampness as the

[74] Petrus Belkmeer, *Disputatio inauguralis physiologico medica de motu ut causa et curatione generali omnium morborum* (Harderwijk, 1735). After his graduation Belkmeer became a physician and teacher at the Baptist church in Enschede.

[75] Johannes de Gorter, *De perspiratione insensibili*, 2nd ed. (Leiden, 1736), 339–42. A multitude of hygrometer designs circulated, all based on the principle of absorption levels of various materials. See Joachim d'Alencé, *Verhandelingen over de barometers, thermometers, en notiometers of hygrometers* (The Hague, 1730).

[76] De Gorter, *De perspiratione insensibili*, 109.

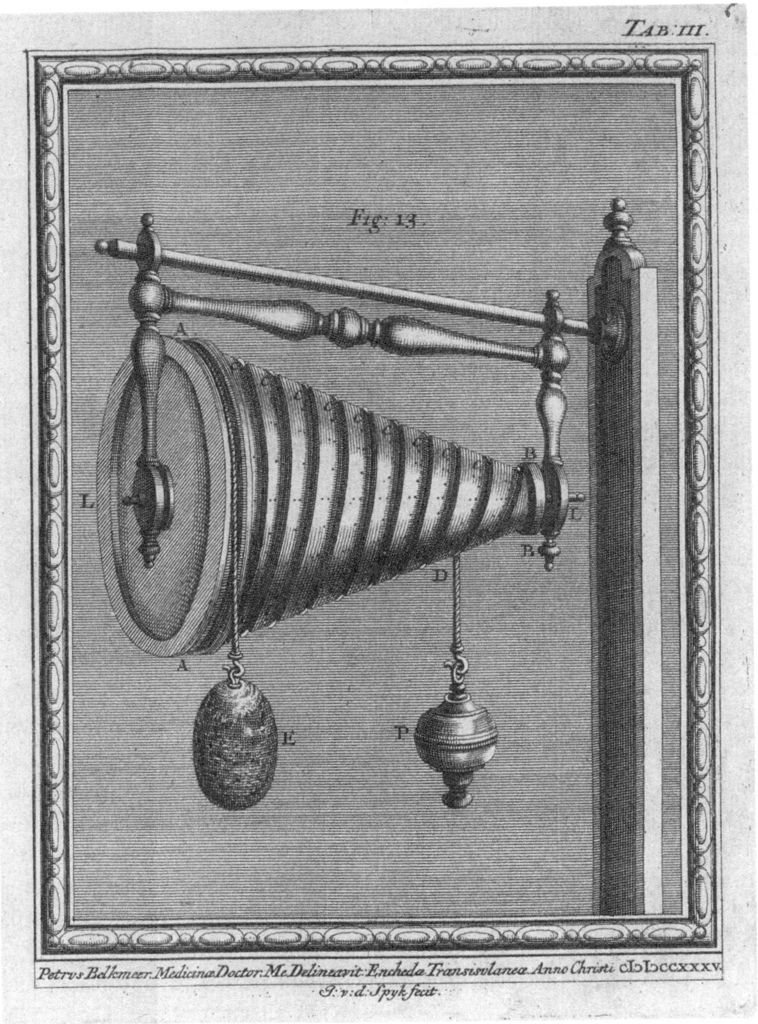

Fig. 6.4 Jacobus van der Spijk, hygrometer by Petrus Belkmeer, published in De Gorter, *De perspiratione insensibili* (Leiden, 1736). Allard Pierson, University of Amsterdam, O 62-6242

root cause of catarrh was supported by the observation that particular regions in the body were especially affected by the cold air: 'the lung and other breathing parts that lay bare in the air', such as the skin, throat, and nose, 'and [the areas] where it predominantly does its harm, are the most affected by this disease'.[77]

Fellow-physicians of the Boerhaave school agreed with De Gorter on the accumulation of nervous juice was the cause of disease, and even death. Referring to the autopsies carried out by the Swiss physician Théophile Bonet (1620–1689), Boerhaave recalled cases of obstructed and overflowing nervous juice in patients who had suffered from catarrh:

> Someone died because of a suffocating catarrh. There was an enormous supply of fluid in between the dura and pia mater. The surface of the windings appeared gelatinous, and when punctured water spewed out. [...] The cold, cough, full-headed catarrh; the head of the dead [patient] was overflowing with water.[78]

Boerhaave ascribed these and other instances in which fluids aggregated to nervous diseases. Gaubius, too, agreed with De Gorter's conclusion that bitterly cold and moist air could prevent the body's natural secretion of perspiration. He argued that, in a freezing and damp atmosphere, 'smaller vessels are contracted, the humours inspissated, the pores shut, the perspiration suppressed'.[79]

Catarrh was not only explained as an excessive accumulation of nervous juice; it was especially in combination with the property of sharpness that the symptoms manifested themselves. De Gorter developed an external and internal theory to explain the presence of sharpness in the nervous juice. External sharpness entered the body by way of harsh and acidic elements such as salt, vinegar, poisons, or—in the case of catarrh—some discrete and malignant particle in the air and wind, which were introduced to the body via the pores. Since much remained unknown about this miasmic sharpness, De Gorter hypothesised that any harmful substance would need to be 'thin and fine, and of such a nature that for its treatment it

[77] De Gorter, *Morbi epidemii*, 3; idem, *Doorgaande ziekten*, 5.

[78] Boerhaave, 'De morbis nervorum', 19 January 1731 in Schulte, *De morbis nervorum*, 88–9; Boerhaave, *De morbis nervorum*, 74.

[79] Hieronymus David Gaubius, *Institutiones pathologiae medicinalis* (Leiden, 1758), 208; idem, *The Institutions of Medicinal Pathology*, trans. Charles Erskine (Edinburgh, 1778), 139.

requires either an improvement or an excretion'.[80] More likely than this was, however, the internal emergence of sharpness in the course of the generation of the nervous juice, which was either not well-prepared in the first place, or retained in the body for too long, having become sharp due to having experienced numerous circulations through the nerves. When the weakened body was unable to excrete these sharp particles they caused serious damage. De Gorter located the sharp material mostly where insensible perspiration ordinarily took place, namely in between the nerves and their enclosing membranes and fibres, and in the ganglia (a network of nerves).[81]

Fortunately for De Gorter's patients, a possible treatment for catarrh was just around the corner. De Gorter aimed to find a workable solution for his patients, and reasoned that drugs could dissolve the thick and slimy mucus, whether caused by internal or external causes.[82] Patients simply had to take sudorific and diaphoretic medicines, which lifted their obstructed perspiration, started a profuse sweat, and thus expelled the sharpness from their nervous systems. Early modern pharmacists generally distinguished sudorifics from diaphoretics: whereas the former promoted the expulsion of sweat, the latter diminished the resistance of the exhaling vessels in the skin and promoted insensible perspiration. In practice, however, these remedies were often the same, and differed only in their degree of action. According to De Gorter, 'sudorifics do not differ much from diaphoretics, except that they are a little more powerful, and therefore they weaken more, and excite several evils, which are not expected from the use of gentle diaphoretics'.[83] Indeed, in the recipes De Gorter provided in his *Formulae medicinales* the sudorific drugs overlap with the diaphoretic drugs.[84] The sudorific drug that caused sweating most effectively, according to De Gorter, was sal ammoniac.

The selection of sudorific and diaphoretic drugs was extensive, but De Gorter specifically chose sal ammoniac, because it agreed with his neurological interpretation of perspiration. Within Galenic medicine all

[80] De Gorter, *Morbi epidemii*, 4; idem, *Doorgaande ziekten*, 7–8.

[81] De Gorter, *De perspiratione insensibili*, 26–35.

[82] De Gorter, *Morbi epidemii*, 7; idem, *Doorgaande ziekten*, 15.

[83] Johannes de Gorter, *Medicinae compendium in usum exercitationis domesticae digestum*, 2 vols (Leiden, 1731–1737), vol. 2, 218.

[84] See *s.v.* 'diaphoretica' and 'sudorifera' in the 'Index formularum medicinalium generalis'. Johannes de Gorter, *Formulae medicinales cum indice virium quo ad inventas indicationes inveniuntur medicamina* (Harderwijk, 1750), [H2v], [X4v].

simplicia (simple ingredients) exhibiting hot and dry qualities were thought to contain the capacity to dissolve. They were, therefore, believed to open the pores, in particular syrup made from blessed thistle (*cardus Benedictus*), devil's bit (*radix scabiosa*), and aromatic angels' root (*radix angelica*).[85] Yet chemical processes of distillation and sublimation introduced medicines made from the most potent essences of plants, animals, and minerals. Chemically produced sweat-inducing drugs included the silvery-white antimony (*antimonium diaphoreticum*), sometimes taken with opium, and the spirits of wine stone, deer horn, and bezoar.[86] De Gorter published over a hundred recipes for decoctions, pastes from powders, plasters, pills, drops, drinks, infusions, soaps, potions, spirits, and scents, all believed to induce perspiration.[87] But despite the existence of this rich assortment of sudorific drugs, De Gorter preferred to prescribe sal ammoniac as the most effective treatment for catarrh. He argued that the ingested sal ammoniac would be absorbed in the blood vessels of the oesophagus and stomach. Once in circulation, it would be discharged as part of the nervous juice and reach the infected skin, lungs, and mucous membrane. It was there that the 'spirit of sal ammoniac' (*spiritus salis ammoniacus*)—combined with steam baths and poultices (a soft, moist mask consisting of bran, flour, and herbs, and applied to the body to relieve soreness and inflammation)—dissolved the mucus. Furthermore, as sal ammoniac carried the chemical property of sublimation, and turned into vapour upon being heated, it promoted insensible perspiration. It freed patients from their symptoms by purging their bodies from all corrupt 'sharp and thin fluid' (*humor tenuis et acris*), and sweating out the 'slimy rawness' of the superfluous nervous juice.[88] Hence, De Gorter's reason for preferring sal ammoniac over numerous other sudorific drugs

[85] De Farvacques, *Medicina pharmaceutica, of Groote algemeene schatkamer der drôgbereidende geneeskonst*, 3 vols (Leiden, 1741), vol. 1, 18. This book was originally published in 1681 but reprinted on Gaubius' recommendation.

[86] For numerous other drugs, see Buisen, *Uitwerpingen*, 184–5; Noël Chomel, *Huishoudelyk woordboek: Vervattende vele middelen om zyn goed te vermeerderen, en zyne gezondheid te behouden, Met verscheiden wisse en beproefde middelen*, trans. Jan Lodewyk Schuer and A.H. Westerhof, 2 vols (Leiden and Amsterdam, 1743), 1455.

[87] De Gorter, *Formulae medicinales*. Under *diaphoretica*, it mentioned *apozema, bolus, cataplasma, electuarium, emplastrum, epithema, fotus, guttulae, haustus, infusio, lavamentum, mixtura, pilulae, potio, pulveres interni, spiritus*, and *suffitus*.

[88] De Gorter, *Morbi epidemii*, 20–2; idem, *Doorgaande ziekten*, 71–7.

was not based on simple trial and error. Rather, his neurological approach to perspiration lay at its basis.

Many physicians agreed that sal ammoniac had a dissolving and opening effect on the human body. Purified salts like sal ammoniac, Boerhaave's students were told, were a 'powerful, stimulating, aperient, diuretic, sudorific, and resolving medicine; capable also of promoting insensible perspiration, and of entring [*sic*] into the lungs, nostrils, and cavities of the head and mouth, in the form of invisible *effluvia*'.[89] To make sal ammoniac, chemists such as Boerhaave and De Gorter mixed one part of an ordinary salt with five parts of urine in a vessel, and gained by sublimation a white, friable substance.[90] Dissolved in water, sufferers simply held a bottle with sal ammoniac up to their noses and drank ten to twenty drops of the sudorific with some beer every morning and evening.[91] De Gorter claimed to have successfully cured his patients from the widespread disease, and made this public in a short treatise, *Morbi epidemii brevis descriptio et curatio per diaphoresin* ('Brief Description of an Epidemic Disease and the Treatment by Sweating', 1733).[92]

* * *

In the early eighteenth century physicians of the Boerhaave school reconceptualised the physiology of perspiration, and debated the important role of the nerves. Although the ancient notion of insensible perspiration continued to be seen as essential to bodily health, the way in which physicians thought about the internal physiological process of perspiration underwent profound changes. First, new weighing measurements and anatomical observations moved physiologists away from Galenic concepts of perspiration, which were bound up with those of digestion and sleeping, and instead drew them towards a neurological theory. Second, the growing emphasis on the role of nervous juice and the presence of *spirits* or *effluvia* in perspiration was largely indebted to the chemical and botanical investigations of plants. Finally, the catarrh epidemic around 1730 offered a unique opportunity to apply the new physiological theory of

[89] Boerhaave, *A New Method*, vol. 2, 224. Emphasis added.

[90] Chambers, *Cyclopaedia*, vol. 1, 140. See also *s.v.* '*Ammoniacum, Sal*', in the 'Index medicamentorum' in De Gorter, *Formulae medicinales*.

[91] As suggested in Henricus Buisen, *Practyk der medicine, ofte Oeffenende geneeskunde*, 4th ed. (Rotterdam, 1743), 1–11.

[92] De Gorter, *Morbi epidemii*; idem, *Doorgaande ziekten*, 22.

perspiration to a pathological case study. Sal ammoniac proved a most effective drug to sweat out the disease.

It should be noted, however, that this historical account of the changes that took place in perceptions of perspiration and the nerves was far from a straightforward advancement in eighteenth-century medicine. During the catarrh epidemics in the 1730s the theory of obstructed insensible perspiration and the efficacy of sal ammoniac were a topic of vigorous debate.[93] In May 1733 Antonius Stochius expressed his disappointment about the ineffectiveness of sal ammoniac, claiming that his patient's condition had, in fact worsened, after the sudorific had been administered. Stochius favoured multiple bloodlettings to treat catarrh instead.[94] His criticism hit a nerve, for it also raised doubts about Johannes de Gorter's neurological interpretation of perspiration. Stochius noted that the dissection of a human body barely produced three drops of nervous juice, so how could just a few drops ever weigh as much as 45 ounces? De Gorter had argued that, when a nerve of a living animal was cut, a large volume of juice would pour out, but Stochius claimed that this fluid was not nervous juice at all, but rather drainage from the lymphatic vessels. He also responded to De Gorter's observation that, when the nerves were laid bare, they felt moist and wet. Stochius wryly observed that 'the renowned author seems not to remember that all the viscera, vessels, and internal parts of the body are always moist externally'. He did not propose an alternative physiology to explain insensible perspiration, but concluded that De Gorter's presentation of nervous juice as the source of perspiration was based on mere assumption. 'Truly, a Professor on his Chair may blazon such things in his lecture hall to his disciples', he wrote, 'but wanting to impose such absurd things on the learned world does not pass'.[95]

[93] The De Gorter-Stochius debate prompted the publication of six pamphlets within a year. These are, in chronological order: Stochius, *De morbo epidemico*; De Gorter, *Doorgaande ziekten*; Stochius, *Algemeene volk-ziekte*; Henricus Vosch van Avesaet, *Lapis lydius animadversionum expertissimi domini A. Stochii … J. Gorteri … brevem ejus descriptionem et curationem morbi epidemii per diaphoresin* (Harderwijk, 1734); Cornelius Belkmeer, *De streelende speel-pop van Dr. Antoni Stochius naakt uytgekleet en ten toon gestelt in de beschouwing rakende syne sogenaamde geneeskundige verhandeling en aanmerkingen over eene algemeene volk-ziekte* (Enkhuizen, 1734); Anthonius Stochius, jnr, *De streelende speel-pop van Dr. Cornelis Belkmeer, door hem eerst naakt uitgekleed, en onder den gewaanden naam van Be-Minnaar der Waarheid, door zynen kwalyk geslepen bril beschouwt, dog nu volkomen ontleed in haare mismaaktheid ten toon gesteld, en den poppemaker weder t'huis gezonden* (Enkhuizen, 1734).

[94] Stochius, *De morbo epidemico*; idem, *Algemeene volk-ziekte*.

[95] Stochius, *De morbo epidemico*, 23–7; idem, *Algemeene volk-ziekte*, 21–6.

Despite this fierce criticism, De Gorter refused to compromise on his neurological theory of perspiration, and he did not change his practice of prescribing sudorific drugs either. Quite to the contrary, he solidified his theory by continuing to lecture on nervous juice and insensible perspiration, and by publishing on the topic. He thus revised and improved his *De perspiratione insensibili* for a second edition which appeared in 1736.[96] In 1737 he also published the second volume of his *Medicinae compendium*, a popular textbook on general diseases and therapies, in which he devoted a separate chapter to insensible perspiration. In that chapter, De Gorter proudly stated that:

> As is described in my book *De Perspiratione Insensibili*, the augmenting medicines, which produce the most excellent and in all respects the least weakening evacuation, both from the arterial extremities as from the nerves, are today called *diaphoretics* by all the Physicians.[97]

Thanks to chemistry, perspiration played a pivotal role in preserving health. But as the next chapter will show, chemistry was about to play an even bigger role in the understanding of disease.

[96] According to De Gorter, his publisher requested a reprint, but he first wanted to improve and elaborate on the text; De Gorter, *De perspiratione insensibili*, [***].

[97] De Gorter, *Medicinae compendium*, vol. 2, 215.

CHAPTER 7

Semen in Flux

In the early modern period venereal disease was notorious, both among sufferers and physicians. As Herman Boerhaave noted, numerous men experienced an involuntary flux or dripping of seed from their penis without an erection or arousal.[1] The proliferation of novel microscopic observations of semen by Antoni van Leeuwenhoek and Nicolaas Hartsoeker revealed ***animalcules*** ('little animals', or spermatozoa) in semen.[2] Yet this discovery did little to resolve the mystery of infertility or the morbid discharges from the urethra known as gonorrhoea ('flux of semen'). Because what exactly ***caused*** venereal disease?[3] Ever since its outbreak in

[1] Herman Boerhaave, 'Tractatus medicus de lue venerea praefixus Aphrodisiaco', in *Aphrodisiacus* (Leiden, 1728); idem, *A Treatise on the Venereal Disease and its Cure in all its Stages and Circumstances*, trans. John Barker (London, 1729).

[2] Much has been written about the discovery of sperm and the debates about its role in procreation. See for example Clara Pinto-Correia, *The Ovary of Eve: Egg and Sperm and Preformation* (Chicago, 1997); Matthew Cobb, *The Egg and Sperm Race: The Seventeenth-Century Scientists Who Unravelled the Secrets of Sex, Life and Growth* (London, 2006); Shirley A. Roe, *Matter, Life, and Generation: Eighteenth-Century Embryology and the Haller-Wolff Debate* (Cambridge, 1981).

[3] In the early modern period, physicians considered venereal disease and gonorrhoea as the same disease. The fundamental distinction was made in the nineteenth century. On the conflation of gonorrhoea and syphilis as merely different degrees of the same disease, see Kenneth M. Flegel, 'Changing Concepts of the Nosology of Gonorrhea and Syphilis', *Bulletin of the History of Medicine*, 48 (1974), 571–588; Ludwik Fleck, *Genesis and*

R. E. Verwaal, *Bodily Fluids, Chemistry and Medicine in the Eighteenth-Century Boerhaave School*, Palgrave Studies in Medieval and Early Modern Medicine,
https://doi.org/10.1007/978-3-030-51541-6_7

the fifteenth century, physicians were puzzled by its rapid spread, and its severe symptoms, and they struggled to consolidate the disease with contemporary medical theory. Diseases like gonorrhoea and syphilis were challenging traditional humoral theory, for they defied a clear explanation as an imbalance of the humours, and they further did not respond to the customary treatments of bloodletting and purgation. And as these diseases were generally tainted with issues of immorality, physicians often left their treatment and study to surgeons.[4] This changed at the turn of the eighteenth century. Hieronymus Gaubius asked his students: 'does the genital liquor, not by the *animalculae*, but by some very singular quality, besides the common affections, contract a peculiar acrimony, by which it induces salacity?' Was perhaps an inertness of semen the cause of sterility?[5] Gaubius, in other words, suggested that physicians ought to investigate the changeable properties of bodily fluids like semen to explain venereal diseases.

This chapter demonstrates that the growing appreciation of chemistry in eighteenth-century medicine motivated physicians to investigate the *cause* and development of disease in terms of the changeable chemical properties of the fluids. Gaubius in particular advocated a better understanding of the bodily fluids in chemical terms in the search for an explanation for disorder and disease. The combination of systematic reasoning and his experimental investigations in the chemical laboratory redefined bodily fluids as active, heterogeneous, and open to change. Morbid semen, as I will explain in more detail, was perceived to be characterised by blandness or sharpness. Gaubius attempted to understand genital disorders and the nature of disease in general by ascribing properties to bodily solids and fluids.

The emphasis on (chemical) causes of diseases led to a new appreciation of pathology, not just as the counterpart of the much more important discipline of physiology, but as a discipline in its own right, with its own textbooks and extensive lecture series. At the turn of the eighteenth

Development of a Scientific Fact, ed. Thaddeus J. Trenn and Robert K. Merton, trans. Fred Bradley (Chicago, 1979).

[4] Jon Arrizabalaga, John Henderson, and Roger French, *The Great Pox: The French Disease in Renaissance Europe* (New Haven, 1997), 56–87. Some surgeons accused physicians of negligence because they refused to research and treat gonorrhoea. See for example Abraham Titsingh, *Cypria tot schrik van haar bondgenooten, en redding der gestruikelden ter eere van de heelkonst en tot dienst der heelmeesteren geschreven*, 2 vols (Amsterdam, 1742).

[5] Hieronymus David Gaubius, *Institutiones pathologiae medicinalis* (Leiden, 1758), 175; idem, *The Institutions of Medicinal Pathology*, trans. Charles Erskine (Edinburgh, 1778), 116.

century disease was either understood as a collection of symptoms that could be perceived by the practitioner, or as lesions in organs that might be observed in post-mortem examinations. The chemical laboratory then stimulated a third way of understanding pathology. Experiments enhanced the chemical understanding of bodily fluids like blood, urine, and milk, both in practice and in theory. Based on experiments, chemists identified basic constituent elements, each with their own properties, which together were thought to make up the entire body. Gaubius went a step further: he understood the chemical properties of fluids as *causes* for bodily disorders, thus allowing for a new coherent and systematic theory of disease, or pathology.[6] In the course of the eighteenth century pathology became a rational system. It was no longer a series of descriptions of particular diseases, but a set of basic principles through which physicians could study and explain the causes and effects of disease.

Medical practitioners had debated the origins and symptoms of disease since antiquity, but the eighteenth century witnessed an intensification of disease theory in two fields of study: nosology and morbid anatomy. The empirical approach to disease based on symptoms was called nosology.[7] Well known is the system from English physician Thomas Sydenham (*c.* 1624–1689), which arranged diseases in the order of their symptoms. Sydenham argued that the causes of diseases could with no certainty be known, so physicians should restrict themselves to the investigation of symptoms. Medical practitioners, Sydenham argued, ought to perceive only the superficies of bodies, not the minute processes in nature's 'abyss of cause', because that would only lead to 'curious and irrelevant speculations'.[8] In a similar vein, the French physician and professor of medicine in Montpellier François Boissier de Sauvages de la Croix (1706–1767) undertook an ambitious project to classify all diseases on the basis of their symptoms. His aim was to reveal the essential character of every disease, and to place the corresponding and identifying symptom in a classificatory

[6] The word pathology derives from the Greek *pathos*: 'suffering, disease' and the Latin *logia*: 'study'.

[7] Nosology derives from the Greek *nosos*: 'disease, sickness'. On nosology, see for example Jacalyn Duffin, *History of Medicine: A Scandalously Short Introduction*, 2nd ed. (Toronto, 2010), 73–76.

[8] Thomas Sydenham, *The Whole Works of That Excellent Practical Physician Dr. Thomas Sydenham Wherein Not Only the History and Cures of Acute Diseases Are Treated of, after a New and Accurate Method: But Also the Shortest and Safest Way of Curing Most Chronical Diseases*, trans. John Pechy (London, 1701).

scheme, to assist practicing physicians in diagnosis and treatment.[9] But these pursuits in nosology did not stop contemporaries from asking the more basic questions of *why* and *how* diseases occurred. It was aetiology, the question of cause, that occupied Gaubius.

A second field of disease theory that developed in the eighteenth century was the anatomy of the diseased body, known as morbid anatomy.[10] Largely spurred by the work of famous Italian professor Giovanni Battista Morgagni (1682–1771), post-mortem examination focused on the so-called seats of disease (*sedibus morborum*) by revealing which bodily organs ordinarily accommodate a certain disease. Numerous corpses were dissected to discern any morbid changes in their organs. Thanks to Morgagni, the opening of bodies and the study of patients' anatomy became immediately relevant to medical theory. The normal and abnormal appearance of organs came to define health and disease.[11] Scholars have paid much attention to the rise of morbid anatomy, for it visually showcased how medical research became truly 'objective'.[12] The borders between morbid anatomy and surgery were ultimately blurred.[13] Furthermore, newly founded hospitals supplied early nineteenth-century researchers with masses of raw material, allowing diseases to be approached in great num-

[9] Roy Porter, *The Greatest Benefit of Mankind: A Medical History of Humanity* (New York, 1999), 260–262; Julian Martin, 'Sauvages's Nosology: Medical Enlightenment in Montpellier', in *The Medical Enlightenment*, ed. Andrew Cunningham and Roger French (Cambridge, 1990), 111–137.

[10] In the late eighteenth century, the terms pathological anatomy and morbid anatomy were synonymous. See for example Eduard Sandifort, *Observationes anatomico-pathologicae*, 4 vols (Leiden, 1777–1781); idem, *Museum anatomicum academiae Lugduno-Batavae descriptum*, 2 vols (Leiden, 1796). Part 2 was titled 'Tabulae anatomico pathologicae'.

[11] Porter, *The Greatest Benefit*, 263–265; Mary Lindemann, *Medicine and Society in Early Modern Europe* (Cambridge, 1999), 86–87; Jürgen Konert and Holger G. Dietrich, 'Giovanni Battista Morgagni und der Beginn der Pathologischen Anatomie', in *Anatomie*, ed. Jürgen Helm and Karin Stukenbrock (Stuttgart, 2003), 127–138; Fabio Zampieri, Alberto Zanatta, Cristina Basso, and Gaetano Thiene. 'Cardiovascular Medicine in Morgagni's De Sedibus: Dawn of Cardiovascular Pathology', *Cardiovascular Pathology*, 25 (2016), 443–452.

[12] Early modern savants were dedicated to 'truth-to-nature'. Applying objectivity to early modern science would be anachronistic. See Lorraine Daston and Peter Galison, *Objectivity* (New York, 2007).

[13] As propagated by, for example, Matthew Baillie in Andrew Cunningham, *The Anatomist Anatomis'd: An Experimental Discipline in Enlightenment Europe* (Farnham, 2010), 219. See also the classic essay by Owsei Temkin, 'The Role of Surgery in the Rise of Modern Medical Thought', *Bulletin of the History of Medicine*, 25 (1951), 248–259.

bers rather than on the basis of individual autopsies.[14] Cunningham has argued that, thanks to the systematic investigation of many corpses, pathology developed as a sub-discipline of anatomy.[15] Yet an analysis of Gaubius's pathology asks for a qualification of that claim. Gaubius ultimately realised that morbid anatomy could not demonstrate all diseases for two reasons. One, quite a few diseases such as the cold involved fluids that cannot be investigated anatomically. And two, morbid anatomy cannot clarify the cause of disease.

While I recognise the development and importance of nosology and morbid anatomy, I argue that we can distinguish a third branch of disease theory in eighteenth-century medicine; one which was inspired by a chemical understanding of the fluids. Examining the works and lectures of Gaubius, and with genital fluxes as a case study, this chapter contributes an additional angle to our common perception of eighteenth-century medical theory as based on symptoms and dissections. It is therefore relevant beyond the case of semen, demonstrating the reconceptualisation of disease as it was disseminated in eighteenth-century medical literature. The possibility of investigating the changing properties of fluids suggested that the nature of a disease could be formulated in terms of chemical elements and be traced back to its cause. In short, the understanding of bodily fluids and seminal fluxes in terms of chemical elements produced a pathology not simply as the knowledge of a particular illness, but as a new kind of disease theory. Whereas the word pathology was first used in the early seventeenth century, this chapter demonstrates how Gaubius redefined the term to refer to a general and rational theory or system of the causes and effects of disease.

Gonorrhoea

Knowledge of successful treatments of venereal disease was much sought after in the early modern period. Many men suffered from *gonorrhoea virulenta* or venereal running, and some experienced excessive pain while urinating. Anxiety about genital health and fertility was widespread, as can be inferred from the innumerable advertisements in Dutch newspapers promoting pills, balms, and other medicaments to cure venereal disease.

[14] Lindemann, *Medicine and Society*, 87; Russell C. Maulitz, *Morbid Appearances: The Anatomy of Pathology in the Early Nineteenth Century* (Cambridge, 1987).

[15] Cunningham, *The Anatomist Anatomis'd*, 196.

In 1748, for example, Andries Leppink from Amsterdam and C. van Zanten from The Hague advertised that 'by a many-year trial, a physician and naturalist invented an agent, certain and infallible against the Weakening and Infertility, both in Men and Women; either arising by Onania or self-pollution [*i.e.* masturbation], or by old and badly treated Venereal inconveniences and other illnesses'. Those who wanted to consult the practitioner and were interested in buying this product had to send a letter with money worth *f*1.20.[16]

Gonorrhoea attracted much attention because it was widespread in early modern Europe and beyond. Ship's surgeons working for the Dutch East India Company frequently noted cases of gonorrhoea among the crewmen in their diaries. In 1699, Abraham Bogaert (1663–1727) listed the medicines he prescribed to four crewmen suffering from inflammatory discharges. Also, the surgeon Christoffel Schabagger noted a case of nine infected crewmen whom he cured within five weeks.[17] In the eighteenth century physicians started to recognise different stages and degrees of this devastating disease.[18] According to Boerhaave, the disease could, after initial contagion, cause the excretion of pus, and spread to other parts of the body. He distinguished five different stages of gonorrhoea, each extending deeper into the male genitalia: the disease could emanate from the glands of the penis in the first stage, to the prostate gland in the last.[19] The dripping of purulent matter from the urethra could be accompanied by a painless genital ulcer, a *chancre* ('creeping ulcer'). This could ultimately lead to

[16] Advertisements in *'s Gravenhaegse Courant*, 16 December 1748 and *Leydse Courant*, 18 December 1748. Others advertised 'Arcane pills' and 'bottle of balsam' against gonorrhoea for *f*3 in *Oprechte Haerlemsche Courant*, 30 May 1754. By comparison, many books in octavo were sold for less than *f*1. See also *De groote zonden van vuile zelfs-bevleckinge, door jonge en oude, mans- en vrouwspersoonen, ontdekt*, trans. J. van Hode (Rotterdam, 1730). Michael Stolberg, 'An Unmanly Vice: Self-Pollution, Anxiety, and the Body in the Eighteenth Century', *Social History of Medicine*, 13 (2000), 1–21; idem, 'Self-Pollution, Moral Reform, and the Venereal Trade: Notes on the Sources and Historical Context of *Onania* (1716)', *Journal of the History of Sexuality*, 9 (2000), 37–61. On the commercialisation of medicinal substances in the eighteenth century, see Frank Huisman, 'Gezondheid te koop: Zelfmedicatie en medische advertenties in de Groninger en Ommelander Courant, 1743–1800', *Focaal*, 21 (1993), 90–130.

[17] See A.E. Leuftink, *Harde heelmeesters: Zeelieden en hun dokters in de 18e eeuw*, 2nd ed. (Zutphen, 2008), 121–128.

[18] See for example Philip K. Wilson, 'Exposing the 'Secret Disease': Recognising and Treating Syphilis' in his *Surgery, Skin and Syphilis: Daniel Turner's London (1667–1741)* (Amsterdam and Atlanta, 1999), 149–189.

[19] Boerhaave, 'Lue venerea'; idem, *Venereal Disease*.

the last stage of the disease, known as the pox or lues, today generally identified as syphilis. By identifying these stages, Boerhaave hinted at distinguishing gonorrhoea from syphilis as different diseases, yet this distinction was not followed up. In fact, the notion that gonorrhoea was a symptom or early stage of syphilis was confirmed by the Scottish surgeon and anatomist John Hunter (1728–1793), who allegedly infected himself with a patient's pus so as to closely observe the development of gonorrhoea over an extended period of time.[20]

Besides those who wrote studies on the symptoms and causes, other physicians were occupied with the history and treatment of syphilis. For centuries the pox was considered as foreign, coming in from the Americas. But when the history and dissemination of the disease were researched by Portuguese-born physician António Sanches, he argued that syphilis was native to Europe.[21] Working as a court physician in St Petersburg, Sanches learned about the value of a corrosive sublimate in combination with steam baths as a remedy against syphilis from a German surgeon who had practised in Tobolsk, Siberia. When Sanches communicated this secret to his friend and fellow-disciple of Boerhaave, Gerard van Swieten, the latter pushed him to write a book on the subject. Sanches started writing in 1746, published in 1750, and would continue publishing on the subject for the rest of his life.[22] Other physicians developing treatments for syphilis included the Amsterdam doctor Johannes Schlichting (1703–1765), who, in 1755, expanded his study on the subject with an analysis of the mineral

[20] John Hunter, *A Treatise on the Venereal Disease* (London, 1786); George Qvist, 'John Hunter's Alleged Syphilis', *Annals of the Royal College of Surgeons of England*, 59 (1977), 205–209.

[21] António Nunes Ribeiro Sanches, *Dissertation sur l'origine de la maladie vénérienne* (Paris, 1750); idem, *Examen historique sur l'apparition de la maladie vénérienne en Europe* (Lisbon [Paris], 1774). These works were published anonymously. The edition published by Gaubius in Leiden in 1777 was the first to mention Sanches by name: António Nunes Ribeiro Sanches, *Dissertation sur l'origine de la maladie vénérienne, pour prouver que ce mal n'est pas venu d'Amérique, mais qu'il a commencé eu Europe par une epidémie. Suivie de l'examen historique sur l'apparition de la maladie vénérienne en Europe*, new ed. (Leiden, 1777).

[22] António Nunes Ribeiro Sanches, *Observations sur les maladies vénériennes* (Paris, 1785), 3. On Sanches, see David Willemse, *António Nunes Ribeiro Sanches, élève de Boerhaave, et son importance pour la Russie* (Leiden, 1966); idem, 'Gerard van Swieten in zijn brieven aan Antonio Nunes Ribeiro Sanches (1739–1754)', *Scientiarum Historia*, 14 (1972), 113–143; Luuc Kooijmans, *De geest van Boerhaave: Onderzoek in een kil klimaat* (Amsterdam, 2014).

waters of Vianen. By chemical experiment these waters were proven to contain salts and minerals that were deemed therapeutically beneficial.[23]

Clearly gonorrhoea was a much-researched disease by professors, practitioners and surgeons alike, and the topic of many dedicated treatises. In what follows I explore eighteenth-century medical reflections on diseases in general, comparing three branches of disease theory: nosology, morbid anatomy, and pathology.

A 'Rational Pathology'

In July 1758 Gaubius wrote to his best friend Sanches in Paris that 'finally, I am pleased to send you my *Pathology* so much completed'.[24] Gaubius's textbook *Institutiones pathologiae medicinalis* (1758) testifies to his long-term preoccupation with the nature of disease in chemical terms over a long period of time. From the outset Gaubius intended his students to be the audience of his *Institutiones*. When Boerhaave passed away in 1738, the professor of botany, Adriaan van Royen (1704–1779), was initially appointed to take over his lectures on the theory and practice of medicine. This included the teaching of pathology. However, it was rumoured that Van Royen's first class caused the English students to flee the lecture hall and implore Gaubius to teach instead.[25] Gaubius took over and his lectures were indeed widely appreciated. The 22-year-old student Johann Heinrich de Brunn (1732–1759) from Schaffhausen enthusiastically wrote that 'Gaubius's lectures are so outstanding, that to be here solely for the sake of this great man, I would never regret it'.[26]

[23] Johannes Daniel Schlichting, *Syphilidos mnemosynon criticon of vrye en oneenzydige gedachten over ongemakken, door 't gebruyk der teeldeelen*, 3rd ed. (Amsterdam, 1755), 649. These elements included *crocus martis* (saffron from steel, or red iron oxide), *sal neutrum* (a salt neither acid nor alkali), and *spiritus mineralis* (the distilled spirit from minerals).

[24] Gaubius to Sanches, Leiden, 21 July 1758, in Sophia W. Hamers-van Duynen, *Hieronymus David Gaubius (1705–1780): Zijn correspondentie met Antonio Nunes Ribeiro Sanches en andere tijdgenoten* (Assen and Amsterdam, 1978), 107.

[25] Johan Frederik Gronovius to Carl Linnaeus, Leiden, 17 March 1739, in *The Linnaean Correspondence*, linnaeus.c18.net/Letter/L0278, Letter L0278 (consulted 28 April 2017). It remains unknown whether it were the contents, or the style and language of Van Royen's lecture that abhorred the students.

[26] De Brunn to Von Haller, Leiden, 30 September 1753, in Albrecht von Haller, ed. *Epistolarum ab eruditis viris ad Alb. Hallerum scriptarum*, 6 vols (Bern, 1773–1775), vol. 3, 419. De Brunn, who was already a medical doctor, matriculated at Leiden on 5 August 1753.

Notably there was something new to appreciate in Gaubius. Sources on Gaubius's teachings reveal a shift in emphasis from physiology to pathology. Extant students' lecture notes demonstrate that Gaubius started out by following Boerhaave's *Institutiones* religiously, reading out specific paragraphs and then explaining and commenting on them (see Fig. 7.1).[27] But a shift can be clearly discerned as time went on. While Boerhaave's compact yet all-encompassing textbook discussed physiology, and in its shadow also the topics of pathology, semiotics, hygiene, as well as therapeutics, Gaubius's lectures were almost exclusively devoted to pathology.[28] To compare, the last edition of Boerhaave's *Institutiones* that appeared during his lifetime (in 1734), included 347 pages on physiology and 62 on pathology. By contrast, Gaubius's lecture notes from 1751 show that he only devoted 157 folios to physiology versus 273 folios to pathology, and an additional 389 to aetiology.[29] Another manuscript confirms this shift in focus, for it summarises and comments on all of Gaubius's pathology lectures, except the section on physiology.[30]

On top of this striking difference of pages dedicated to the different branches of medicine, further far-reaching, and contentious,

[27] See for example 'Dictata Gaubii in Pathologiam Boerhavii', 2 vols, 1745. Copenhagen, Royal Library, Thott 691, 692 kvart; 'H.D. Gaubius, Institutiones pathologicae', 4 vols, Leiden, University Library, BPL 850; 'Viri Celeberrimi H.D.Gaubii M.D. [...] Praelectionum Theoreticarum Secundum Institutiones Medicas H. Boerhaave', 6 vols, Leiden: Philip Bonk, 1751. Leiden, University Library, BPL 1475. According to a contemporary, Boerhaave's *Institutiones* was an 'intelligible, regular, and rational System'. Cited in Andrew Cunningham, 'Medicine to Calm the Mind: Boerhaave's Medical System, and Why It Was Adopted in Edinburgh', in *The Medical Enlightenment*, ed. Andrew Cunningham and Roger French (Cambridge, 1990), 40–66, here 41.

[28] Boerhaave had defined pathology as 'The second Branch of Physic [*i.e.*, medicine] treating of Diseases, their Differences, Causes and Effects, or Symptoms; by which the human Body is known to vary from its healthy State'. Albrecht von Haller, ed. *Dr. Boerhaave's Academical Lectures on the Theory of Physic: Being a Genuine Translation of his Institutes and Explanatory Comment*, 6 vols (London, 1742–1746), vol. 1, 77.

[29] Herman Boerhaave, *Institutiones medicae in usus annuae exercitationis domesticos digestae*, 5th ed. (Leiden and Rotterdam, 1734). Gaubius, 'Praelectionum Theoreticarum', Leiden, University Library, BPL 1475, vols 1–3.

[30] 'Compendium Praelectionum H.D. Gaubii Med. Prof. in Pathologiam Nosologiam & Aetiologiam', Leiden, *c.* 1750, 1768. Leiden, University Library, BPL 1476. This manuscript follows exactly the same structure as BPL 1475, except that it leaves out the sections on physiology and symptomatology, and starts with a definition of pathology instead. Interleaved sheets of paper preserve notes in another hand, and some are dated 1768, giving proof of the continued use of these manuscripts by different persons.

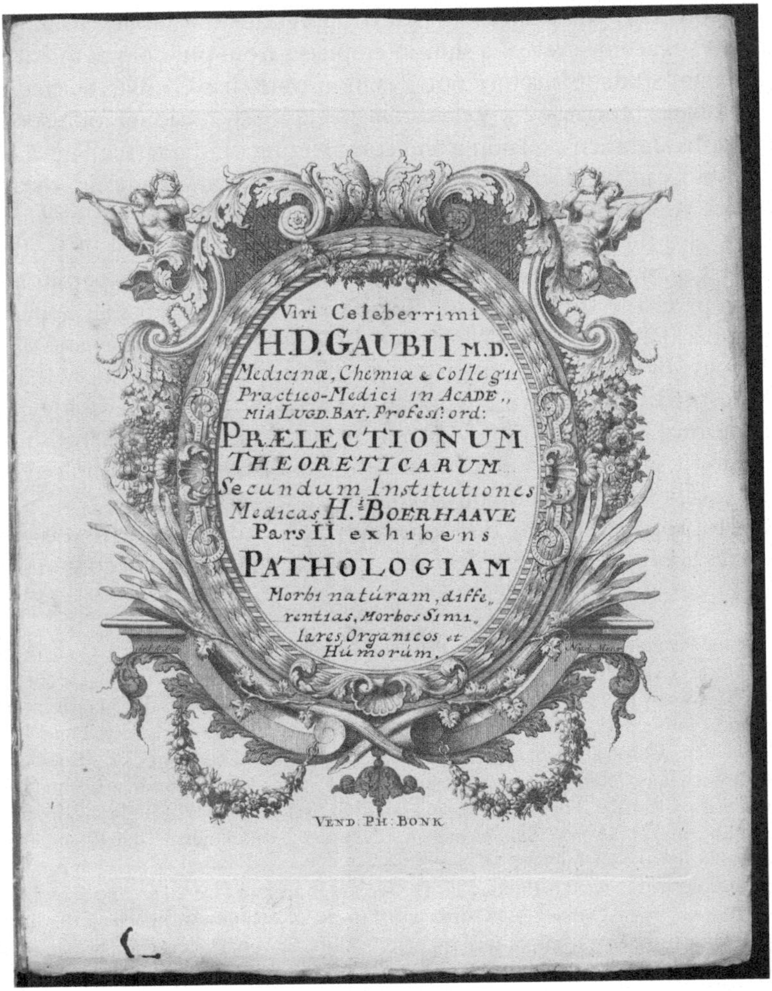

Fig. 7.1 Title page of Gaubius's pathology lecture dictates based on Boerhaave's *Institutiones* (Leiden, 1751). Leiden University Library, BPL 1475

developments occurred in physicians' concept of disease. In the 1750s Gaubius decided to no longer follow his predecessor's *Institutiones*, and wrote his own textbook instead. This attests to a transition from general knowledge of medicine towards a more specialised knowledge of disease theory, which he called 'rational Pathology'.[31] As Gaubius wrote in 1754 to Boerhaave's nephew, the professor of anatomy in St Petersburg, Abraham Kaau-Boerhaave, 'I am still busy creating a *System of Pathology* for my home lessons, because I am overwhelmed by the variety of subjects that I have to discuss and that are increasing every year'.[32] Notions of systems of medicine, or physic, were very common in the eighteenth century. These consisted mainly of systematically structured lectures on the human body as a whole, which professors read out to their students and then publish as textbooks. But flaws in old systems and the discovery of new observations necessitated new systems: the goal of every system was to incorporate all known facts and doctrines in a single and comprehensive structure with explanatory power. For example, William Cullen's (1710–1790) own textbook *First Lines of the Practice of Physic* (1784) presents the Scottish professor's reflections on past systems:

> An Humoral Pathology still continued to make a great part of every system. In these circumstances, it was soon perceived, that chemistry promised a much better explanation than the Galenic or Aristotelian philosophy had done [...] there appeared in the writings of Stahl, of Hoffman, and of Boerhaave three new, and considerably different, Systems of Physic.[33]

The development of chemistry as a tool for medicine had, therefore, spurred the development of new systems of physiology and pathology in the eighteenth century. Cullen highly praised Boerhaave's efforts, because Boerhaave's system included both modern and ancient writings, producing a system 'superior to any that had ever before appeared'.[34] And yet, advancements and new insights revealed flaws in Boerhaave's system, ultimately leading to new systems by Gaubius and Cullen alike.

[31] Gaubius, *Institutiones*, 337.

[32] Gaubius to Kaau, Leiden, 14 August 1754 in Hamers-van Duynen, *Hieronymus David Gaubius*, 219–222. Emphasis added.

[33] William Cullen, *First Lines of the Practice of Physic*, 4th ed., 4 vols (Edinburgh, 1784), x–xi.

[34] Ibid., xxv.

Like other systematic thinkers, Gaubius wrote his book first and foremost for his students, as is evident from a number of sources. In November 1755 Jakob Christoph Ramspeck (1722–1797) reported that 'at two o'clock I again listen to Gaubius teaching Pathology according to his own dictates, which now they sweat to get under the press, and of which the first part already circulates in the hands of students'.[35] In the preface the professor addressed his students directly: 'It is right, gentlemen, that I offer you, and that I also dedicate these Institutions of Pathology to you, which I have endeavoured to compose for your use, and which I finally finished after three years of delay in printing'.[36] That the project was undertaken at the request of the students did not mean that it was any easier for Gaubius to write in comparison to writing a treatise with colleagues in mind. Quite the opposite, in fact. Gaubius confided his struggles to Sanches, saying that writing such a treatise is a difficult job: 'One finds many thorns and rough stones on the road, especially, if one is not writing for masters but for pupils'. Worse even, 'if I had not started it, I would not finish it'.[37] Although developing a new *Institutiones* had cost him blood, sweat and tears, Gaubius's book successfully fulfilled didactic and pedagogic purposes in teaching pathology in a systematic manner.

For Gaubius, his new pathology was essential to rescue medicine from 'doubtful and fallacious' speculations and bad practice. This was a serious problem at the time, because incomplete or inaccurate theories could, by virtue of eloquence, still be convincingly communicated as trustworthy knowledge. Yet the consequences could be grave, Gaubius sighed:

> How often health is ruined by an improper regimen, diseases protracted by a preposterous method of cure, and finally, how many untimely deaths are thus occasioned; it is not without concern, I must say, that I am in doubt whether mankind benefits more than it is hurt by the medical art'.[38]

If he was to devise a new pathology, it had to follow a rigorous methodology. Gaubius argued that 'only such theorems be proposed, concerning the truth of which there is the clearest evidence, from *faithful observation*

[35] Ramspeck to Von Haller, Leiden, 4 November 1755 in Von Haller, *Epistolae*, vol. 3, 525.

[36] Gaubius, *Institutiones*, [*2r].

[37] Gaubius to Sanches, Leiden, 29 December 1756 in Hamers-van Duynen, *Hieronymus David Gaubius*, 102.

[38] Hieronymus David Gaubius, *Institutiones pathologiae medicinalis*, 2nd ed. (Leiden, 1763), [*4v]; idem, *Institutions*, vi.

and sound reasoning, the only sure foundation of medicine instituted upon rational principles'.[39] For Gaubius and his students this meant that only thorough and systematic observation – as opposed to observation undertaken in a casual and cursory way – could test and falsify hypotheses.[40] And only studious attention to nature, as opposed to creative inventions and preconceived ideas, Gaubius argued, could lead to a perfect history of diseases. In other words, it was only through rigorous observation and reasoning that medical students would be armed against fanciful hypotheses, and would learn to distinguish falsehood from truth. According to Gaubius's discussion on methodology and verisimilitude, rational pathology would be truly useful to teaching the practice of medicine.[41]

So what exactly did this 'rational pathology' entail? In Gaubius's eyes, pathology as disease theory went beyond the detailed description of the most common symptoms, beyond the visual depiction of morbid phenomena, and beyond the narration of interesting case studies. Instead, Gaubius investigated the *causes* of disorders and placed them in a systematic order. He launched an account of latent forces or 'noxious powers' (*potentiis nocentibus*), and predisposing causes or 'seeds of disease' (*seminiis morborum*).[42] Together they could bring about morbid conditions of the body, and were hence responsible for the outbreak of illness. The noxious powers were inspired by the Galenic *res non-naturam*, or things non-natural, which supported health when used correctly, but could also be abused and cause disorder.[43] Food and drink, airs and places, and sleeping

[39] Gaubius, *Institutiones*, 2nd ed., [*3r]; idem, Institutions, iv. Emphasis added. This rationalist attitude was similar to Boerhaave's, who had also argued for observation in combination with 'rigorous and disciplined reasoning'. Cited in Lester S. King, 'Rationalism in Early Eighteenth Century Medicine', *Journal of the History of Medicine and Allied Sciences*, 18 (1963), 257–271, here 257.

[40] On how the method of observation changed in the early modern period, see Gianna Pomata, 'Sharing Cases: The Observationes in Early Modern Medicine', *Early Science and Medicine*, 15 (2010), 193–236; ead. 'Observation Rising: Birth of an Epistemic Genre, 1500–1650', in *Histories of Scientific Observation*, ed. Lorraine Daston and Elizabeth Lunbeck (Chicago, 2011), 45–80.

[41] Roger French has argued how the genre of *Institutes* was necessarily a rationalist account to be useful in education in *Medicine before Science: The Rational and Learned Doctor from the Middle Ages to the Enlightenment* (Cambridge, 2003), 225–229.

[42] Gaubius discussed the noxious powers in *Institutiones*, 200–312. The seeds of disease were discussed at 312–320. See also Thomas H. Broman, 'The Medical Sciences', in *The Cambridge History of Science*, ed. Roy Porter (Cambridge, 2003), 463–484, here 478.

[43] Lelland J. Rather, 'The "Six Things Non-Natural": A Note on the Origins and Fate of a Doctrine and a Phrase', *Clio Medica*, 3 (1968), 337–347; Peter H. Niebyl, 'The Non-

and waking were examples of these non-naturals. In Gaubius's new system, noxious powers either flowed from internal sources, such as strong emotions or the inordinate excretion of bodily fluids, or were caused by external factors, for example, the climate or the abuse of medicines. Gaubius's seeds of disease, by contrast, were inspired by all things natural (*res naturam*), such as age, sex, or state of bodily solids and fluids, and those things against nature (*res preter-naturam*), such as hereditary medical conditions or the vitiation of fluids.

Gaubius's rational pathology was especially innovative because of his inclusion of chemical elements and their properties in his explanation of how illnesses came about. Based on an organisation of the bodily solids and fluids as a permutation of four chemical elements (water, phlogiston, salt, and earth) Gaubius suggested that properties inherent to these elements made the body liable to a specific vitiation or pathological condition. For example, an excess of water taken in or retained could give rise to an increased tenuity to the fluids. In combination with an acrimony or other noxious power, the bitter and thin fluid could easily penetrate and corrode other body parts, potentially causing all kinds of harm, such as haemorrhage and diarrhoea.

Gaubius considered blood the mother of all fluids, including semen. In the chemical laboratory, through the process of distillation, Gaubius had discovered that healthy blood consisted of water, bitter oil, volatile salt, and black earth – as I have described in the chapter on blood.[44] Gaubius extended these results to the entire body, arguing that the bones, muscles, organs, and fluids were all essentially combinations of these four principles. The substance of water formed the greatest part of the liquids, and had the inherent chemical properties of fluidity and easy dissipation when subjected to heat. The element of phlogiston or 'the principle of heat' was the inflammable part of the body, believed to exist in all combustible bodies and to be released in the process of combustion. Water and phlogiston were both connected to the third, saline part, because salt was supposedly covered by phlogiston and was diluted in water.[45] Finally, earth formed the

Naturals', *Bulletin of the History of Medicine*, 43 (1971), 486–492; Sandra Cavallo and Tessa Storey, *Healthy Living in Late Renaissance Italy* (Oxford, 2013); James Kennaway and Rina Knoeff, eds., *Lifestyle and Medicine in the Enlightenment: The Six Non-Naturals in the Long Eighteenth Century* (London, 2020).

[44] Gaubius, *Institutiones*, 155–160; idem, *Institutions*, 103–106.

[45] On phlogiston, see Hasok Chang, *Is Water H2O? Evidence, Realism and Pluralism* (Dordrecht, 2012), 1–70.

basis of the body as a whole, because it lent the solids and liquids their required cohesion and firmness.[46] In the case of semen, the main constituents were phlogiston and salt and only little water. Indeed, Boerhaave's translator Ephraim Chambers recorded that 'by chemical analysis, [semen] is found to consist almost intirely [*sic*] of oil, and volatile salts, blended together by the mediation of a little phlegm'.[47] In Gaubius's system, therefore, semen also was understood as a combination of these constituent parts.

It is clear that the chemical configuration of the body was essentially principlist. Whereas compositionist explanations of matter were based on the presumption of immutable and basic chemical substances, which could be combined and separated while remaining intact and unchangeable in nature, the principlist understanding was based on the concept of the changeability of properties.[48] This chemical analysis of the body in terms of basic principles was crucial to the explanation of pathological phenomena. The simplest diseases, for example, were described in terms of the four principles and their varying degrees of cohesion, a property of the element. If earth or water were out of proportion, a vitiated 'force of cohesion' (*vis cohaerendi*) could cause stiffness or frailty in the solids, and dilution or a glutinous consistency in the fluids. For example, the excessive force of cohesion was considered a natural seed of disease which, in combination with a noxious power such as a cold atmosphere or the consumption of too much glutinous food, might result in stagnation and obstruction, infarction, an impediment of the secretions, and tumours.

Gaubius's pathology was also apt to explaining venereal diseases. Some male genital disorders included priapism (a persistent and painful erection) and satyriasis (excessive sexual desire). In the new rational pathology, the source for these disorders was traced back to the chemical composition of semen. Gaubius maintained that far from being a static arrangement of elements, the constituent principles of semen were subject to change. Although semen could retain a recognisable identity in different circumstances, an imbalance in the elements could nevertheless cause it to acquire morbid properties. Gaubius explained that bodily fluids such as semen

[46] Gaubius, *Institutiones*, 56–58; idem, *Institutions*, 36–38.

[47] See *s.v.* 'Seed, semen' in Ephraim Chambers, *Cyclopaedia, or, An Universal Dictionary of the Arts and Sciences*, 2 vols (London, 1728), 2: 46.

[48] On principlism and compositionism, see Robert Siegfried, *From Elements to Atoms: A History of Chemical Composition* (Philadelphia, 2002); Hasok Chang, 'Compositionism as a Dominant Way of Knowing in Modern Chemistry', *History of Science*, 49 (2011), 247–268.

could turn from bland to sharp, adopting an 'acrid, inciding, and corrosive quality, [which] is able to dissolve the cohesion of the solids, to affect with pain the parts endowed with sensation, and to stimulate to contraction those that are animated with the vital power'.[49] Hence, priapism and satyriasis were caused by a particular sharpness of semen, or in Gaubius's words, by the 'redundance, acritude, and virulency of the seminal liquor'.[50] In general, bodily fluids would by merit of their chemical properties sustain health. Yet a change in their constituent elements could make them predisposed to disease.

In short, Gaubius aimed at understanding diseases through a coherent and systematic disease theory or pathology, explaining the causes of disease, or the aetiology, in terms of the chemical properties of bodily solids and fluids. Pathology, in other words, went beyond the study of individual cases and symptoms, and was rather concerned with understanding the nature of disease. Building on Boerhaave's medical and chemical lectures, this focus showcases Gaubius's appreciation of chemistry *within* medicine, not only in the practical analysis of bodies, but also in theoretical and rational thinking. This was the defining characteristic of the discipline of pathology, which set it apart from the practices of identifying symptoms or observing the changes in injured tissue.

Classifying Symptoms

The pathology of Hieronymus Gaubius was not only innovative because of its incorporation of chemistry, his emphasis on the causes of disease was also contrary to other physicians' inclinations to focus solely on symptoms. Indeed, in the early modern period, competing medical systems had caused many physicians to doubt the necessity of finding the cause of disease. Whereas the ancients had propagated a humoral system of four bodily humours, chymists such as Paracelsus instead propagated a so-called *tria prima* of sulphur, mercury, and salt. Amidst the confusion about the causes of disease, physicians of the turn of the eighteenth century decided to shift their attention to symptoms. The evaluation of symptoms and signs was that they were, first of all, detectable and observable by the senses – not speculations of the mind. Second, they were useful for evaluating a patient's condition, making a prognosis, and deciding upon a

[49] Gaubius, *Institutiones*, 136; idem, *Institutions*, 90.
[50] Gaubius, *Institutiones*, 452; idem, *Institutions*, 314.

treatment. The medical authors considered in this section had, therefore, taken the initiative to collect and compare all known symptoms and signs of diseases in order to construct a classification by symptoms – a field called nosology. But, as this section shows, it was this approach that Gaubius later rejected: he argued that one would never come to truly understand the nature of disease by solely studying symptoms.

One of these reform-minded physicians was Thomas Sydenham, who revolutionised the way physicians looked at disease.[51] Sydenham had collected many case histories during epidemic outbreaks, but instead of reproducing the single case studies he produced a generalised history of the symptoms and courses of individual diseases, without speculating about their causes.[52] Sydenham had described the symptoms of gonorrhoea, for example, as 'an uncommon pain in the parts of generation and a kind of rotation of the testicles'. In those men who were circumcised, however, 'a spot not unlike the measles appears on the glans [*sic*]'. This was followed by the morbid discharge flowing from the urethra and 'resembling semen'. When the disease turned virulent, 'this matter becomes green, and is mixed with a watery humour streak'd with blood'.[53] Sydenham then listed a number of treatments without indulging in explanations about the workings or effectiveness of these medicaments. Various pills, purging potions, balms, courses of bloodletting, liniments, and ointments could be prescribed to treat gonorrhoea or relieve the pain. The practical-minded Sydenham therefore gathered established practices and methods of curing, collected via numerous careful observations of each disease.

Similar ambitions formed the foundation of the work of several other university professors in the course of the eighteenth century. François Boissier de Sauvages, physician in Paris and later professor of medicine at Montpellier, wanted to expand beyond the symptoms of a single disease, and aimed for a comprehensive system that would include all symptoms. Sauvages's ambitious nosological project received some high-profile endorsement: 'I praise the excellent plan to arrange the diseases into

[51] With epidemics, 'diseases acquired a weightier ontology'. See French, *Medicine before Science*, 191–192.

[52] Andrew Cunningham, 'Thomas Sydenham: Epidemics, Experiment and the "Good Old Cause"', in *The Medical Revolution*, ed. Roger French and Andrew Wear (Cambridge, 1989), 164–190.

[53] Thomas Sydenham, *The Entire Works of Dr Thomas Sydenham, Newly Made English from the Originals*, ed. John Swan (London, 1742), 596–599.

classes', Boerhaave wrote in a letter that was reproduced as a preface to Sauvages's book. 'It seems to me that the work will be of a very high level, when you will be capable to finish it as happily as you have made a well-balanced design to it'.[54] By keeping elaborate notes of his clinical observations in Montpellier hospitals, Sauvages's aim was to determine which observable sign revealed the essential character of a disease, and to place all these signs into an encyclopaedic, classificatory scheme, intended to assist physicians in diagnosis and treatment.[55] The fruits of his first endeavour appeared in 1731, and arranged diseases in classes, genera, and species.[56] After thirty further years of clinical observation, Sauvages published his definitive work, *Nosologia methodica sistens morborum classes, genera et species* ('Methodological Nosology, in which Diseases are Arranged by Classes, Genera, and Species'). Taking into account the various circumstances in which symptoms might appear, Sauvages ultimately distinguished an astounding 2400 species, divided into 295 genera, 44 orders, and 10 classes.[57]

In this system, gonorrhoea was classified in the ninth class on fluxes, in the third order on serous or watery fluxes, and was defined as a genus divided into six species. If a patient with some venereal distemper approached a physician for treatment, the diagnosis could be performed according to Sauvages's checklist of the characteristic signs. The first two species, namely *gonorrhoea pura* and *gonorrhoea libidinosa*, involved the flux of a humour from the urethra either with or without sexual inclination. If a cloudy fluid was discharged during libidinous dreams, this was regarded as an involuntary pollution, and hence classified as the third species. *Gonorrhoea virulenta* or *gonorrhoea syphilitica* constituted an impure copulation of a purulent matter, and often presented with dysuria, that is, excessive pain during peeing. When there was no pain, it was commonly known as a gleet, or a 'dripper' (*druipert* in Dutch). The last two species

[54] Boerhaave to Boissier de Sauvages, Leiden, 24 April 1731 in G.A. Lindeboom, ed. *Boerhaave's Correspondence*, 3 vols (Leiden, 1962–1979), vol. 3, 166–167.

[55] Volker Hess and J. Andrew Mendelsohn, 'Sauvages' Paperwork: How Disease Classification Arose from Scholarly Note-Taking', *Early Science and Medicine*, 19 (2014), 471–503.

[56] François Boissier de Sauvages, *Nouvelles classes de maladies, qui dans un ordre semblable à celui des botanists, comprennent les genres & les espèces de toutes les maladies, avec leurs signes & leurs indications* (Avignon, 1731).

[57] François Boissier de Sauvages, *Nosologia methodica sistens morborum classes, genera et species, juxta Sydenhami mentem et botanicorum ordinem*, 3 vols (Amsterdam, 1763).

were excretions of pus or leaks of urine due to an infection of the foreskin or ulceration of the glands. Sauvages classified these not as actual gonorrhoea but as spurious, since the flux was not actually discharge from the urethra.[58] In Sauvages's eyes, then, the observation of visible signs was sufficient to recognise the genera of disease, and to determine a classification into species: 'Symptoms are easier to know than the causes, as they affect the *senses*, so they are infinitely more correct, especially if they are constant, in allowing us to know the true character of the diseases, and to distinguish them'. Sauvages warned his students against medical dogma and the most fanciful hypotheses about the cause of disease. Instead, physicians ought to rely on experience, 'confirmed by novel experiments [to] produce some physical certainty'.[59] In the field of nosology, then, only the observation and experience of symptoms allowed for the definition of diseases and the distinction between different diseases.

Whereas Boissier de Sauvages was thoroughly enmeshed in his endeavour to describe and differentiate thousands of species of disease, Gaubius was occupied with teaching to his students a clear and coherent system of pathology. He was intimately familiar with Sauvages's work, as he owned copies of *Pathologia methodica* (1752) and *Nosologia methodica* (1763).[60] But instead of attending to detailed descriptions of individual symptoms, Gaubius preferred a joint assessment of the various kinds of effects, in a systematic way and related to the causes of disease. After all, Gaubius stated, a physician needs to understand the causes and seats of disease to assess the symptoms clearly, and to make a right diagnosis.[61] He divided all symptoms into three general categories. The first were the 'alienated sensible qualities', or those qualities observed by the physician in the patient's body which did not necessarily cause any inconvenience. These qualities may therefore be considered signs, not symptoms, because signs were

[58] François Boissier de Sauvages, *Nosologie methodique, dans laquelle les maladies sont rangées par classes, suivant le systême de Sydenham, & l'ordre des Botanistes*, trans. Pierre François Nicolas, 3 vols (Paris, 1771), vol. 3, 186–191. For a different arrangement, see for example William Cullen, *Nosology: Or, a Systematic Arrangement of Diseases, by Classes, Orders, Genera, and Species* (Edinburgh, 1800), 174.

[59] As translated by S.E. Starkstein and G.E. Berrios, 'The "Preliminary Discourse" to Methodical Nosology, by Francois Boissier de Sauvages (1772)', *History of Psychiatry*, 26 (2015), 477–491. See also Elizabeth A. Williams, *A Cultural History of Medical Vitalism in Enlightenment Montpellier* (Aldershot, 2003), 80–111.

[60] 'Bibliotheca Gaubiana sive Catalogus librorum viri celeberrimi Hieronymi Davidis Gaubii', (Leiden, 1783), 112.

[61] Gaubius, *Institutiones*, 337; idem, *Institutions*, 236.

generally understood to be indications of a disease in a patient's complexion without causing the patient any physical or mental discomfort. Examples included the colour of the skin, smell and heat of the body, and hardness in body parts. The other two categories of symptoms included the 'vitiations of the excretions' and 'impeded actions'.[62] Gaubius's tripartite division of symptoms may perhaps appear to be yet another, more abstract nosology, yet in contrast to Sauvages's system, the symptoms were here linked to the causes of disease.

The way Gaubius perceived semen in various disorders is illustrative of his system of pathology. Faulty fluids were categorised in two groups: first, all those cases in which the body discharged fluids which, in normal circumstances, would be retained; second, all those cases in which the body retained fluids which ought to be secreted. And indeed, the inordinate secretion and retention of semen was thought to explain a wide variety of symptoms. Too much sexual intercourse – and hence too much ejaculation – would drain one's vital strength or energy, for the copious loss of semen meant that bodily fluids had to be drawn from other parts of the body:

> From too much venery [comes] lassitude, debility, immobility, feebleness of the loins, pains of the encephalos [*i.e.* brain], heats, convulsions, hebetude of all the senses, especially the sight, blindness, fatuity, a febrile circulation, exsiccation [*i.e.* dehydration], leanness, pulmonic and dorsal tabefaction [*i.e.* deterioration], effemination.[63]

The actual occurrence of these symptoms depended on the proper 'seeds of disease', such as difference in age, temperament, habit, and vital power.[64] This explained exceptional cases when some men grew 'languid even with moderate venery, other[s] scarce to be exhausted by the greatest excess'.[65] Abstinence from intercourse and the retention of one's precious semen rarely proved hurtful, the professor observed. Yet the combination of sexual intercourse with the avoidance of seminal emission was not recommended: 'A restrained ejection of the seminal liquor at the termination

[62] Gaubius, *Institutiones*, 338–461.
[63] Ibid., 288–289; idem, *Institutions*, 195.
[64] Gaubius, *Institutiones*, 312; idem, *Institutions*, 220.
[65] Gaubius, *Institutiones*, 289; idem, *Institutions*, 195.

of excitement, induces spermatocele, cirsocele, scirrhosity and cancer of the testicles'.[66]

Fellow physicians of the Boerhaave School agreed with Gaubius that excessive evacuations of semen lead to a general weakness. Johannes de Gorter, for example, wrote extensively on the retention and evacuation of bodily fluids, reasoning that ejaculation equated energy loss. Whenever a man discharged too much semen, or lose his testicles (like eunuchs), 'the vigour of the whole body perishes'.[67] Albrecht von Haller likewise stated that 'semen is discharged with a considerable loss of strength throughout the whole body'.[68] Excessive sexual activity would, indeed, lead to much disturbance, because the vigour of the body was too much exhausted, Van Swieten argued. He reminded his readers that the insensible perspiration would be diminished by immoderate venery, causing 'concoctive faculties' to weaken and the worst kind of fevers to arise. In some cases, 'a great quantity of liquid semen itself is also discharged when he [the patient] goes to urine and stool, nor does he retain the semen in his sleep, but loses it whether he sleeps with his wife or not'.[69]

Gaubius's third and final group of symptoms were the so-called impeded actions. The morbid secretion of semen was an essential part of male genital disorders. One such symptom concerned erectile dysfunction, or, 'when the male [member] is languid or incapable for an erection'. As we may expect from Gaubius's rational pathology, he linked this symptom of a limp penis directly to its causes, which he found primarily in the solid parts of the phallus. The muscular motion and postures of the penis were numbed and sluggish because of inactivity. He also mentioned semen as a potential cause of erectile dysfunction: the 'paucity, or vapid inertness of the seed' caused the patients to suffer from an inability to maintain an erection. Both the insufficient quantity of genital liquor and its properties of blandness, wateriness, and chemically inactivity would fail to provide the necessary stimulus. Furthermore, another side effect of insufficient

[66] Gaubius, *Institutiones*, 290; idem, *Institutions*, 196.

[67] Johannes de Gorter, *De perspiratione insensibili Sanctoriana-Batava tractatus experimentis propriis in Hollandia* (Leiden, 1725), 179–180.

[68] Albrecht von Haller, ed. *Praelectiones academicae in proprias institutiones rei medicae*, 6 vols (Göttingen, 1739–1744), vol. 5, pt. 1, 401; idem, *Academical Lectures*, vol. 5, 82.

[69] Gerard van Swieten, ed. *Commentaria in Hermanni Boerhaave Aphorismos de cognoscendis et curandis morbis*, 5 vols (Leiden, 1742–1772), vol. 2, 46; idem, *The Commentaries upon the Aphorisms of dr. Herman Boerhaave*, 18 vols (London, 1744–1773), vol. 5, 106.

and faulty semen was the 'impotency to emit the seed'.[70] As this example illustrates, Gaubius proposed a threefold division of symptoms, giving examples for each category, and discussing them in direct relation to their potential causes – as opposed to simply scrutinising the symptoms as Sydenham and Sauvages did.

The reason why Gaubius was opposed to an exclusive focus on symptoms was that these could be easily confused with the disease itself. Although a symptom could be detected by the senses, this observation did not reveal its origin. Some symptoms were, indeed, brought about by the disease, and therefore were directly related to it, but other symptoms could be the result of bodily peculiarities that were present prior to or independently from the outbreak of the disease. For example, convulsions, fevers, regurgitations, diarrhoea, sweats, and many other disorders could be symptoms of particular diseases, but they could also be brought about by ingesting the wrong kind of food, excessive physical exercise, or other causes. Also, the relationship between the different body parts, and between the body and the soul, might result in a phenomenon which Gaubius called the 'symptom of a symptom'. A multitude of morbid effects might, therefore, successively occur in a patient, but not help the physician determine the true nature of a particular disease. Fever, vomiting, diarrhoea, and sweat, for example, often accompanied each other in various diseases, but were 'by no means to be reckoned the constant effect flowing from the cause'. The opposite was also true: the natural powers in the body might suppress or even remove certain symptoms, which would otherwise be seen as characteristic of disease, and restore health. Gaubius established a hierarchy, distinguishing essential or primary symptoms from secondary or non-necessary symptoms.[71] The direct or indirect relation between symptoms and the nature and cause of disease, in other words, must both be studied. In Gaubius's view, then, sensing and studying symptoms alone would simply not suffice.

[70] Gaubius, *Institutiones*, 451–452; idem, *Institutions*, 313–314. Torpidity inferred a 'diminished sensibility of the living solid'.

[71] Gaubius, *Institutiones*, 36–51; idem, *Institutions*, 24–33.

Morbid Anatomy

The nosology debate continued throughout the eighteenth century, mainly because many authors considered symptoms too ambiguous to serve as a reliable explanation for disease. Andrew Cunningham has argued that, in response, anatomists increasingly delved into morbid anatomy, thereby changing the nature of pathology 'from a largely non-anatomical to an exclusively anatomical study'.[72] Indeed, many prominent physicians such as Steven Blankaart, John Hunter, and Frederik Ruysch dissected bodies and observed hundreds of curious cases.[73] Yet Gaubius contested the validity of an anatomical approach just as much as he did that of nosology. Pimples, warts, and wounds were perhaps easily observed in an autopsy, but the varieties of morbid secretions – such as gonorrhoea – were barely discernible in a corpse. In this section, therefore, I wish to nuance Cunningham's claim that pathology formed a sub-discipline of anatomy.

One of the figureheads of morbid anatomy was the Italian Giovanni Battista Morgagni. On his dissection table, disease was cut open to the bone, as he argued that disease was associated with perceivable structural changes. Over his 60-year-long career Morgagni combined his position as a university professor at Padua with the performance of public dissections of patients who had died in the public hospitals of Bologna and Padua. Morgagni was well known in the medical community for his *Adversaria anatomica*, first published in 1706. Volumes 4 to 6 of the *Adversaria*, as well as his *Epistolae anatomicae*, appeared in Leiden in 1723 and 1728 thanks to the mediation of Boerhaave.[74] Morgagni used his experience with dissecting in his promotion of an approach to disease that was quintessentially anatomical, as is evident from his magnum opus, *De sedibus et causis morborum per anatomen indagatis* ('On the Seats and Causes of

[72] Cunningham, *The Anatomist Anatomis'd*, 186. In Leiden, Eduard Sandifort (1742–1814) was mainly responsible for introducing pathological anatomy to the medical curriculum. See Marieke M.A. Hendriksen, *Elegant Anatomy: The Eighteenth-Century Leiden Anatomical Collections* (Leiden, 2015), 63–65.

[73] Steven Blankaart, *Anatomia practica rationalis, sive Rariorum cadaverum morbis denatorum anatomica inspectio* (Amsterdam, 1688); Hunter, *Venereal Disease*; Frederik Ruysch, *Alle de ontleed- genees- en heelkundige werken*, trans. Ysbrand Gysbert Arlebout, 3 vols (Amsterdam, 1744).

[74] Herman Boerhaave to Giovanni Battista Morgagni, Leiden, 1718–1737 in Lindeboom, *Boerhaave's Correspondence*, vol. 2, 78–91.

Diseases Investigated by Anatomy', 1761).[75] Modern physicians and scientists sometimes make the anachronistic claim that Morgagni was the pioneer of nineteenth-century experimental pathology.[76] Yet Morgagni himself considered his work the continuation of the predecessors' work. Not once did he use the term pathology to describe his work, but exclusively spoke of morbid anatomy.[77]

Morgagni investigated more than 700 corpses to uncover their 'seats of disease' by distinguishing the variations in the appearance of organs. He did not simply attempt to locate diseases in various parts of the body, in which case he would have been able to use the terms 'place' (*locus*) or 'position' (*situs*). Rather, he intentionally used the term 'seat' (*sedes*) to denote the particular location or usual place of residence of a disease. Furthermore, Morgagni merged the concept of the seat with that of the cause of disease. Lesions discovered post-mortem in an organ, in other words, were the source of the disease, its progress, and symptoms.[78] This anatomical approach to disease enabled Morgagni to differentiate between levels of pathological decomposition and to make generalisations about different species of disease. The dissection of a 30-year-old man who appeared to have had three testicles, for example, revealed only two testes and a hernia as the cause of epiplocele and enterocele, meaning that a part of the omentum that normally surrounded the abdominal organs had prolapsed and descended into his scrotum.[79] Morbid anatomy, in short, seemed to provide a new and solid basis for nosology, since the analysis

[75] Giovanni Battista Morgagni, *De sedibus et causis morborum per anatomen indagatis* (Venice, 1761). The first three books in Morgagni's magnum opus, printed in folio and spanning almost a thousand pages, grouped all diseases by body part: diseases of the head, thorax, and stomach. The fourth book concerned diseases treated with surgery across the entire body; the fifth and final book contained additions relating to all preceding books.

[76] See for example François Paoli, *Jean-Baptiste Morgagni, ou, La naissance de la médecine moderne* (Paris, 2013).

[77] An 'old' versus 'new' pathology can be considered to run parallel to the 'old' and 'new' physiology, as argued by Andrew Cunningham, 'The Pen and the Sword: Recovering the Disciplinary Identity of Physiology and Anatomy before 1800 – I: Old Physiology–the Pen', *Studies in History and Philosophy of Biological and Biomedical Sciences*, 33 (2002), 631–666.

[78] Fabio Zampieri, Alberto Zanatta, and Gaetano Thiene, 'An Etymological "Autopsy" of Morgagni's Title: *De sedibus et causis morborum per anatomen indagatis* (1761)', *Human Pathology*, 45 (2014), 12–16.

[79] Morgagni, *De sedibus*, 180; idem, *The Seats and Causes of Diseases Investigated by Anatomy*, trans. Benjamin Alexander, 3 vols (London, 1769), vol. 2, 545–546.

was not based on external appearances, but on internal and perceptible bodily structures.

Yet not all seats and causes of disease were as easily situated. Morgagni's very first statement on gonorrhoea already highlighted the main drawback to observation: 'I never, or scarcely ever, saw those evident marks of disease'.[80] Morgagni's anatomical approach to finding the seat of gonorrhoea involved the dissection of numerous corpses, which he dissected literally from head to toe. One case history he recounted was that of a carpenter, who turned out to have had an enlarged prostate gland in proportion to the penis. Yet there were 'no peculiar appearances' of the disease, 'except that the internal surface of the urethra seem'd to be somewhat moister, and more red, than usual'. The other genital organs such as the Cowper's gland appeared normal. The case of a 25-year-old man who had suffered from virulent gonorrhoea did not reveal much either. Besides Morgagni's observation of a whitish line in the urethra, there were 'very few things differing from the natural structure'.[81] Morbid anatomy, in short, disappointed in the search for a satisfactory observation and location of seminal flux in the solid parts of the body.

This was precisely why Gaubius – who was familiar with Morgagni's work – propagated the chemical method, which enabled him to closely examine variations in bodily fluids, and hence gain more insight into the cause of abnormal discharges of blood and other fluids.[82] In the *Institutiones*, Gaubius argued that a fluid's properties were contingent and could therefore move from the healthy to the morbid. In the case of gonorrhoea, the continuous and inconvenient discharge moved between several chemical properties, from being mild and tenuous, to abounding and turgescent, and even acrid and virulent. This chemical sharpness was not supposed to be envisioned as a particle with a predetermined figure, size or shape – as would have been the case in a corpuscular or Cartesian understanding of sharpness. Rather, chemical experiments in the laboratory had clearly indicated a property of sharpness in fluids. When, for example, medical students mixed fresh urine with quicklime or chalk in a glass vessel, and distilled it above a great fire, they learned that there

[80] Morgagni, *De sedibus*, vol. 2, 194; idem, *The Seats*, vol. 2, 592.

[81] Morgagni, *De sedibus*, vol. 2, 196; idem, *The Seats*, vol. 2, 597–598.

[82] Gaubius owned Giovanni Battista Morgagni, *Adversaria anatomica omnia*, 6 vols (Leiden, 1723); and *Epistolae anatomicae duae novas observationes, et animadversiones complectentes* (Leiden, 1728). See 'Anatomici' in 'Bibliotheca Gaubiana', 54.

ensued a clear liquor 'of an exceeding fiery and pungent taste and odour'. This salt from the human body had a 'violent degree of acrimony, or corrosiveness'.[83] Unfortunately, there is no evidence that this experiment was often conducted with semen or gonorrhoeal discharge – possibly because it was considered improper to do so. Nevertheless, all fluids were considered to have an inherent rigidity and sharpness. Chemical acrimony, according to Gaubius, neither could nor needed to be envisioned as a sharp particle, a tiny solid of a certain configuration. Instead, he regarded it as a stingingly bitter property inherent to all bodily fluids.[84] This understanding could be translated to gonorrhoea, when a sharp discharge combined with morbid states of the male anatomy, such as 'irritability of the genitals, and too great laxity of the ducts', resulted in copious flows.[85] In short, defining bodily fluids in chemical terms allowed for a coherent theory of pathology that emphasised dynamic and contingent properties.

Important differences between Gaubius's and Morgagni's approaches can now be clearly discerned. Morgagni's morbid anatomy was based on the dissection of bodies who had ceased living as a result of disease, and on individual cases, and was constructed according to the seats of diseases in different parts of the body. Gaubius, on the other hand, built a systematic pathology, which did not rely on individual case histories, but instead drew conclusions about the nature and cause of disease, explaining variations in the chemical condition of both bodily solids and fluids. Just like Boerhaave taught his students the first principles of physiology, on the basis of which his students could draw their own conclusions and devise their own experiments, so Gaubius formulated the first principles of pathology, which could be used for the explanation of many different kinds of diseases.

Chemistry and anatomy were not mutually exclusive, however. In fact, when Gaubius's colleague Johannes de Gorter discussed the disease of gonorrhoea in 1761, he combined the anatomy of male genitalia with Gaubius's chemical notion of sharpness. In the same year in which Gaubius published his *Institutiones*, De Gorter resigned from his appointment as court physician to Elisabeth, Empress of Russia (1709–1762). Upon his

[83] Herman Boerhaave, *A New Method of Chemistry: Including the Theory and Practice of that Art: Laid down on Mechanical Principles, and Accommodated to the Uses of Life*, trans. Peter Shaw and Ephraim Chambers, 2 vols (London, 1727), vol. 2, 191.

[84] Gaubius, *Institutiones*, 137; idem, *Institutions*, 90–91.

[85] Gaubius, *Institutiones*, 453; idem, *Institutions*, 315.

return to the Dutch Republic, publisher Isaak Tirion requested another print of the sold-out *Gezuiverde geneeskonst* ('Medicine Purified', 1744; second edition 1750). For the third edition, De Gorter added new chapters, including one on venereal disease, which mostly dealt with gonorrhoea.[86] De Gorter defined a dripper as the 'unnatural, continuous, and excessive discharge of Mucilaginous or Seminal substances, good or putrefied, through the urethra, without any erection of the penis nor lustfulness'.[87] Anatomy of the genitalia enabled a classification of and distinction between genuine and deceptive flux. The genuine seminal flux was a leak of fluid prepared in the testicles; deceptive flux was not semen at all, but a fluid originating from the prostate, Cowper's gland, or lubricating glands in the penis. The malignant form of gonorrhoea started, De Gorter explained, with an 'impure mixing of a thin *sharp* liquid leaking from the urethra with a not unpleasant irritation'.[88] Slowly this fluid would develop a high viscosity, growing thicker and stickier, eventually causing twinges and heat in the urethra. After emitting a stench and swelling considerably – or promoted by a treatment with mercury – the affected area would ultimately discharge the seminal substance in a large quantity, and the genitalia would be restored to health. Like Gaubius, then, De Gorter used the chemical properties of semen in his description of the morbid alterations in the fluid.

The conception of bodily fluids in terms of their chemical properties enabled physicians to emphasise the critical role of variable circumstances, such as nutrition and life style. In the case of seminal flux, Gaubius believed that the properties of fluids were contingent depending on circumstance. The evil of gonorrhoea could not only be induced by sexual overindulgence and a venereal virus, but also by 'the abuse of things promoting venery and diuretics', that is, aphrodisiacs and drugs causing an increased urination.[89] Furthermore, different kinds of food could influence the balance between blandness and sharpness in the body. Everyday foods such as fruit, corn, and fish were particularly appropriate for the human body because of their mild and bland nature. But Gaubius warned that luxury foods would 'introduce acrid, aromatic, spiritous [*sic*], saline, pungent,

[86] Johannes de Gorter, *Gezuiverde geneeskonst, of kort onderwys der meeste inwendige ziekten*, 3rd ed. (Amsterdam, 1761), [*2v].

[87] Ibid., 85.

[88] Ibid., 85–86. Emphasis added.

[89] Gaubius, *Institutiones*, 453; idem, *Institutions*, 315.

and heating things, as being grateful to the palate, [but] which are fraught with innumerable evils'.[90]

In short, the case of gonorrhoea illustrates how eighteenth-century Boerhaavian pathology was rooted in chemistry. Understanding body parts in terms of chemical elements and their inherent chemical properties made them active and susceptible to react with internal and external circumstances, for better or for worse. The fact that the body and the causes of disease could be understood in chemical terms provided physicians with the predisposing and occasional causes to explain disease: semen could be sharp or bland, but external circumstances, such as excessive sexual activity, the venereal virus, and acrid foods could also influence the varying degrees of sharpness in semen, and hence could lead to disorders such as gonorrhoea. This exercise not only put the practice of chemistry centre stage, it also reshaped medical perceptions of disease, and enabled a reinvention of the field of pathology.

Disseminating the New Pathology

Hieronymus Gaubius's pathology textbook represented a new approach to disease theory, which was fundamentally different from other eighteenth-century enterprises in nosology and morbid anatomy. Yet analysing the contents of his book tells us very little about the reception and dissemination of this knowledge. Was this publication solely used in Leiden, or did professors at other universities also teach Gaubius's pathology? Did the *Institutiones* solely fulfil the educational role of a textbook, or did it spur further research in this field? Whereas most modern medical historians do not acknowledge Gaubius's contribution to medicine, the famous nineteenth-century German pathologist and so-called Pope of Medicine, Rudolf Virchow (1821–1902), certainly thought otherwise. Looking back at the history of his discipline of pathology, he pointed out that 'the *Institutiones Pathologiae Medicinales* of Hieronymus Davides Gaubius was the first textbook of general pathology which the world had seen; it remained the standard until far into our own century'. In fact, when the young Virchow was a student at the Friedrich Wilhelm Institute in Berlin in 1839, Gaubius's work was still part of the students' reading

[90] Gaubius, *Institutiones*, 138; idem, *Institutions*, 91.

list.[91] To understand how and why Gaubius's pathology was appreciated across the eighteenth and nineteenth centuries, we need to explore the dissemination of the book through its readers and users.

When the *Institutiones* was first published in 1758, Gaubius sent some copies to his friends. Writing to António Sanches, for example, he asked that his friend read 'this small work' – of almost 500 pages nonetheless – and 'judge it and let me have your sincere remarks in due time the way I know you: as a fair and capable judge, and a friend who does not flatter'.[92] Gaubius repeated his request a few years later when a second edition was planned.[93] The Leiden booksellers Samuel and Johannes Luchtmans published this edition in 1763, which was not much different from the first, except for the new, much longer preface. If reprints are any indication of popularity, the book did very well: numerous editions appeared in Leiden, Leipzig, Edinburgh, and Venice between 1759 and 1787, and translations into French, Dutch, English, German, and Russian were published, respectively, in Paris, Leiden, Edinburgh, Zurich and Berlin, and St Petersburg.[94]

In the second half of the eighteenth century, professors of medicine across Europe recognised the merits of the *Institutiones* for teaching pathology. When Groningen physician Matthias van Geuns was appointed professor of medicine at the University of Harderwijk in 1776, he used the book in his lectures.[95] In line with Gaubius's systematic reasoning, Van

[91] Rudolf Virchow, *Hundert Jahre allgemeiner Pathologie* (Berlin, 1895); idem, *Disease, Life and Man: Selected Essays*, trans. and ed. by Lelland J. Rather (Stanford, CA, 1958), 174, 93.

[92] Gaubius to Sanches, Leiden, 21 July 1758 in Hamers-van Duynen, *Hieronymus David Gaubius*, 107.

[93] Gaubius to Sanches, Leiden, 23 September 1761 in ibid., 124.

[94] The translations were: Hieronymus David Gaubius, *Pathologie de M. Gaubius*, trans. Pierre Seu (Paris, 1770); idem, *Pathologie de M. Gaubius*, trans. Pierre Seu, rev. ed. (Paris, 1788); Alexander Balthazaar, *Pathologia chirurgicalis of heelkundige ziektekunde* (Leiden, 1772–1776); idem, *Pathologia chirurgicalis of heelkundige ziektekunde*, 2 vols (Leiden, 1777). Gaubius, *Institutions*. Idem, *Anfangsgründe der Krankheitenlehre des Menschen*, trans. Daniel Andreas Diebold (Zurich, 1781); idem, *Anfangsgründe der Medicinischen Krankheitslehre*, trans. Christian Gottfried Gruner (Berlin, 1784). A revised and expanded edition of Gruner's translation appeared in 1797. Finally, a Russian translation appeared by Petr Martynovich Gofman as idem, *Начальныя основанія врачебныя пафологіи, то есть науки о свойствѣ, причинахъ, припадкахъ и различіяхъ болѣзней, въ человѣческомъ тѣлѣ случающихся*, trans. Petr Martynovich Gofman (St Petersburg, 1792).

[95] J.H. Sypkens Smit, *Leven en werken van Matthias van Geuns M.D., 1735–1817* (Assen, 1953), 27. As a student Van Geuns had followed Gaubius's lectures on pathology and kept

Geuns's pathology lecture started with a schematic outline of all the branches of pathology.[96] Van Geuns separated the 'study of man' (*anthropologia*), which entailed physiology and pathology, from the 'study of remedies' (*iamatologia*), which included therapy and dietetics – the latter extending beyond advice on healthy nutrition to include rules or regimen for healthy living. Pathology was subsequently divided into the fields of nosology, aetiology, and symptomatology – in other words, the kinds, causes, and effects of disease. Van Geuns continued to teach the main classification of diseases and their causes as he saw fit, but consistently with reference to Gaubius's *Institutiones*. Van Geuns's successor Jan Bleuland (1756–1838) equally recommended that his students study the works of Gaubius. In his inaugural lecture, Bleuland stressed how honourable it would be to follow such an example through emulation.[97]

Outside the Dutch Republic the *Institutiones* also enjoyed much popularity as a student textbook. At the University of Altdorf near Nuremberg, the professor of chemistry and medicine Johann Christian Gottlieb Ackermann (1756–1801) adopted it as a textbook for his courses. At the request of bookseller E.C. Grattenauer, Ackermann prepared a Nuremberg edition that contained minor additions.[98] But the new pathology was perhaps most eagerly appropriated in Scotland. When in 1766 William Cullen succeeded to the chair of theory of medicine, which had previously been filled by his colleague Robert Whytt (1714–1766), he became responsible for teaching the institutes of medicine. After lecturing on physiology and health, Cullen's 1767 spring lectures on pathology consisted of readings from and a commentary on Gaubius's textbook.[99] This was a continuation of Whytt's practice, who had already adopted Boerhaave's *Institutiones*,

notes. See Matthias van Geuns, 'Ad Gaubii Instit. patholog'. Leiden, 1759. Amsterdam, University Library, MS 1377 I F 66.

[96] Matthias van Geuns, 'Schema institutionem pathologiae medicinalis Cl. Gaubii'. Harderwijk, *c.* 1775. Amsterdam, University Library, MS 1382 I E 61, ff. 1r–16v.

[97] Jan Bleuland, *Oratio, qua memoria Hieronymi Davidis Gaubii cum omnibus, tum praesertim medicinae studiosis commendatur* (Harderwijk, 1792), 7. On Bleuland, see Ruben E. Verwaal and Reina de Raat, 'Verzameldrift: De anatomische collectie van professor Jan Bleuland', *Geschiedenis der geneeskunde*, 14 (2010), 138–145.

[98] Hieronymus David Gaubius, *Institutiones pathologiae medicinalis*, ed. Johann Christian Gottlieb Ackermann, 3rd ed. (Nuremberg, 1787). By this time, Gaubius's textbook had already been reprinted twice in Leipzig in 1759 and 1781. Ackermann's edition included his commentaries and Hahn's short biography of Gaubius.

[99] Charles Blagden, 'Cullen's Comment on Gaubius's Pathology'. Edinburgh, 1767. Wellcome Library, London, MS 1927.

and later Gaubius's pathology textbook for his teachings.[100] When students complained that Cullen was spending too much time commenting on Gaubius, Cullen stressed the usefulness of the book to the practice of medicine. He replied that once they were treating patients, they ought to 'consult especially Dr Gaubius. He is infinitely the best writer upon the subject; he is more full and comprehensive, discovers a greater erudition, and a greater knowledge of medical opinions and facts, than any other pathologist'.[101]

In 1778 an English translation by surgeon Charles Erskine, sewn in wrappers and in octavo format, was sold in Edinburgh for 4 shillings.[102] When reviewed in *The Critical Review*, the new pathology received high praise:

> Of all the branches of medical enquiry, *pathological disquisitions* afford the understanding the greatest pleasure; for while they depend on theoretical principles deducible from the laws of the animal oeconomy, they likewise serve to lay the surest and most satisfactory foundation for rational practice.[103]

Praising Gaubius as one of 'the most eminent professors in the modern schools of physic', the reviewer was of the opinion that the author 'has judiciously avoided all speculative disquisitions that tend not to the solid improvement of science. The original work is unquestionably one of the most valuable productions in this department of medical knowledge'.[104] One could not wish for a more laudatory review.

Gaubius's new pathology was not only adopted across Europe, but some medical authors even attempted to emulate his system of pathology. In the early 1770s Irish surgeon and accoucheur David MacBride (1726–1778) published *A Methodical Introduction to the Theory and Practice of Physic*, a book which shared many similarities with Gaubius's

[100] Roger French, *Robert Whytt, the Soul, and Medicine* (London, 1969), 9.

[101] Cited in John Thomson, *An Account of the Life, Lectures and Writings of William Cullen, M.D.*, 2 vols (Edinburgh, 1832–1859), vol. 1, 325–326.

[102] A reprint of the Latin textbook had already appeared as Hieronymus David Gaubius, *Institutiones pathologiae medicinalis*, 2nd ed. (Edinburg, 1762).

[103] Tobias George Smollett, 'The Institutions of Medicinal Pathology by H.D. Gaubius', *The Critical Review: Or, Annals of Literature*, 47 (1779), 183–186, here 183.

[104] Ibid., 184, 6.

Institutiones.[105] MacBride had been appointed secretary to the Medico-Philosophical Society in Dublin, and started teaching a course of medical lectures in his home in the winter of 1766–67. His book based on these lectures, was clearly inspired by Gaubius's pathology, for it often referred to and summarised parts of the *Institutiones.* On the topic of the morbid sharpness of fluids, for example, MacBride gave 'a short sketch of it [a necessary part of the pathology], as laid down by Gaubius, who is one of the latest, and supposed to be the best writer on this branch'. MacBride distinguished his work from Gaubius's by also considering the practice of medicine. He remarked, for instance, that while 'all this is perfectly systematical, and does very well to read; but when we come to look for these different species of acrimony in our patients, we shall not often be able either to find out or distinguish them'.[106]

The Scottish physician James Makittrick (aka Makittrick Adair, 1728–1801) was of the same mind as MacBride. Makittrick was in the process of making commentaries on Gaubius's textbook in preparation for his teaching responsibilities in Dublin, when MacBride had gone ahead of him. Makittrick nevertheless decided to publish his work, for he thought of it as 'of considerable use to young gentlemen who study physic regularly, as it may facilitate the attainment of those principles, by which their practice, as *rational physicians*, must be regulated'.[107] Makittrick's book adopted its own unique structure, but the general assumptions about the branch of pathology were exactly the same as Gaubius's: Makittrick recognised the essential role of chemistry in medicine, opened with a discussion of the constituent principles of the body, and classified diseases according to their predisposing, preternatural, and proximate causes. In the case of semen, for example, similarities with Gaubius's work can easily be discerned: 'semen makes considerable changes in the body'. In case of disease, the seminal liquor 'may be either too thin and aqueous; or they may be acrid [by which] their secretion or excretion may be impeded, or preternaturally increased'.[108]

[105] David MacBride, *A Methodical Introduction to the Theory and Practice of Physic* (London, 1772). A Latin translation appeared as *Introductio methodica in theoriam et praxin medicinae*, trans. Johannes Fredericus Clossius, 2 vols (Utrecht, 1774). The original was reissued in an enlarged edition in Dublin in 1776.

[106] MacBride, *Physic*, 85–87.

[107] James Makittrick, *Commentaries on the Principles and Practice of Physic* (London, 1772), iii. Emphasis in original.

[108] Ibid., 75, 233.

These examples of the readership, use, and emulation of Gaubius's work provide sufficient evidence for the wide dissemination of the new pathology across Europe. Many universities adopted the *Institutiones* for educational purposes, while the works of MacBride and Makittrick testify to the continuing enterprise of matching, perhaps even surpassing Gaubius's endeavour in pathology.

* * *

This chapter has demonstrated that the eighteenth century saw the development of a new field of disease theory. This branch of medicine was not primarily based on the body's organs, such as morbid anatomy, nor on the classification of external characteristics of patients' bodies, such as nosology, but on a chemical analysis of the bodily fluids. On the basis of Gaubius's *Institutiones pathologiae medicinales* this chapter has thus argued against the notion of pathology as a sub-discipline, but presented it as a field in its own right. Pathology developed into a discipline for which chemistry was a guiding principle, because it allowed for an examination of changes in bodily fluids in living patients, and hence providing a deeper examination of varying states of the living body.

How, then, were these approaches to disease understood to relate to each other? On the one hand, seeing how medical experts were coming to grips with venereal disease, these systems worked alongside each other rather than against one another.[109] Different branches of disease theory were extensively explored through bedside observation, anatomical dissection, and chemical experiment. Whereas medical writers drew from a pool of shared assumptions and ideas about observation and experience, hypothesis and theory, they all sought to improve and reform medical theory, academic education, and bedside practice. Notably, however, the eighteenth-century discipline of pathology was the overarching disease theory, which included nosology and morbid anatomy, because it aimed for a knowledge of the nature, cause, and effect of disease. The branches of nosology, aetiology, and therapeutics were seen as different parts of the general study of pathology, a domain in which Gaubius specialised. In the words of Boissier de Sauvages, for example, '[p]athology, as we hear from

[109] Broman, 'The Medical Sciences', 481.

the Arabs, is that theoretical part of Medicine, which treat of things contrary to nature, that is, of disease, the causes and symptoms of disease'.[110]

Throughout the eighteenth century, medical professors debated gonorrhoea and the ontological question of disease in general. Through this process, disease was increasingly perceived as an identifiable phenomenon on its own, ultimately reaching ontological status in the nineteenth century.[111] And as such, disease was susceptible to re-examination and redefinition within the discipline of pathology.

[110] François Boissier de Sauvages, *Pathologia methodica, seu de cognoscendis morbis* (Amsterdam, 1752), 1. Emphasis in original. With 'Arabs', Sauvages mainly referred to Ibn Sina.

[111] French, *Medicine before Science*, 231.

CHAPTER 8

Conclusion

The case studies on saliva, blood, urine, milk, sweat, and semen ultimately boil down to three main conclusions. They concern the impact of the Boerhaave school, the role of chemistry of the fluids for physiology and pathology, and how these preceded many aspects of modern medicine in the nineteenth century. These conclusions, I hope, reinforce further studies on medicine and natural philosophy in the eighteenth century.

The first conclusion I draw concerns the importance of the Boerhaave school in the long eighteenth century. In October 1818, for example, the immense library and art collection of the late physician and professor of medicine, Matthias van Geuns (1735–1817), was auctioned in Utrecht.[1] Among the rich collection of books, theses, and manuscripts on matters medical, chemical, botanical, theological, philosophical, and historical sat a noteworthy picture gallery of 12 portraits, mounted in gilt frames of ebony. The first five portraits were those of Hippocrates, Thomas Sydenham, Friedrich Hoffmann, Georg Ernst Stahl, and Herman Boerhaave. The final seven comprised of portraits of Johannes de Gorter, Bernhard Siegfried Albinus, Gerard van Swieten, Albrecht von Haller, Hieronymus Gaubius, Johannes Oosterdijk Schacht, and finally an

[1] 'Bibliotheca Geunsiana, sive, Catalogus librorum quibus usus est vir celeberrimus Matthias van Geuns' (Utrecht, 1818). On the life and works of Matthias van Geuns see J.H. Sypkens Smit, *Leven en werken van Matthias van Geuns M.D., 1735–1817* (Assen, 1953).

R. E. Verwaal, *Bodily Fluids, Chemistry and Medicine in the Eighteenth-Century Boerhaave School*, Palgrave Studies in Medieval and Early Modern Medicine,
https://doi.org/10.1007/978-3-030-51541-6_8

engraving of a bust of Boerhaave in a black frame.[2] Such a collage of deceased yet very distinguished physicians and chemists memorialised the Boerhaave school. It was an international school of which Van Geuns had surely envisioned himself a part, even though he had never met its originator.

At first glance Van Geuns's intriguing collage of portraits may appear an ode to a bygone age, but on closer inspection, it also poignantly demonstrates the importance and longevity of the medical research developed by the members of the Boerhaave school. For indeed long after Boerhaave's death in 1738, the Leiden professor continued to inspire the works of and bonds between his disciples. When in 1775 a 70-year-old Gaubius delivered his speech in Peter's Church to celebrate the bicentenary of the founding of Leiden University, he reflected on medical history and observed that 'people have eventually seen the rise of *the Boerhaavian school*, not a new school, but one very old, based on the Hippocratic one'.[3] This school of physicians, Gaubius argued, was imbued with the spirit and system of Boerhaave, grounded in indisputable discoveries and pure reasoning. Gaubius retired that same year, but he continued to work, overseeing a revised edition of his pathology textbook as well as the publication of António Sanches's seminal study on venereal disease. In correspondence with his old-time friend Sanches, Gaubius wrote to him he was impressed by his 'vivacity of spirit', joking that his must be proof of the distinction between body and soul.[4]

Some 20 years earlier, Albrecht von Haller retired from the chair of medicine, yet continuing to be one of the most prolific writers. In 1753 he left the university in Göttingen and returned to his native city of Bern. But despite retreating from teaching and increasingly experiencing

[2] 'Bibliotheca Geunsiana', 326. On the memorial importance of picture-galleries and collections in early modern Europe, see David van der Linden, 'Memorializing the Wars of Religion in Early Seventeenth-Century French Picture Galleries: Protestants and Catholics Painting the Contested Past', *Renaissance Quarterly*, 70 (2017), 132–178; Lieke van Deinsen, *The Panpoëticon Batavûm: The Portrait of the Author as a Celebrity* (Amsterdam, 2016).

[3] Hieronymus David Gaubius, *Oratio panegyrica in auspicium seculi terti Academiae Batavae* (Leiden, 1775), 53; idem, *Feestrede van Hieronymus David Gaubius by den heugelyken aanvang der derde eeuwe van Hollands Hooge Schoole te Leyden, den agsten van Sprokkelmaand 1775*, trans. Pieter van den Bosch (Leiden, 1775), 89–93. Emphasis added.

[4] Gaubius to Sanches, Leiden, 25 November 1777, in Sophia W. Hamers-van Duynen, *Hieronymus David Gaubius (1705–1780): Zijn correspondentie met Antonio Nunes Ribeiro Sanches en andere tijdgenoten* (Assen and Amsterdam, 1978), 107.

short-sightedness, his fluent pen never slowed. While in Switzerland he wrote philosophical novels as well as an 8-volume magnum opus, *Elementa Physiologiae Corporis Humani* ('Physiological Elements of the Human Body').[5] The hefty volumes narrated all physiological phenomena, supported by elaborate anatomical descriptions and experimental results. Some parts in the book departed from the way Boerhaave had originally thought about them—particularly in the case of the concepts of sensibility and irritability.[6] This was not so much a departure from Boerhaave, but a logical consequence of Boerhaave's anti-authoritarian outlook, wanting his students to think for themselves and improve upon the knowledge previously obtained. And still, up to his late 70s, Von Haller continued to think about his former teacher, as he wrote to his daughter:

> Fifty years have almost elapsed since I was the disciple of the immortal Boerhaave; but his image is continually present to my mind. I have always before my eyes the venerable simplicity of that great man, who possessed, in an eminent degree, the talent of persuading.[7]

Beyond people's minds and letters, the image of Boerhaave and his school was advanced and promoted in the material and visual culture of the eighteenth-century society. Commissioned by Boerhaave's daughter, Johanna Maria de Thoms (1712–1791), a monument was erected in honour of her father in Peter's Church in Leiden in 1762 (see Fig. 8.1). Designed by Frans Hemsterhuis (1721–1790) and constructed in the workshop of master mason Antoni Wapperom (1729–1781), the Boerhaave memorial was an oval urn of white marble placed on a rectangular pedestal of black marble. The funerary vase was modestly adorned with a flame on top and six heads linked together by drapery, representing the four stages of life in men and women: childhood, youth, maturity, and old age. The six heads may also implicitly have been a symbol of the virtue of Prudentia of the present, past, and future: *intelligentia*, *memoria*, and

[5] Albrecht von Haller, *Elementa physiologiae corporis humani*, 8 vols (Lausanne, 1757–1766).

[6] See Hubert Steinke, *Irritating Experiments: Haller's Concept and the European Controversy on Irritability and Sensibility, 1750–1790* (Amsterdam, 2005), 93–124.

[7] Albrecht von Haller, *Letters from Baron Haller to his Daughter, on the Truths of the Christian Religion* (London and Edinburgh, 1780), 64. See also Rina Knoeff, 'Herman Boerhaave at Leiden: Communis Europae praeceptor', in *Centres of Medical Excellence?*, ed. Ole Peter Grell, Andrew Cunningham, and Jon Arrizabalaga (Farnham, 2010), 269–286.

Fig. 8.1 Boerhaave memorial. Engraving by Abraham Delfos, in Gerard van Swieten, *Commentaria in Hermanni Boerhaave Aphorismos de cognoscendis et curandis morbis* (Leiden, 1742–1772). (Credit: Rijksmuseum, Amsterdam. CC0 1.0)

providentia.[8] The pedestal was decorated with Boerhaave's profile portrait suitably set against a staff of Asclepius crossed with a burning torch. The inscription read 'dedicated to Boerhaave's health-bringing talents'.

The monument came to represent the continued flourishing of the Boerhaave school and medical research. For when Paulus 's-Graeuwen (1715–1779) was appointed the new professor of medicine and chemistry at the University of Groningen in 1771, he commissioned his new house to be adorned with lavishly ornamented wallpaper depicting flowers, ribbons as well as the white marble vase of Boerhaave's memorial statue.[9] This was no coincidence, for as a student 's-Graeuwen had been part of the last cohort attending Boerhaave's lectures in Leiden in 1736 and 1737. Following a career as a medical practitioner in the city of Zutphen, 's-Graeuwen succeeded Johannes de Gorter as the new professor of medicine and chemistry at Harderwijk in 1755.[10] Thanks to the intervention of Gaubius—who played a pivotal role in the appointment of medical professors at Dutch universities, and thereby secured Boerhaave's legacy—'s-Graeuwen was subsequently appointed to Groningen in 1770.[11] Once in the north, 's-Graeuwen inherited the chemistry course by his predecessor Wouter van Doeveren and extended it from three days to four days a week.[12] In his inaugural lecture, 's-Graeuwen explicitly aligned himself with Boerhaave, praising him for having combined the works of the ancients with those of the moderns, verifying both by his own experience and observations. And on a more personal note, 's-Graeuwen stated that 'I will keep in memory that I myself had been a student of his, and as I recall his most dearest admirer'.[13] The memorial and wallpaper, in short, perfectly illustrate that the community of medical researchers following in Boerhaave's footsteps was widespread and continued to thrive until the

[8] Frits Scholten, 'Frans Hemsterhuis's Memorial for Herman Boerhaave: A Monument of Wisdom and Simplicity', *Simiolus*, 35 (2011), 199–217.

[9] R.H. Alma et al., *Het Hinckaertshuis: Zeven eeuwen bouwhistorie en bewoners*, ed. T.J. Hoekstra et al. (Groningen, 2012).

[10] De Gorter had left for St Petersburg the year before, to join Abraham Kaau-Boerhaave and António Sanches at the court of Elizabeth, Czarina of Russia. See Aegidius W. Timmerman, 'Johannes de Gorter: Een schets van zijn leven en werk', *Nederlands Tijdschrift voor Geneeskunde*, 112 (1968), 35–41.

[11] Cornelis Hendricus Velse (1719–1808) to Paulus 's-Graeuwen, The Hague, 16 October 1770. Zeeland Library, Middelburg, Handschrift 1940, f. 1r.

[12] See *Series Lectionum*, University of Groningen, 1771–1779.

[13] Paulus 's-Graeuwen, *Oratio inauguralis de anatomiae pathologicae utilitate ac necessitate* (Groningen, 1771), 12.

end of the eighteenth century. And as this book has shown, Boerhaave's fame was continuously spread by his students, who played an important role in the transmission of his ideas long after his passing.

By keeping Boerhaave's image alive members of the Boerhaave school also consolidated their own public image, simultaneously establishing and legitimising the central place of chemistry in medicine. Gerard van Swieten, for example, used Abraham Delfos's engraving of the Boerhaave memorial as the frontispiece to his magnum opus, the *Commentaria* (see Fig. 8.1).[14] This made the book in itself a monument to Boerhaave, and at the same time, it aimed to establish Van Swieten as his intellectual successor—a title which, arguably, might just as well apply to Gaubius, Von Haller, and others. Van Swieten institutionalised the Boerhaave school by completely reorganising the medical faculty in Vienna. He established a botanical garden, built a chemical laboratory, and expanded the anatomical theatre. The bases for teaching the theory and practice of medicine were, of course, Boerhaave's textbooks: the *Institutiones medicae* for physiology, Van Swieten's commentaries on Boerhaave's *Aphorismi* for medical practice, Boerhaave's *Materia medica* for pharmacy, and finally, Boerhaave's *Elementa chemiae* for chemistry. It was not until 1785 that Pascal Joseph von Ferro (1753–1809) observed that Boerhaave's textbooks were no longer considered the best for use in medical education.[15] By analysing the works of De Gorter, Gaubius, Von Haller, Van Swieten, and many other Boerhaave disciples, this book has shown that the Boerhaave school had been a medical and a cultural phenomenon, which involved multiple generations of physicians.

* * *

The second conclusion I want to highlight concerns the central role of chemistry of the fluids played in the conceptualisation and formalisation of physiology and pathology in the eighteenth century. From the sixteenth century onwards, anatomical investigations uncovered the structure of the human body and anatomists continued to do so throughout the

[14] Delfos's engraving was included in the fourth edition of Van Swieten's *Commentaria* in 1766.

[15] Pascal Joseph von Ferro, *Einrichtung der medizinischen Fakultät zu Wien* (Vienna, 1785); Max Neuburger, 'Boerhaave's Einfluß auf die Entwicklung der Medizin in Oesterreich', *Janus*, 23 (1918), 215–222.

eighteenth century.[16] But in order to better account for the functioning of the living body, an understanding of physiology was necessary. This explains the reaffirmed interest in the bodily fluids, their properties, and variability depending on certain conditions. By means of new methods and instruments, Gaubius and Thomas Schwencke were able to examine the nature of blood more thoroughly. In laboratories, heating, evaporation, distillation and fermentation, putrefaction, and crystallisation, as well as other chemical processes, made it possible to see how the fluids behaved under various circumstances. These *in vitro* observations of saliva, blood, and milk were then used to better understand *in vivo* processes of digestion, nutrition, and secretion. In short, with the help of chemistry, medical professors and students gained new understandings of physiology.

In addition to the healthy body, chemistry of the fluids assisted in explaining the causes and effects of diseases in new ways. Based on chemical analyses, the comparison of bodily fluids such as sweat and semen with morbid secretions like phlegm and inflammatory discharges was a way to gain new pathological knowledge. A concrete example of this, as I have shown, was the elaborate experimentations performed with urine, enabling new explanations for the cause of bladder stones. Johannes de Gorter, likewise, developed a chemico-pathological theory of the disease of catarrh, rooted in a neurological description of obstructed perspiration and stagnating particles turning sharp. This theory directed De Gorter's preferred treatment, namely the prescription of the ammonia, which would induce sweating and dissolve the sharpness of the fluids. Gaubius ultimately undertook the Herculean task of describing all possible causes of disease in terms of the changeable chemical properties of the fluids. His systematic theory of disease or pathology recognised the importance of symptoms and morbid anatomy but was innovative in that it proposed a set of basic principles through which the aetiology or the causes of disease could be explained.

Although the bodily fluids as materials of study were at the forefront in this book, it is undeniable that scientific tools and instruments, their development and application, significantly accelerated advances in physiology

[16] Andrew Cunningham, *The Anatomist Anatomis'd: An Experimental Discipline in Enlightenment Europe* (Farnham, 2010); Marieke M.A. Hendriksen, *Elegant Anatomy: The Eighteenth-Century Leiden Anatomical Collections* (Leiden, 2015).

and pathology.[17] The introduction of various measuring instruments enabled physicians to probe the fluids beyond their first-hand sensory experiences. From 1718 onwards, Boerhaave, Gaubius, and Schwencke measured the temperature and weight of blood by means of thermometers and hydrometers.[18] De Gorter rebuilt Santorio's weighing chair to measure his insensible perspiration and correlated it with the humidity of the air by using Petrus Belkmeer's design of a hygrometer. Often requiring some skill and assistance, these measuring instruments allowed for repetition and a common ground for comparison with other measurements taken. While measuring instruments were designed to passively record, others had a more modifying impact. The facilities of the chemical laboratory equipped physicians to chemically manipulate the nature and behaviour of the fluids in various conditions. Medical faculties across Europe founded chemical laboratories where professors and students, with the help of assistants, performed chemical procedures for research and educational purposes. The invention of Boerhaave's little furnace, in particular, enabled the slow and controlled heating and distilling of bodily fluids, as well as other vegetal and animal materials, making it the ideal instrument for students to learn the practice of chemistry.[19] Therapeutic devices like breast pumps appeared in the medical marketplace, too, promoting the flow of milk and other fluids. Although contemporaries debated the results produced by these instruments, their usage provided a framework for thinking about nature. Instruments and chemical practices helped produce and refine key concepts, such as chemical principles. As such, the various configurations and actions of the instruments applied to bodily fluids significantly contributed to the formation of a new medicine in the eighteenth century.

* * *

[17] Insightful reflections on the impact of instruments in history of science include Frans van Lunteren, 'Clocks to Computers: A Machine-Based "Big Picture" of the History of Modern Science', *Isis*, 107 (2016), 762–776; John Tresch and Emily I. Dolan, 'Toward a New Organology: Instruments of Music and Science', *Osiris*, 28 (2013), 278–298.

[18] On Boerhaave and Fahrenheit's thermometer, see John C. Powers, 'Measuring Fire: Herman Boerhaave and the Introduction of Thermometry into Chemistry', *Osiris*, 29 (2014), 158–177.

[19] Marieke M.A. Hendriksen and Ruben E. Verwaal, 'Boerhaave's Furnace: Exploring Early Modern Chemistry through Working Models', *Berichte zur Wissenschaftsgeschichte*, 43 (2020), 385–411.

Finally, in identifying the Boerhaave school as an important community of researchers who established an eclectic medical theory based on the analysis of the bodily fluids, this book nuances some longstanding assumptions about revolutionary breakthroughs in medicine and chemistry at the turn of the nineteenth century. In recent years scholars have started to move away from the dominant focus on the chemical revolution around 1800, arguing that earlier developments in the fields of agriculture, pharmacy, and mineralogy were of crucial importance to chemistry.[20] This book strengthens this line of argument, and additionally details the importance of Boerhaavian chemistry in eighteenth-century medicine, arguing that it preceded and formed the basis of many of the revolutionary developments of the early nineteenth century.

The traditional perception of revolutionary changes taking place in the nineteenth century has long been the dominant view, and one to which many nineteenth-century scientists themselves contributed.[21] For example, in a paper read at the monthly meeting of the Medical Society of Victoria in Melbourne, Thomas Shearman Ralph (1813–1891) stated: 'Up to a recent date, it has been remarked, that the history of the life of the red corpuscle of the blood had yet to be written; to this I add, that the organic, chemical, physiological, and pathological history of the blood as a whole, has yet to be furnished to the medical student'.[22] While Ralph certainly made valuable observations, medical students had been applying microscopes and chemical experiments to the analysis of blood for more than a century at that stage. But it is on the basis of such observations that scholars have previously characterised the chemistry of the early eighteenth

[20] See, for example, Lawrence M. Principe, ed. *New Narratives in Eighteenth-Century Chemistry* (Dordrecht, 2007); Matthew D. Eddy, Seymour Mauskopf, and William R. Newman, eds., *Chemical Knowledge in the Early Modern World* (2014); Hjalmar Fors, *The Limits of Matter: Chemistry, Mining, and Enlightenment* (Chicago, 2015); John Stewart, 'Chemical Affinity in Eighteenth-Century Scottish Physiology and Agriculture' (Doctoral Thesis, The University of Oklahoma, 2013); Joppe van Driel, 'The Filthy and the Fat: Oeconomy, Chemistry and Resource Management in the Age of Revolutions, 1700–1850' (Doctoral Thesis, University of Twente, 2016).

[21] Revolution-based narratives of the birth of the clinic and the chemical revolution are indebted to Michel Foucault, *Naissance de la clinique: une archéologie du regard médical* (Paris, 1963); Thomas S. Kuhn, *The Structure of Scientific Revolutions*, 2nd ed. (Chicago, 1974).

[22] Thomas Shearman Ralph, 'Observations and Experiments with the Microscope on the Effects of Various Chemical Agents on the Blood', *Australian Medical Journal*, 11 (1866), 230–241, here 230.

century as a discipline suffering from theoretical stagnation and fallaciousness, and this theory was supposed to be proven by the misguided acceptance of the existence of phlogiston as an element. It was only from the nineteenth century onwards, they have argued that scientists came to realise the value of chemical methods to medical practice and diagnosis. Animal chemistry or clinical chemistry, as it was called, explored physiological processes as chemical reactions, and it was predominantly executed by chemists who were not physicians.[23]

But having traced blood, urine, and other bodily fluids in the laboratory, this book has emphasised the wide-spread recognition among eighteenth-century medical men—physicians, professors, and students alike—of chemistry as a pillar of medicine.[24] The Boerhaave school established and disseminated theoretical innovation in physiology and pathology by means of innovative scientific instruments and chemical experiments. As time progressed, the level of abstraction and quantification improved thanks to the investigation of larger research samples and the application of more accurate measurements. Researchers isolated the elementary substances and chemical properties of bodily fluids, and they made their findings relevant to public health concerns. In other words, the Boerhaave school established chemistry as an academic discipline with an influence beyond the confines of the chemical laboratory, from the early eighteenth century onwards. Doing so, members of the Boerhaave school anticipated many aspects of modern medicine at the turn of the nineteenth century.

By analysing Boerhaave as the beginning of a new and international research programme, this book opens up new avenues for research on medicine and natural philosophy in this period. Boerhaave's students may be traced in other fields of expertise, most notably in botany and natural history, as is the case for Martinus Houttuyn (1720–1798) and Johannes le Francq van Berkhey (1729–1812).[25] Also, the role of chemistry for the appreciation of mineral waters or the study of tropical diseases in the

[23] The main publication is Justus von Liebig, *Animal Chemistry or Organic Chemistry in Its Application to Physiology and Pathology*, trans. William Gregory (Cambridge, 1842). See for example Louis Rosenfeld, 'Justus Liebig and Animal Chemistry', *Clinical Chemistry*, 49 (2003), 1696–1707.

[24] Scholars tend to picture eighteenth-century chemistry as subordinate to medicine. For example, chemistry is presented as a 'handmaid' to medicine in F.R. Jevons, 'Boerhaave's Biochemistry', *Medical History*, 6 (1962), 343–362, here 346.

[25] Martinus Houttuyn, *Natuurlyke historie of uitvoerige beschryving der dieren, planten en mineraalen, volgens het samenstel van den heer Linnaeus*, 37 vols (Amsterdam, 1761–1785);

Americas and Asia may be re-evaluated by taking the Boerhaavians into account.[26] If we consider Boerhaave's death in 1738 not the end of an era, but rather a new beginning, the study of erudite individuals and academic institutions, of chemical practices and the production of medical knowledge in the Dutch Republic and far beyond may gather new momentum.

Johannes le Francq van Berkhey, *Natuurlyke historie van Holland*, 4 vols (Amsterdam, 1769–1778).

[26] On mineral water see, for example, Philippus Ludovicus de Presseux, *Dissertatio medica inauguralis de aquis Spadanis* (Leiden, 1736); and Jean Philippe de Limbourg, *Dissertatio physico-medica inauguralis de hominis principiis, morbisque a causis internis animatis in genere et verminosis in specie* (Leiden, 1746). On the use of Boerhaave's work by practitioners in the tropical colonies see Mark Harrison, *Medicine in an Age of Commerce and Empire: Britain and its Tropical Colonies, 1660–1830* (Oxford, 2010), 47–63.

Bibliography

Manuscripts and Objects

Amsterdam, University Library

MS 1343 I E 9–10 Matthias van Geuns, 'Ex collegio chemico H.D. Gaubii annotata', Leiden, 1758–1759.

MS 1377 I F 66 Matthias van Geuns, 'Ad Gaubii Instit. patholog'. Leiden, 1759.

MS 1382 I E 61 Matthias van Geuns, 'Schema institutionem pathologiae medicinalis Cl. Gaubii'. Harderwijk, c. 1775.

MS 1403 P IC 38 'Aenteekeninge, uijt ordre en volgens instructie van Haare Edele Grootagtbaare de Heeren Burgermeesteren, van die geene, die van den steen syn gesneden, 't zeedert den 23en April 1700 tot den eersten Januarij 1801. Gehouden door de stads doctoren en overluijden van het Chirurgijns gilde', Amsterdam, 1700–1801.

Bordeaux, University Library of Arts, University Michel de Montaigne

Bdx.U3.Ms.4 'Cl. Hier. Gaubii, in Academia Lugduno-Batava professoris extraordinarii et lectoris chemiae, PRAELECTIONES PUBLICAE CHEMICAE annorum 1731, 1732 et 1733, sive examen chemicum humorum corporis humani, habitum in auditorio chemico', 2 vols, 1731–1733.

R. E. Verwaal, *Bodily Fluids, Chemistry and Medicine in the Eighteenth-Century Boerhaave School*, Palgrave Studies in Medieval and Early Modern Medicine,
https://doi.org/10.1007/978-3-030-51541-6

COPENHAGEN, ROYAL LIBRARY

Thott 691, 692 kvart 'Dictata Gaubii in Pathologiam Boerhavii', 2 vols, 1745.

GRONINGEN, UNIVERSITY LIBRARY

HS 98 Anton Gabriel Meder, 'Tractatus collegii chymici', Leiden, 1710–1713.

HAARLEM, NOORD-HOLLANDS ARCHIEF

NHA 444, 373 Hollandsche Maatschappij der Wetenschappen te Haarlem, 1752–1975. Prijsvraag 8, Zog, 1760–1761.

LEEUWARDEN, TRESOAR

s 66 AW 'Bibliotheca Boerhaaviana, sive Catalogus librorum … Hermanni Boerhaave', Leiden, 1739.

LEIDEN, RIJKSMUSEUM BOERHAAVE

V25790 Little Boerhaave stove, 1800–1900.
V05517 Breast pump made of tin and glass, G. Hoepe, Berlin, 1800–1850.

LEIDEN, MUSEUM DE LAKENHAL

3589 Model of Sepulchre of Herman Boerhaave in the Pieterskerk, Leiden, late 18th century.

LEIDEN, UNIVERSITY LIBRARY

BPL 850 'H.D. Gaubius, Institutiones pathologicae', 4 vols, c. 1750.
BPL 1475 'Viri Celeberrimi H.D. Gaubii M.D. Medicinae, Chemiae, & Collegii Practico-Medici in Academia Batava, qua Leidae est, Professoris Ordinarii. PRAELECTIONUM THEORETICARUM Secundum Institutiones Medicas H. Boerhaave', 6 vols, Leiden: Philip Bonk, 1751.
BPL 1476 'Compendium Praelectionum H.D. Gaubii Med. Prof. in Pathologiam Nosologiam & Aetiologiam', Leiden, *c.* 1750, 1768.
BPL 1477 'Viri Clarissimi Hieronymi Davidis Gaubii Med. Doctoris, Ejusdem ac Chemiae et Collegii Practico-Medici in Academia Batavia, quae Leidae est, Professoris Ordinarii, DICTATA IN CHEMIAM', 3 vols, Leiden: Philip Bonk, *c.* 1750.

BPL 1478 Gerhardus Vermeer, 'Comentaria ex ore Clarissimi Viri J. De Gorter excerpta', Harderwijk, c. 1740.

F 699 Herman Boerhaave, 'Praelectiones publice habitae de morbis nervorum', Leiden, 1730–1735. Microfiche copy of MS XIII 11.

UBL 042, 348–350 Archieven van Senaat en Faculteiten, 1575–1877; Universiteit Leiden (ASF); D. Studenten.

London, British Library

Sloane MS 3183–3184 'Descripta quaedam ab Ore Hermanni Boerhaavii inter pronunciandum lectiones ejus de Chemia in 2 tomis. Lugd. Bat. Anno 1719', 2 vols, Leiden, c. 1719–1730.

London, Royal College of Surgeons of England

MS0114/1/1 'Collegii Chemici ex D.D. Boerhaaven'. Leiden, 1702.

London, Science Museum

A124874 Breast pump of brass and glass, Europe, 1771–1830.

A606855 Breast reliever for self use, made of glass, c. 1850–1880.

A606862 Breast reliever for self use, made of glass, place and date unknown.

London, Wellcome Library

A641255 Nipple shield of sterling silver with the maker's initials and George III's head, 1786–1821.

A606830 Nipple shield of glass, c. 1800.

A606829 Nipple shield of ivory, c. 1800.

MS 1927 Charles Blagden, 'Cullen's Comment on Gaubius' Pathology'. Edinburgh, 1767.

MS 2480–2485 George Paterson, 'Gaubii Chemia', Leiden, c. 1755.

Lyon, Inter-University Library, Diderot

Mss et R 63 'Dictata cl. Gaubii in Ætiologiam, seu Commentarium a §o 737 et sequentes usque ad § 800 Institutionum Medicarum Boerhaave', Leiden, 1745.

Mss et R 64 'Dictata clariss. Gaubii in Symptomatologiam Pathologiae', Leiden, 1745.

MIDDELBURG, ZEELAND LIBRARY

Handschrift 1940 Letter from Cornelis Hendricus Velse to Paulus 's-Graeuwen, The Hague, 16 October 1770.

Handschrift 4940 Jacobus de Puyt, 'Observatie van een verlossinge van drie kinders', Middelburg, 1777.

Handschrift 6270–6271 'Hieron. David. Gaubii Med. Doct. Med. & Chemio Professoris in Academia Leijdensi DICTATA IN CHEMIAM', 2 vols.

MONTPELLIER, INTER-UNIVERSITY LIBRARY, SECTION MEDICINE

H 166 'Hieronymi Gaubii prælectiones publicæ chemicæ, annis 1731-32-33 sive Examen Chemicum humorum Corporis humani habitum in auditorio chemico', 1731–1733.

OXFORD, BODLEIAN LIBRARY

MS. Rawl. C 535 'Hermanni Boerhaave Chemia', Rawlinson Collection Mss. 12382.

ST PETERSBURG, S.M. KIROV MILITARY MEDICAL ACADEMY

MS XIII 11 Herman Boerhaave, 'Praelectiones publice habitae de morbis nervorum', Leiden, 1730–1735.

THE HAGUE, ROYAL LIBRARY

424 B 2, nr. 14 Hieronymus Gaubius to Anthony Adriaan van Iddekinge, Leiden, 8 February, 15 May, 24 August, 2 September 1753.

UPPSALA, UNIVERSITY LIBRARY

Waller MS cod-00068 Georg Michael Wepfer, 'Collegium Chymicum experimentale', Schaffhausen, 1714.

VIENNA, ONB

Cod. 12713–12714 Han 'Epistolae ad Antonium Ribeiro Sanches', 2 vols, 1735–1783.

YALE, BEINECKE RARE BOOK AND MANUSCRIPT LIBRARY

Mellon MS 117 'Annotationes ex praelectionibus chemicis', Leiden, 1772–1773.

Primary Sources

Arnold, Godfried. *Historie der kerken en ketteren van den beginne des Nieuwen Testaments tot aan het jaar onses Heeren 1688.* 3 vols Amsterdam: Sebastiaan Petzold, 1701–1729.

Balthazaar, Alexander. *Pathologia chirurgicalis of heelkundige ziektekunde.* Leiden: Cornelis de Pecker, 1772–1776.

———. *Pathologia chirurgicalis of heelkundige ziektekunde.* 2 vols Leiden: J. Meerburg and J.J. Thyssens, 1777.

Barlow, G.H., and George Owen Rees. 'An Account of Observations Made under the Superintendance of Dr Bright on Patients Whose Urine Was Albuminous'. *Guy's Hospital Reports 2nd Series,* 1 (1843), 189–316.

Bartsch, Johannes. *Dissertatio medica inauguralis sistens calorem corporis humani effectum hygraulicum.* Leiden: Coenraad Wishoff, 1737.

Behrens, Konrad Barthold. *Ob das Wasserbeschawen in Krankheiten etwas nuze, und wie weit demselben zu trauen.* Hildesheim: [s.n.], 1688.

Belkmeer, Cornelius. *De streelende speel-pop van Dr. Antoni Stochius naakt uytgekleet en ten toon gestelt in de beschouwing rakende syne sogenaamde geneeskundige verhandeling en aanmerkingen over eene algemeene volk-ziekte.* Enkhuizen: Meynard Semeyns, 1734.

Belkmeer, Petrus. *Disputatio inauguralis physiologico medica de motu ut causa et curatione generali omnium morborum.* Harderwijk: Johannes Rampen, 1735.

à Bergen, Johannes Christophorus. *Dissertatio medica inauguralis de lithiasi.* Leiden: Dirk Haak, 1765.

Bergman, Tobern. 'Tilläggning Om Blåse-Stenen'. *Kongliga Vetenskaps Academiens Handlingar,* 37 (1776), 333–8.

van Beverwijck, Johan. *Wercken der genees-konste, bestaende in den Schat der gesontheyt, Schat der ongesontheyt, Heel-konste.* Amsterdam: Jan Jacobsz Schipper, 1680.

'Bibliotheca Boerhaaviana sive catalogus librorum instructissimae bibliothecae viri summi D. Hermanni Boerhaave'. Leiden: Samuel Luchtmans, 1739.

'Bibliotheca Gaubiana sive Catalogus librorum viri celeberrimi Hieronymi Davidis Gaubii'. Leiden: Samuel and Johannes Luchtmans, 1783.

'Bibliotheca Geunsiana, sive, Catalogus librorum quibus usus est vir celeberrimus Matthias van Geuns'. Utrecht: Johannes Altheer, 1818.

Bicker, Lambertus. *Verhandeling van het zog der vrouwen, ter beantwoordinge op de vraage by de Hollandsche Maatschappye der Weetenschappen opgegeeven in de jaaren 1760 en 1761.* Leiden: Samuel en Johannes Luchtmans, 1763.

Bidloo, Govard. *Anatomia humani corporis.* Amsterdam: Johannes van Someren, Jan van Dyck, Hendrick and Dirk Boom, 1685.

Blankaart, Steven. *Verhandelinge van de opvoedinge en ziekten der kinderen.* Amsterdam: Hieroymus Sweerts, 1684.

———. *Anatomia practica rationalis, sive Rariorum cadaverum morbis denatorum anatomica inspectio*. Amsterdam: Cornelis Blankaart, 1688.

Bleuland, Jan. *Oratio, qua memoria Hieronymi Davidis Gaubii cum omnibus, tum praesertim medicinae studiosis commendatur*. Harderwijk: Johan Moojen, 1792.

Boerhaave, Herman. *Institutiones medicae, in usus annuae exercitationis domesticos digestae*. Leiden: Johannes van der Linden, 1708.

———. *Institutiones medicae in usus annuae exercitationis domesticos digestae*. alt. ed. Leiden: Johannes van der Linden, 1713.

———. *Boerhaave's Institutions in Physick*. Translated by Joseph Browne. 2nd ed. London: Jonas Browne, 1715.

———. *Sermo academicus de chemia suos errores expurgante*. Leiden: Pieter van der Aa, 1718.

———. *Institutiones et experimenta chemiae*. 2 vols 'Paris': [s.n.], 1724.

———. *Institutiones et experimenta chemiae*. 2 vols Venice: Sebastian Coleti, 1726.

———. *A New Method of Chemistry: Including the Theory and Practice of that Art: Laid down on Mechanical Principles, and Accommodated to the Uses of Life*. Translated by Peter Shaw and Ephraim Chambers. 2 vols London: John Osborn and Thomas Longman, 1727.

———. 'Tractatus medicus de lue venerea praefixus Aphrodisiaco'. In *Aphrodisiacus*. Leiden: Johannes Arnold Langerak, Johan and Herman Verbeek, 1728.

———. *A Treatise on the Venereal Disease and its Cure in all its Stages and Circumstances*. Translated by John Barker. London: T. Cox and J. Clarke, 1729.

———. *Elementa chemiae, quae anniversario labore docuit in publicis, privatisque scholis*. 2 vols Leiden: Isaac Severinus, 1732.

———. *Institutiones medicae in usus annuae exercitationis domesticos digestae*. 5th ed. Leiden and Rotterdam: Dirk Haak, Samuel Luchtmans, Johannes and Herman Verbeek, and Johan Daniel Beman, 1734.

———. *Elements of Chemistry: Being the Annual Lectures*. Translated by Timothy Dallowe. 2 vols London: J. and J. Pemberton; J. Clarke; A. Miller; and J. Gray, 1735.

———. *A New Method of Chemistry: Including the History, Theory, and Practice of the Art*. Translated by Peter Shaw. 2nd ed. 2 vols London: Thomas Longman, 1741.

———. *Kortbondige spreuken wegens de ziektens, te kennen en te geneezen*. Translated by Kornelis Love Jakobsz. Amsterdam: Johannes Gysius, 1741.

———. *Praelectiones academicae de morbis nervorum*. Edited by Jacob van Eems. Leiden: Pieter van der Eyk and Cornelis de Pecker, 1761.

Boerhaave, Herman, and Frederik Ruysch. *Opusculum anatomicum de fabrica glandularum in corpore humano*. Leiden: Pieter van der Aa, 1722.

Boissier de Sauvages, François. *Nouvelles classes de maladies, qui dans un ordre semblable à celui des botanists, comprennent les genres & les espèces de toutes les maladies, avec leurs signes & leurs indications*. Avignon: B. d'Avanville, 1731.

———. *Pathologia methodica, seu de cognoscendis morbis*. Amsterdam: De Tournes, 1752.

———. *Nosologie methodique, dans laquelle les maladies sont rangées par classes, suivant le systême de Sydenham, & l'ordre des Botanistes*. Translated by Pierre François Nicolas. 3 vols Paris: Hérissant, 1771.

de Bordeu, Théophile. *Recherches sur le pouls, par rapport aux crises*. Paris: De Bure l'aîné, 1756.

———. *Inquiries Concerning the Varieties of the Pulse: And the Particular Crisis Each More Especially Indicates*. London: T. Lewis and G. Kearsly, 1764.

Bourgeois, Pieter. *Dissertatio pathalogico-medica inauguralis de calculo et remediis cum solventibus*. Leiden: Johan Heyligert, 1744.

Bright, Richard. 'Tabular View of the Morbid Appearances in 100 Cases Connected with Albuminous Urine'. *Guy's Hospital Reports,* 1 (1836), 380–402.

Buisen, Henricus. *Verhandelinge van de uitwerpingen des menschelyke lighaams, bestaande in pis, afgang, zweet, kwyl; en braaking*. Rotterdam: Hermanus Kentlink, 1731.

———. *Practyk der medicine, ofte Oeffenende geneeskunde*. 4th ed. Rotterdam: Herman Kentlink, 1743.

Burton, William. *An Account of the Life and Writings of Herman Boerhaave*. London: Henry Lintot, 1743.

Cadogan, William. *An Essay upon Nursing, and the Management of Children, from their Birth to Three Years of Age*. London: J. Roberts, 1750.

Canisius, Cornelius. *Dissertatio medica inauguralis de calculo in genere et preacipue renum et vesicae*. Leiden: Coenraad Wishoff, 1757.

Cats, Jacob. *Houwelyck: Dat is de gansche gelegentheyt des Echten staets*. 7 vols Middelburg: Jan Pietersz van de Venne, 1625.

Chambers, Ephraim. *Cyclopaedia, or, An Universal Dictionary of the Arts and Sciences*. 2 vols London: J. and J. Knapton, 1728.

Chomel, Noël. *Huishoudelyk woordboek: Vervattende vele middelen om zyn goed te vermeerderen, en zyne gezondheid te behouden, Met verscheiden wisse en beproefde middelen*. Translated by Jan Lodewyk Schuer and A.H. Westerhof. 2 vols Leiden and Amsterdam: S. Luchtmans and H. Uytwerf, 1743.

———. *Algemeen huishoudelijk-, natuur-, zedekundig- en konstwoordenboek*. Edited by Jacques Alexandre de Chalmot. 2nd ed. Leiden and Leeuwarden: Johannes le Mair and Hendrik Amelius de Chalmot, 1778.

Crauford, Jacobus. *Disputatio medica inauguralis de scorbuto*. Leiden: Abraham Elzevier, 1707.

Cullen, William. *First Lines of the Practice of Physic*. 4th ed. 4 vols Edinburgh: C. Elliot and T. Cadell, 1784.
———. *Nosology: Or, a Systematic Arrangement of Diseases, by Classes, Orders, Genera, and Species*. Edinburgh: William Creech, 1800.
d'Alencé, Joachim. *Verhandelingen over de barometers, thermometers, en notiometers of hygrometers*. The Hague: Jacobus de Jongh, 1730.
David, Jean-Pierre. *Dissertation sur ce qu'il convient faire pour diminuer ou supprimer le lait des femmes*. Paris: Vallat-Lachapelle, 1763.
———. 'Wat 'er behoore gedaan te worden, om het Zog der Vrouwen te vermeerderen, te verminderen, of te doen verdwynen'. *Verhandelingen uitgegeeven door de Hollandse Maatschappy der Weetenschappen te Haarlem*, 7 (1763), 2–76.
De groote zonden van vuile zelfs-bevleckinge, door jonge en oude, mans- en vrouwspersoonen, ontdekt. Translated by J. van Hode. Rotterdam: [*s.n.*], 1730.
Diderot, Denis, and Jean le Rond d'Alembert, eds. *Encyclopédie, ou dictionnaire raisonné des sciences, des arts et des métiers*. 17 vols Paris: Briasson, 1751–1765.
Dionis, Pierre. *Traite general des accouchemens*. Paris: Charles-Maurice d'Houry, 1718.
Doorschodt, Henricus. *Dissertatio medica inauguralis de lacte*. Leiden: Johannes Hasebroek, 1737.
van Dueren, Johan. *De ontdekking der bedriegeryen van de gemeene pis-bezienders*. Amsterdam: Timotheus ten Hoorn, 1688.
Egeling, Jan. *Dissertatio chemico-medica inauguralis de lacte*. Utrecht: Joannis Broedelet, 1759.
Ellis, John. *An Essay towards a Natural History of the Corallines, and other Marine Productions of the Like Kind, Commonly Found on the Coasts of Great Britain and Ireland*. London: A. Millar, J. and J. Rivington, and R. and J. Dodsley, 1755.
Everwijn, Joannes Gulielmus. *Specimen medicum inaugurale de lithogenesia in vesica urinaria*. Leiden: Johannes Hasebroek, 1758.
de Farvacques, Robert. *Medicina pharmaceutica, oft drôgh-bereydende gheneeskonste: met besondere aenmerckingen op verscheyde misbruycken, die soo wel in de medecyne als chymie zyn voor-vallende*. Brussels: F. Foppens, 1681.
———. *Medicina pharmaceutica, of Groote algemeene schatkamer der drôgbereidende geneeskonst*. 3 vols Leiden: Isaac Severinus, 1741.
Farwell, Christopher. *Dissertatio medica inauguralis de calculo renum et vesciae urinariae*. Leiden: Johannes and Hermanus Verbeek, 1729.
von Ferro, Pascal Joseph. *Einrichtung der medizinischen Fakultät zu Wien*. Vienna: R. Gräffer, 1785.
van Foreest, Pieter. *De incerto, fallaci, urinarum iudicio*. Leiden: Plantin, 1589.
de Fourcroy, Antoine François, and Louis Nicolas Vauquelin. 'Premier mémoire pour servir à l'histoire naturelle, chimique et médicale, de l'urine humaine'. *Annales de Chimie*, 31 (1798), 48–71.

———. 'Second mémoire pour servir à l'histoire naturelle, chimique et naturelle de l'urine humaine dans lequel on s'occupe spécialement des propriété de la matière particulière qui la caractérise'. *Annales de Chimie*, 32 (1799), 80–162.

le Francq van Berkhey, Johannes. *Natuurlyke historie van Holland.* 4 vols Amsterdam: Yntema and Tieboel, 1769–1778.

Gallandat, Isaac Henri. 'De zogpomp, voorgesteld en aangeprezen'. *Verhandelingen van het Zeeuwsch Genootschap der Wetenschappen*, 13 (1786), 538–58.

Gaubius, Hieronymus David. *Oratio inauguralis qua ostenditur chemiam artibus academicis jure esse inserendam.* Leiden: Coenraad Wishoff, 1731.

———. *Oratio de vana vitae longae, a chemicis promissae, exspectatione.* Leiden: Isaac Severinus, 1734.

———. *Sermo academicus de regimine mentis quod medicorum est.* Leiden: Boudewijn Janssoon van der Aa, 1747.

———. *Institutiones pathologiae medicinalis.* Leiden: Samuel and Johannes Luchtmans, 1758.

———. *On the Passions: or a Philosophical Discourse Concerning the Duty and Office of Physicians in the Management and Cure of the Disorders of the Mind.* Translated by J. Taprell. London: W. Flexney, 1760?

———. *Institutiones pathologiae medicinalis.* 2nd ed. Edinburg: A. Donaldson and J. Reid, 1762.

———. *Institutiones pathologiae medicinalis.* 2nd ed. Leiden: Samuel and Joannes Luchtmans, 1763.

———. *Sermo academicus de regimine mentis quod medicorum est.* Leiden: Samuel and Johannes Luchtmans, 1763.

———. *Pathologie de M. Gaubius.* Translated by Pierre Seu. Paris: Vincent, 1770.

———. *Feestrede van Hieronymus David Gaubius by den heuglyken aanvang der derde eeuwe van Hollands Hooge Schoole te Leyden, den agsten van Sprokkelmaand 1775.* Translated by Pieter van den Bosch. Leiden: Samuel and Johannes Luchtmans, 1775.

———. *Oratio panegyrica in auspicium seculi terti Academiae Batavae.* Leiden: Samuel and Johannes Luchtmans, 1775.

———. *The Institutions of Medicinal Pathology.* Translated by Charles Erskine. Edinburgh: C. Elliot and T. Cadell, 1778.

———. *Anfangsgründe der Krankheitenlehre des Menschen.* Translated by Daniel Andreas Diebold. Zurich: Orell, Gessner, Füsslin and Comp, 1781.

———. *Anfangsgründe der Medicinischen Krankheitslehre.* Translated by Christian Gottfried Gruner. Berlin: Sigismund Friedrich Hesse, 1784.

———. *Institutiones pathologiae medicinalis.* Edited by Johann Christian Gottlieb Ackermann. 3rd ed. Nuremberg: E.C. Grattenauer, 1787.

———. *Pathologie de M. Gaubius.* Translated by Pierre Seu. rev. ed. Paris: Théophile Barrois, 1788.

———. *Начальныя основанія врачебныя пафологіи, то есть науки о свойствѣ, причинахъ, припадкахъ и различіяхъ болѣзней, въ человѣческомъ тѣлѣ случающихся*. Translated by Petr Martynovich Gofman. St Petersburg: Nauk, 1792.

Gebeurtenissen, voorgevallen in de maanden January en February, anno 1740 in, en veroorzaakt door de nooit meer gehoorde vehemente en strenge winter. Enkhuizen: Hendrik Callenbach, 1740.

de Genlis, Félicité. *Adèle en Theodoor, of Brieven over de opvoeding* Translated by Elisabeth Wolff-Bekker. 3 vols The Hague: Isaac van Cleef, 1782–1783.

de Gorter, Herman Boerhaave. *Disputatio medica inauguralis de lacte et lactatione*. Harderwijk: Johannes Moojen, 1751.

de Gorter, Johannes. *De perspiratione insensibili Sanctoriana-Batava tractatus experimentis propriis in Hollandia*. Leiden: Janssoon van der Aa, 1725.

———. *Medicinae compendium in usum exercitationis domesticae digestum*. 2 vols Leiden: Pieter and Boudewijn Janssoon van der Aa, 1731–1737.

———. *Morbi epidemii brevis descriptio et curatio per diaphoresin*. Harderwijk: Willem Brinkink, 1733.

———. *Korte beschryving van een algemene doorgaande ziekten, in deze tijd nog woedende, en desselfs genezing door sweetinge*. Translated by Amos Lambrechts. Amsterdam: Gerrit Bouman, 1733.

———. *De perspiratione insensibili*. 2nd ed. Leiden: Janssoon van der Aa, 1736.

———. *Formulae medicinales cum indice virium quo ad inventas indicationes inveniuntur medicamina*. Harderwijk: Johannes Wigmans, 1750.

———. *Gezuiverde geneeskonst, of kort onderwys der meeste inwendige ziekten*. 3rd ed. Amsterdam: Isaak Tirion, 1761.

———. *Praxis medicae systema*. Edited by David de Gorter. 2nd ed. Harderwijk: Johannes Moojen, 1767.

de Graaf, Reinier. *De succi pancreatici natura et usu exercitatio anatomico-medica*. Leiden: [s.n.], 1664.

———. *Alle de wercken, so in de ontleed-kunde, als andere deelen der medicyne*. Amsterdam: Abraham Abrahamse, 1686.

Groen, J. van der. *Den Ervaren huys-houder; zijnde het III. deel van het Vermakelyck landt-leven*. Amsterdam: Marcus Doornick, 1683.

de Haen, Antonius. *Ratio medendi in nosocomio practico*. Vienna: Johann Thomas Trattner, 1756–1773.

———. *Opuscula quaedam inedita*. Edited by Joseph Eyerel. 2 vols Vienna: Joseph Camesma, 1795.

von Haller, Albrecht. *Dissertatio inauguralis sistens experimenta et dubia circa ductum salivalem novum Coschwizianum*. Leiden: Coenraad Wishoff, 1727.

———, ed. *Praelectiones academicae in proprias institutiones rei medicae*. 6 vols Göttingen: Abram Vandenhoeck, 1739–1744.

———, ed. *Dr. Boerhaave's Academical Lectures on the Theory of Physic: Being a Genuine Translation of his Institutes and Explanatory Comment.* 6 vols London: W. Innys, 1742–1746.

———. *Primae lineae physiologiae.* 2nd ed. Göttingen: Abram Vandenhoeck, 1751.

———. *Physiology: Being a Course of Lectures upon the Visceral Anatomy and Vital Oeconomy of Human Bodies.* Translated by Samuel Mihles. 2 vols London: W. Innys and J. Richardson, 1754.

———. *Elementa physiologiae corporis humani.* 8 vols Lausanne: Bousquet, d'Arnay, Grasset, 1757–1766.

———, ed. *Epistolarum ab eruditis viris ad Alb. Hallerum scriptarum.* 6 vols Bern: Societas typographica, 1773–1775.

———. *Letters from Baron Haller to his daughter, on the truths of the Christian religion.* London and Edinburgh: J. Murray and William Creech, 1780.

Harvey, William. *Exercitationes de generatione animalium.* Amsterdam: Johannes Janssonius, 1651.

———. *Anatomical Exercitations Concerning the Generation of Living Creatures.* London: Octavian Pulleyn, 1653.

von Hellwig, Ludwig. *Valentin Kräutermanns Curieuser und vernünfftiger Urin-Artzt.* Arnstadt: Johann Jacob Beumelburg, 1735.

van Helmont, Jan Baptist. *Ortus medicinae: id est, initia physicae inaudita. Progressus medicinae novus, in morborum ultionem ad vitam longam.* Edited by Franciscus Mercurius van Helmont. Amsterdam: Lowijs Elzevier, 1652.

———. *Ortus medicinae.* Edited by Franciscus Mercurius van Helmont. Amsterdam: Lowijs Elzevier, 1652.

———. *Opera omnia.* Frankfurt: Hieronymus Christian Paulli, 1707.

Hoffmann, Friedrich. *Observationum physico-chymicarum.* Halle: Rengeriana, 1722.

Horsman, Samuel. *Dissertatio medica inauguralis de calculo renum et vesicae.* Leiden: C. Boutesteyn & Fil., 1721.

Houttuyn, Martinus. *Natuurlyke historie of uitvoerige beschryving der dieren, planten en mineraalen, volgens het samenstel van den heer Linnaeus.* 37 vols Amsterdam: Frans Houttuyn, 1761–1785.

Hulshof, Hermannus. *Dissertatio medica inauguralis sistens febrem diariam benignam ex suppressa Sanctoriana perspiratione ortam.* Groningen: Hajo Spandaw, 1740.

Humphrey, Thomas. *Dissertatio medica inauguralis de calculi urinosi generatione et incremento.* Leiden: Johannus Arnoldus Langerak, 1736.

Hunter, John. *A Treatise on the Venereal Disease.* London: sold at No. 13, Castle-Street, Leicester-Square, 1786.

———. *A Treatise on the Blood, Inflammation and Gun-Shot Wounds.* London: Sherwood, Gilbert and Piper, 1794.

Ingenhousz, Jan. *Experiments upon Vegetables.* London: Peter Elmsly and H. Payne, 1779.

von Jacquin, Joseph Franz. *Leerboek der algemeene en artsenijkundige scheikunde.* Translated by Gerardus Plaat. 2 vols Leiden: Abraham and Jan Honkoop, 1794.

Kaau, Abraham. *Perspiratio dicta Hippocrati per universum corpus anatomice illustrata.* Leiden: Samuel Luchtmans, 1738.

de Kedts, Daniel Dolegius. *Disputatio medica inauguralis de calculo in genere, sed praesertim renum et vesicae.* Leiden: Coenraad Wishoff, 1717.

Keill, James. *Tentamina medico-physica ad quasdam quaestiones quae oeconomiam animalem spectant, accomodata: Quibus accessit Medicina statica Britannica.* London: George Strahan, William and John Innys, 1718.

Kind, Joannes Petrus. *Dissertatio medica physico-pathologica inauguralis de remediis calculum in renibus diffringentibus.* Leiden: Johannes van Kerckhem, 1724.

de la Fosse, Frederik. *Specimen medicum inaugurale de aëre vitae et morborum causa.* Leiden: Jean Luzac, 1743.

de La Mettrie, Julien Offray. *L'homme machine.* Leiden: Elie Luzac, 1748.

Lacaze, Louis de. *Idée de l'homme physique et moral, pour servir d'introduction à un traité de médecine.* Paris: Guérin and Delatour, 1755.

Lennartson, Anders. *The Chemical Works of Carl Wilhelm Scheele.* SpringerBriefs in History of Chemistry. Dordrecht: Springer, 2017.

von Liebig, Justus. *Animal Chemistry or Organic Chemistry in Its Application to Physiology and Pathology.* Translated by William Gregory. Cambridge: John Owen, 1842.

de Limbourg, Jean Philippe. *Dissertatio physico-medica inauguralis de hominis principiis, morbisque a causis internis animatis in genere et verminosis in specie.* Leiden: Cornelis Haak, 1746.

Lotichius, Petrus Nicolaus. *Dissertationem physico-medicam inauguralem de phosphoris et phosphoro urinae.* Leiden: Gerard Potvliet, 1757.

Luiscius, Anthonius Ludovicus. *Disputatio medica inauguralis anatomico chemico practica de calculo renum et vesicae.* Leiden: Coenraad Wishoff, 1720.

MacBride, David. *A Methodical Introduction to the Theory and Practice of Physic.* London: W. Strahan, T. Cadell, A. Kincaid, W. Creech and J. Balfour, 1772.

———. *Introductio methodica in theoriam et praxin medicinae.* Translated by Johannes Fredericus Clossius. 2 vols Utrecht: Johannes Schoonhoven, 1774.

Magenis, Arthurus. *Dissertatio medica inauguralis de urina.* Leiden: Gerard Potvliet, 1753.

Major, Johann Daniel. *Dissertatio medica de lacte lunae.* 1667.

Makittrick, James. *Commentaries on the Principles and Practice of Physic.* London: T. Becket and J. Balfour, 1772.

Mauriceau, François. *Traité des maladies des femmes grosses.* Paris: Chez l'auteur, 1681.

Mesnard, Jacques. *Le guide des accoucheurs, ou Le maistre dans l'art d'accoucher les femmes.* Paris: De Bure, Le Breton, Durand, 1743.

Molhuysen, P.C. *Bronnen tot de geschiedenis der Leidsche Universiteit*. Rijks geschiedkundige publicatiën. 7 vols The Hague: Martinus Nijhoff, 1913–1924.
Morgagni, Giovanni Battista. *Adversaria anatomica omnia*. 6 vols Leiden: Johan Arnold Langerak, 1723.
———. *Epistolae anatomicae duae novas observationes, et animadversiones complectentes*. Leiden: Johannes van Kerckhem, 1728.
———. *De sedibus et causis morborum per anatomen indagatis*. Venice: Remondini, 1761.
———. *The Seats and Causes of Diseases Investigated by Anatomy*. Translated by Benjamin Alexander. 3 vols London: A. Millar and T. Cadell, Johnson and Payne, 1769.
le Mort, Jacob. *Chymia, rationibus et experimentis auctioribus, iisque demonstrativis superstructa, in qua malevolorum calumniae modestè simul diluuntur* Leiden: Pieter van der Aa, 1688.
'Musei Gaubiani pars sive Catalogus partis supellectilis, qua usus est vir celeberrimus H. D. Gaubius'. Leiden: Samuel and Johannes Luchtmans, 1783.
Musgrave, Samuel. *Dissertatio medica inauguralis sive apologia pro medicina empirica*. Leiden: Samuel and Johannes Luchtmans, 1763.
Nieuwe Nederduitsche apotheek: Op eene klaare en verstaanbaare wyze onderwys gevende omtrent de beste dagelyks gebruikt wordende geneeskundige bereidingen. Leiden: Pieter van der Eyk, 1753.
Paracelsus. *De urinarum ac pulsuum judiciis*. Cologne: Arnold Birckmann, 1568.
———. *Opera medico-chemica*. 3 vols Frankfurt: Palthenius, 1603.
———. *Opera: Bücher und Schrifften*. Strassburg: Lazarus Zetzner, 1616.
———. *Opera omnia medico-chemico-chirurgica*. 3 vols Geneva: Jean Antoine & Samuel de Tournes, 1658.
Pemberton, Henry. *Dissertatio physico-medica inauguralis de facultate oculi, qua ad diversas rerum conspectarum distantias se accommodat*. Leiden: Pieter van der Aa, 1719.
Peyer, Bernhardus. *Dissertatio medica inauguralis de tussi*. Leiden: Coenraad Wishoff, 1750.
de Presseux, Philippus Ludovicus. *Dissertatio medica inauguralis de aquis Spadanis*. Leiden: Jean Luzac, 1736.
Ralph, Thomas Shearman. 'Observations and Experiments with the Microscope on the Effects of Various Chemical Agents on the Blood'. *Australian Medical Journal*, 11 (1866), 230–41.
Rau, Johannes Jacobus. *Oratio inauguralis de methodo anatomen docendi et discendi*. Leiden: Samuel Luchtmans, 1713.
Rhode, Ulricus Wilhelmus. *Dissertatio inauguralis chirurgico-medica de nervi punctura*. Leiden: Coenraad Wishoff, 1720.
von Richter, Wilhelm Michael. *Geschichte der Medicin in Russland*. 3 vols Moscow: N.S. Wsewolojsky, 1813–1817.

Robidé van der Aa, C.P.E. *Frederika Louise Wilhelmina van Pruissen, eerste koningin der Nederlanden: als een voorbeeld ter navolging, der Nederlandsche meisjes aangeprezen*. Amsterdam: G.J.A. Beijerinck, 1838.

de Roode, Matthaeus. *Disputatio medica inauguralis de lithiasi*. Leiden: Pieter van de Aa, 1720.

Rosén von Rosenstein, Nils. *Handleiding tot de kennis en geneezing van de ziekten der kinderen*. Translated and edited by Eduard Sandifort. The Hague: Pieter van Cleef, 1768.

Rouelle, Hiliare Marie. 'Observations sur l'urine humaine, & sur celles de vache & de cheval, comparés ensemble'. *Journal de médecine, chirurgie, pharmacie, &c.*, 40 (1773), 451–68.

Ruysch, Frederik. *Alle de ontleed- genees- en heelkundige werken*. Translated by Ysbrand Gysbert Arlebout. 3 vols Amsterdam: Janssoons van Waesberge, 1744.

's-Graeuwen, Paulus. *Oratio inauguralis de anatomiae pathologicae utilitate ac necessitate*. Groningen: Hajo Spandaw, 1771.

Sanches, António Nunes Ribeiro. *Dissertation sur l'origine de la maladie vénérienne*. Paris: Durand and Pissot, 1750.

———. *Examen historique sur l'apparition de la maladie vénérienne en Europe*. Lisbon [Paris]: [*s.n.*], 1774.

———. *Dissertation sur l'origine de la maladie vénérienne, pour prouver que ce mal n'est pas venu d'Amérique, mais qu'il a commencé eu Europe par une epidémie. Suivie de l'examen historique sur l'apparition de la maladie vénérienne en Europe*. new ed. Leiden: Andries Koster, 1777.

———. *Observations sur les maladies vénériennes*. Paris: Théophile Barrois jr., 1785.

Sandifort, Eduard. *Natuur- en genees-kundige bibliotheek*. 11 vols The Hague: Pieter van Cleef, 1765–1775.

———. *Observationes anatomico-pathologicae*. 4 vols Leiden: Pieter van der Eyk and Daniël Vygh, 1777–1781.

———. *Museum anatomicum academiae Lugduno-Batavae descriptum*. 2 vols Leiden: Samuel and Johannes Luchtmans, 1796.

Santori, Santorio. *Commentaria in Artem medicinalem Galeni*. Venice: Giacomo Antonio Somasco, 1612.

———. *Ars de statica medicina, sectionibus aphorismorum septem comprehensa*. Venice: Nicolaum Polum, 1614.

———. *Commentaria in primam fen primi libri Canonis Avicennae*. Venice: Giacomo Sarcina, 1625.

———. *Commentaria in primam sectionem Aphorismorum Hippocratis*. Venice: Marco Antonio Brogiolo, 1629.

———. *De statica medicina et de responsione ad staticomasticem*. Leiden: David Lopez de Haro, 1642.

———. *De ontdekte doorwaasseming of de leidstar der genees-heeren*. Translated by Philippe La Grue. Amsterdam: Jan Rieuwertsz, 1683.

———. *De ontdekte doorwaasseming des menschen lichaams*. Translated by Philippe La Grue. Edited by Steven Blankaart. 2nd ed. Amsterdam: Jacob van Royen and Abraham Uittenbogaart, 1684.

———. *De ontdekte doorwaasseming des menschen lichaams*. Edited by Heydentryck Overkamp. Amsterdam: Timotheus ten Hoorn, 1686.

Scheele, Carl Wilhelm. 'Undersökning om Blåse-stenen'. *Kongliga Vetenskaps Academiens Handlingar*, 37 (1776), 327–32.

———. 'Analysis of the Calculus Vesicae'. Translated by Thomas Beddoes. In *The Chemical Essays*. 145–56. London: J. Murray, W. Gordon and C. Elliot, 1786.

Scheffelius, Christianus Stephanus. *Dissertatio medico-practica inauguralis de lithiasi fellea sive calculo vesiculae biliariae, cujus occasione traditur simul brevis historia lapidis porcini Malacensis et fellis bovis* Leiden: Coenraad Wishoff, 1721.

Schlichting, Johannes Daniel. *Syphilidos mnemosynon criticon of vrye en oneenzydige gedachten over ongemakken, door 't gebruyk der teeldeelen*. 3rd ed. Amsterdam: Jan Morterre, 1755.

Schlosser, Johannes Albertus. *Specimen chemico medicum inaugurale de sale urinae humanae nativo*. Leiden: Georgius Wishoff and Quirinus Visser, 1753.

Schutte, Johann Hendrik. *De wel onderwezene vroedvrouw, of grondig en beknopt onderwys, van het geen een vroedvrouw weeten, en by alle voorkomende natuurlyke en tegennatuurlyke zwaare geboortens door vaardige handgreepen, verrichten moet*. The Hague: Hendrik Bakhuysen, 1771.

Schwencke, Thomas. *Haematologia, sive Sanguinis historia, experimentis passim superstructa*. The Hague: Johannes Martinus Husson, 1743.

———. *Haematologia, ofte Verhandeling van het bloed*. Translated by Abraham Westerhoff. The Hague: Johannes Martinus Husson, 1744.

———. *Schets van de heelmiddelen en haar uytwerkingen op het lichaam*. The Hague: Johannes Martinus Husson, 1745.

———. 'Aanmerkingen over verscheide manieren van bloedstelpen, en de voornaamste bloedstelpende middelen in de heelkunde'. *Verhandelingen van de Hollandse Maatschappy der Weetenschappen*, 2 (1755), 225–50.

Secker, Thomas. *Disputatio medica inauguralis de medicina statica*. Leiden: Hendrik Mulhovius, 1721.

Séwel, Willem. *A Compleat Dictionary English and Dutch* Edited by Egbert Buys. 2 vols Amsterdam: Kornelis de Veer, 1766.

Sluim, Daniel. *Dissertatio medica inauguralis de lacte*. Leiden: Coenraad Wishoff, 1716.

Smollett, Tobias George. 'The Institutions of Medicinal Pathology by H.D. Gaubius'. *The Critical Review: Or, Annals of Literature*, 47 (1779), 183–6.

Snellen, Henricus. *Disputatio chymico-medica inauguralis de urinarum inspectione*. Leiden: Abraham Elzevier, 1701.

Spallanzani, Lazzaro. *Dissertazioni di Fisica Animale, e Vegetabile*. 2 vols Modena: Soc. Tipografica, 1780.

———. *Dissertations Relative to the Natural History of Animals and Vegetables.* Translated by Thomas Beddoes. London: J. Murray, 1784.

Stahl, Georg Ernst. *Fundamenta chymiae dogmaticae et experimentalis.* Nuremberg: Wolfgang Moritz Endter, 1723.

Stalpart van der Wiel, Cornelius. *Dissertatio medica inauguralis de urinae calculo.* Leiden: Abraham Elzevier, 1712.

Stegmann, Johann Gottlieb. *Kurze Beschreibung einer Saug- und Drukpumpe.* Kassel: Waisenhaus-Buchdrukkerey, 1774.

Stein, George Wilhelm. *Kurze Beschreibung einer Brust- oder Milchpumpe.* Kassel: Henrich Schmiedt, 1773.

———. *Korte beschryving eener borst- of zog-pomp.* Utrecht: Joannes van Schoonhoven, 1775.

Stochius, Anthonius, jnr. *De streelende speel-pop van Dr. Cornelis Belkmeer, door hem eerst naakt uitgekleed, en onder den gewaanden naam van Be-Minnaar der Waarheid, door zynen kwalyk geslepen bril beschouwt, dog nu volkomen ontleed in haare mismaaktheid ten toon gesteld, en den poppemaker weder t'huis gezonden.* Enkhuizen: Hendrik Callenbach, 1734.

Stochius, Antonius. *Geneeskundige verhandeling over eene algemeene volk-ziekte, die deze winter gewoedt heeft en nog niet ophout.* Translated by Antonius Stochius, jnr. Enkhuizen: Hendrik Callenbach, 1733.

———. *Dissertatio medico-practico de morbo epidemico hac hyeme grassato, necdum cessante cum animadvers. in cel. J. Gorteri brevem descriptionem et curationem morbi epidemii per diaphoresin.* Enkhuizen: Hendrik Callenbach, 1733.

Stook, Joannes. *Dissertatio practico-medica inauguralis de febri lactea.* Leiden: Jean Luzac, 1746.

Sturm, Johann Christoph. *Collegium experimentale, sive curiosum.* Nuremberg: Wolfgang Moritz Endter, 1676.

Swaving, Christianus. *Dissertatio inauguralis physiologica-medica de excrementis secundae coctionis.* Leiden: Johannes van Abkoude, 1744.

van Swieten, Gerard, ed. *Commentaria in Hermanni Boerhaave Aphorismos de cognoscendis et curandis morbis.* 5 vols Leiden: Johannes and Herman Verbeek, 1742–1772.

———, ed. *The Commentaries upon the Aphorisms of dr. Herman Boerhaave.* 18 vols London: Robert Horsfield and Thomas Longman, 1744–1773.

Swildens, Johan Hendrik. *Vaderlandsch A-B boek voor de Nederlandsche jeugd.* Amsterdam: Willem Holtrop, 1781.

Sydenham, Thomas. *The Whole Works of That Excellent Practical Physician Dr. Thomas Sydenham Wherein Not Only the History and Cures of Acute Diseases Are Treated of, after a New and Accurate Method: But Also the Shortest and Safest Way of Curing Most Chronical Diseases.* Translated by John Pechy. London: Richard Wellington, 1701.

———. *The Entire Works of Dr Thomas Sydenham, Newly Made English from the Originals.* Edited by John Swan. London: Edward Cave, 1742.

Sylvius, Franciscus. *Disputationum medicarum: pars prima primarias corporis humani functiones naturales ex anatomicis, practicis & chymicis experimentis deductas complectens.* Amsterdam: Johannes van den Bergh, 1663.

———. *Praxeos medicae idea nova.* Leiden: Jean le Carpentier, 1671.

———. *A New Idea of the Practice of Physic.* Translated by Richard Gower. London: Brabazon Aylmer, 1675.

———. *Opera medica.* Amsterdam: Daniel Elzevier & Abraham Wolfgang, 1679.

Tersier, Abraham. *Dissertatio physico-medica inauguralis de vita et morte, quam annuente deo ter optimo maximo.* Leiden: Georgius Wishoff, 1739.

———. *Responsa ad rogata bina medica, unus de morbo bubulo, &c., alternus de lactis conformatione, &c., ab Collegis scientiarum Harlemensi proposita.* Dordrecht: Adriaan Walpot, 1762.

———. *Brief aan den wel edelen heere N.N. behelzende een vertoog over de melk* Dordrecht: Adriaan Walpot, 1768.

Tersier, Bartholomeus. *Dissertatio inauguralis medica de intempestivo remediorum usu.* Leiden: Johannes Bos, 1769.

de Thoméze, Leonardus Henricus. *Dissertatio medica inauguralis de calculo biliari.* Leiden: Pieter van der Eyk and Daniel Vijgh, 1773.

Thorvarth, Joannes Guilielmus. *Dissertatio practico-medica inauguralis de lactis defectu.* Leiden: Dirk Haak, 1764.

Thurneisser zum Thurn, Leonhart. *Protokatalepsis oder Praeoccupatio durch zwölff verscheidenlicher Tractaten gemachter Harm Proben.* Frankfurt (Oder): Johann Eichorn, 1571.

———. *Bebaiosis agonismou: Das ist Confirmatio concertationis, oder ein Bestettigung desz jenigen so streittig, häderig, oder zenckisch ist, wie dann ausz Unverstandt die neuwe und vor unerhörte Erfindung der aller nützlichesten und menschlichem Geschlecht der notturftigesten Kunst desz Harnnprobirens ein Zeitlang gewest ist.* Berlin: Im Grauwen Closter, 1576.

Titsingh, Abraham. *Cypria tot schrik van haar bondgenooten, en redding der gestruikelden ter eere van de heelkonst en tot dienst der heelmeesteren geschreven.* 2 vols Amsterdam: Johannes Gysius, 1742.

———. *Diana, ontdekkende het geheim der dwaazen die zig vroedmeester noemen: ter eeren van chirurgia geschreeven.* Amsterdam: [s.n.], 1750.

von Uffenbach, Zacharias Conrad. *Merkwürdige Reisen durch Niedersachsen, Holland und Engelland.* 3 vols Frankfurt & Leipzig: [s.n.], 1753–1754.

Voltelen, Floris Jacobus. *Observationes chemico-medicae de lacte humano, ejusque cum asinino et ovillo comparatione.* Utrecht: Abraham van Paddenburg, 1775.

———. *Oratio aditialis de chemiae hodiernae pretio rite constituendo.* Leiden: Samuel and Johannes Luchtmans, 1784.

Voorduin, J.C. *Proeve eener karakterschets van nu wijlen Hare Majesteit, de Koningin der Nederlanden, Frederika Louisa Wilhelmina.* Utrecht: Robert Natan, 1837.

Vosch van Avesaet, Henricus. *Lapis lydius animadversionum expertissimi domini A. Stochii … J. Gorteri … brevem ejus descriptionem et curationem morbi epidemii per diaphoresin.* Harderwijk: Willem Brinkink, 1734.

Vullyamoz, Marc-Louis. *Dissertatio chemico-medica inauguralis de sale lactis essentiali.* Leiden: Dirk Haak, 1756.

Weeks, Andrew, ed. *Paracelsus (Theophrastus Bombastus von Hohenheim, 1493–1541): Essential Theoretical Writings*, Aries book series. Leiden: Brill, 2008.

Willis, Thomas. *Cerebri anatome: Cui accessit nervorum descriptio et usus.* Amsterdam: Gerbrandt Schagen, 1664.

Wolff-Bekker, Elisabeth. *Proeve over de opvoeding, aan de Nederlandsche moeders.* Amsterdam and The Hague: Johannes Allart and Isaac van Cleef, 1779.

———. *Essai sur l'éducation. Dédié aux Mères de Famille Hollandoises.* The Hague: Isaac van Cleef, 1785.

Wolff-Bekker, Elisabeth, and Agatha Deken. *Economische liedjes.* The Hague: Isaac van Cleef, 1781.

Zehenphenningh, Henricus Petrus Sigismundus. *Dissertatio medico therapeutica inauguralis sistens quaedam therapiae specialis notamina circa abusus remediorum vomitoriorum, laxantium et sudoriferorum.* Leiden: Cornelis de Pecker, 1750.

Secondary Sources

Ackerknecht, Erwin H. *A Short History of Medicine.* rev. ed. Baltimore, London: Johns Hopkins University Press, 1982.

Alma, R.H., L.W. Barneveld, G.L.G.A. Kortekaas, J. Pietersma, C. Rogge, B.G.N. Schuilenga, and T. Tel. *Het Hinckaertshuis: Zeven eeuwen bouwhistorie en bewoners.* Edited by T.J. Hoekstra, *et al.* Groningen: Stiching Het Hinckaertshuis, 2012.

Altena, Peter, and Myriam Everard, eds. *Onbreekbare burgerharten: De historie van Betje Wolff en Aagje Deken.* Nijmegen: Vantilt, 2004.

Anderson, Robert G.W., ed. *Cradle of Chemistry: The Early Years of Chemistry at the University of Edinburgh.* Edinburgh: John Donald, 2015.

Arikha, Noga. *Passions and Tempers: A History of the Humours.* New York: Ecco, 2007.

Arrizabalaga, Jon, John Henderson, and Roger French. *The Great Pox: The French Disease in Renaissance Europe.* New Haven: Yale University Press, 1997.

Aucante, Vincent. *La philosophie médicale de Descartes.* Science, histoire et société. Paris: Presses Universitaires de France, 2006.

Backer, H.J. 'Boerhaave's ontdekking van het ureum'. *Nederlands Tijdschrift voor Geneeskunde*, 87 (1943), 1274–8.

Baggerman, Arianne. *Publishing Policies and Family Strategies: The Fortunes of a Dutch Publishing House in the 18th and Early 19th Centuries*. Leiden: Brill, 2014.

Baldassarri, Fabrizio. 'Botany and Medicine'. In *Encyclopedia of Early Modern Philosophy and the Sciences*, edited by Dana Jalobeanu and Charles T. Wolfe. Cham: Springer, 2020.

Baumann, E.D. *François dele Boe Sylvius*. Leiden: E.J. Brill, 1949.

van Berkel, Klaas. *Universiteit van het noorden: Vier eeuwen academisch leven in Groningen: Deel I: De oude universiteit, 1614–1876*. Studies over de geschiedenis van de Groningse universiteit. Hilversum: Verloren, 2014.

Bertoloni Meli, Domenico. 'The Color of Blood: Between Sensory Experience and Epistemic Significance'. In *Histories of Scientific Observation*, edited by Lorraine Daston and Elizabeth Lunbeck. 117–34. Chicago: The University of Chicago Press, 2011.

———. *Mechanism, Experiment, Disease: Marcello Malpighi and Seventeenth-Century Anatomy*. Baltimore: John Hopkins University Press, 2011.

Beukers, Harm. 'Het laboratorium van Sylvius'. *TGGNWT*, 3 (1980), 28–36.

———. 'Acid Spirits and Alkaline Salts: The Iatrochemistry of Franciscus dele Boë, Sylvius'. *Sartoniana*, 12 (1999), 39–59.

Bhushan, Nalini. 'What is a Chemical Property?'. *Synthese*, 155 (2007), 293–305.

van Biema, Eduard. *Les Huguetan de Mercier et de Vrijhoeven: histoire d'une famille de financiers huguenots de la fin du XVIIe jusqu'à la moitié du XVIII siècle*. The Hague: Martinus Nijhoff, 1918.

Bierens de Haan, Johan Abraham. *De Hollandsche Maatschappij der Wetenschappen, 1752–1952*. Haarlem: J. Enschedé en Zonen, 1952.

Bigotti, Fabrizio. 'A Previously Unknown Path to Corpuscularism in the Seventeenth Century: Santorio's Marginalia to the *Commentaria in Primam Fen Primi Libri Canonis Avicennae* (1625)'. *Ambix*, 64 (2017), 1–14.

———. 'The Weight of the Air: Santorio's Thermometers and the Early History of Medical Quantification Reconsidered'. *Journal of Early Modern Studies*, 7 (2018), 73–103.

———. *Physiology of the Soul: Mind, Body, and Matter in the Galenic Tradition of the Late Renaissance (1550–1630)*. The Age of Descartes. Turnhout: Brepols, 2019.

———. 'Santorio, Sanctorius'. In *Encyclopedia of Early Modern Philosophy and the Sciences*, edited by Dana Jalobeanu and Charles T. Wolfe. Cham: Springer, 2020.

Bigotti, Fabrizio, and David Taylor. 'The Pulsilogium of Santorio: New Light on Technology and Measurement in Early Modern Medicine'. *Societate si Politica*, 11 (2017), 55–114.

Boantza, Victor D. *Matter and Method in the Long Chemical Revolution: Laws of Another Order*. Science, Technology and Culture, 1700–1945. Farnham: Ashgate, 2013.

Boantza, Victor D., and Leslie Tomory. 'The "subtile Aereal Spirit of Fountains": Mineral Waters and the History of Pneumatic Chemistry'. *Early Science and Medicine*, 21 (2016), 303–31.

Boersma, Jacob. *Antonius de Haen, 1704–1776: Leven en werk*. Assen: Van Gorcum, 1963.

Boissier de Sauvages, François. *Nosologia methodica sistens morborum classes, genera et species, juxta Sydenhami mentem et botanicorum ordinem*. 3 vols Amsterdam: De Tournes, 1763.

Bots, J.A.H., M. Evers, W.Th.M. Frijhoff, F. Keverling Buisman, J. Kuys, and M.V.T. Tenten, eds. *Het Gelders Athene: Bijdragen tot de geschiedenis van de Gelderse universiteit in Harderwijk (1648–1811)*, Werken uitgegeven door Gelre. Hilversum: Verloren, 2000.

Bouras-Vallianatos, Petros, and Barbara Zipser, eds. *Brill's Companion to the Reception of Galen*. Leiden: Brill, 2019.

Brechka, Frank T. *Gerard van Swieten and His World 1700–1772*. International Archives of the History of Ideas. The Hague: Nijhoff, 1970.

Broman, Thomas H. 'The Medical Sciences'. In *The Cambridge History of Science*, edited by Roy Porter. 463–84. Cambridge: Cambridge University Press, 2003.

Brown, Theodore M. 'From Mechanism to Vitalism in Eighteenth Century English Physiology'. *Journal of the History of Biology*, 7 (1974), 179–216.

———. 'Physiology and the Mechanical Philosophy in Mid-seventeenth Century England'. *Bulletin of the History of Medicine*, 51 (1977), 25–54.

de Bruijn, J.G., ed. *Inventaris van de prijsvragen uitgeschreven door de Hollandsche Maatschappij der Wetenschappen 1753–1917*. Groningen: Tjeenk Willink, 1977.

Burns, Chester R. 'The Nonnaturals: A Paradox in the Western Concept of Health'. *The Journal of Medicine and Philosophy*, 1 (1976), 202–11.

Bylebyl, Jerome J. 'Nutrition, Quantification and Circulation'. *Bulletin of the History of Medicine*, 51 (1977), 369–85.

Carlyle, Margaret. 'Breastpump Technology and 'Natural' Motherly Milk in Enlightenment France'. *Women's Studies International Forum*, 60 (2017), 89–96.

Cavallo, Sandra, and Tessa Storey. *Healthy Living in Late Renaissance Italy*. Oxford: Oxford University Press, 2013.

———, eds. *Conserving Health in Early Modern Culture: Bodies and Environments in Italy and England*. Manchester: Manchester University Press, 2017.

Chang, Hasok. *Inventing Temperature: Measurement and Scientific Progress*. Oxford studies in philosophy of science. Oxford: Oxford University Press, 2004.

———. 'Compositionism as a Dominant Way of Knowing in Modern Chemistry'. *History of Science*, 49 (2011), 247–68.

———. *Is Water H_2O? Evidence, Realism and Pluralism*. Boston Studies in the Philosophy of Science. Dordrecht: Springer, 2012.

Chang, Ku-ming. 'Fermentation, Phlogiston and Matter Theory: Chemistry and Natural Philosophy in Georg Ernst Stahl's *Zymotechnia Fundamentalis*'. *Early Science and Medicine*, 7 (2002), 31–63.

———. 'Collaborative Production and Experimental Labor: Two Models of Dissertation Authorship in the Eighteenth Century'. *Studies in History and Philosophy of Biological and Biomedical Sciences*, 41 (2010), 347–55.

Christie, J.R.R. 'Historiography of Chemistry in the Eighteenth Century: Hermann Boerhaave and William Cullen'. *Ambix*, 41 (1994), 4–19.

Cislo, Amy Eisen. *Paracelsus's Theory of Embodiment: Conception and Gestation in Early Modern Europe*. The Body, Gender and Culture. London: Pickering and Chatto, 2010.

Clericuzio, Antonio. 'The Internal Laboratory: The Chemical Reinterpretation of Medical Spirits in England (1650–1680)'. In *Alchemy and Chemistry*, edited by Piyo Rattansi and Antonio Clericuzio. Dordrecht: Kluwer Academic Publishers, 1994.

———. 'Chemical and Mechanical Theories of Digestion in Early Modern Medicine'. *Studies in History and Philosophy of Biological and Biomedical Sciences*, 43 (2012), 329–37.

Cobb, Matthew. *The Egg and Sperm Race: The Seventeenth-Century Scientists Who Unravelled the Secrets of Sex, Life and Growth*. London: Free Press, 2006.

Coley, Noel G. 'Animal Chemists and Urinary Stone'. *Ambix*, 18 (1971), 69–93.

———. 'Early Blood Chemistry in Britain and France'. *Clinical Chemistry*, 47 (2001), 2166–78.

———. 'Medical Chemists and the Origins of Clinical Chemistry in Britain (circa 1750–1850)'. *Clinical Chemistry*, 50 (2004), 961–72.

Conrad, Lawrence I., Michael Neve, Vivian Nutton, Roy Porter, and Andrew Wear. *The Western Medical Tradition: 800 BC to AD 1800*. Cambridge: Cambridge University Press, 1995.

Cook, Harold J. *Trials of an Ordinary Doctor: Joannes Groenevelt in Seventeenth-Century London*. Baltimore: The Johns Hopkins University Press, 1994.

———. 'Boerhaave and the Flight from Reason in Medicine'. *Bulletin of the History of Medicine*, 74 (2000), 221–40.

———. *Matters of Exchange: Commerce, Medicine, and Science in the Dutch Golden Age*. New Haven: Yale University Press, 2007.

Corbin, Alain. *Le miasme et la jonquille: l'odorat et l'imaginaire social, 18e–19e siècles*. Paris: Aubier Montaigne, 1982.

———. *The Foul and the Fragrant: Odor and the French Social Imagination*. Leamington Spa: Berg, 1986.

Cunningham, Andrew. 'Thomas Sydenham: Epidemics, Experiment and the "Good Old Cause"'. In *The Medical Revolution*, edited by Roger French and Andrew Wear. 164–90. Cambridge: Cambridge University Press, 1989.

———. 'Medicine to Calm the Mind: Boerhaave's Medical System, and Why It Was Adopted in Edinburgh'. In *The Medical Enlightenment*, edited by Andrew Cunningham and Roger French. 40–66. Cambridge: Cambridge University Press, 1990.

———. *The Anatomical Renaissance: The Resurrection of the Anatomical Projects of the Ancients*. Aldershot: Scolar Press, 1997.

———. 'The Pen and the Sword: Recovering the Disciplinary Identity of Physiology and Anatomy before 1800 – I: Old Physiology–the Pen'. *Studies in History and Philosophy of Biological and Biomedical Sciences*, 33 (2002), 631–66.

———. 'The Pen and the Sword: Recovering the Disciplinary Identity of Physiology and Anatomy before 1800 – II: Old Anatomy–the Sword'. *Studies in History and Philosophy of Biological and Biomedical Sciences*, 34 (2003), 51–76.

———. *The Anatomist Anatomis'd: An Experimental Discipline in Enlightenment Europe*. History of Medicine in Context. Farnham: Ashgate, 2010.

———. 'The Principality of Blood: William Harvey, the Blood, and the Early Transfusion Experiments'. In *Blood – Symbol – Liquid*, edited by Catrien Santing and Jetze Touber. 193–205. Leuven: Peeters, 2012.

Dacome, Lucia. 'Living with the Chair: Private Excreta, Collective Health and Medical Authority in the Eighteenth Century'. *History of Science*, 39 (2001), 467–500.

———. 'Resurrecting by Numbers in Eighteenth-Century England'. *Past and Present*, 193 (2006), 73–110.

———. 'Balancing Acts: Picturing Perspiration in the Long Eighteenth Century'. *Studies in History and Philosophy of Biological and Biomedical Sciences*, 43 (2012), 379–91.

Dannenfeldt, Karl H. 'Sleep: Theory and Practice in the Late Renaissance'. *Journal of the History of Medicine and Allied Sciences*, 41 (1986), 415–41.

Darrow, Floyd L. *The Story of an Ancient Art: From the Earliest Adhesives to Vegetable Glue*. Lansdale, PA and South Bend, IN: Perkins Glue Company, 1930.

Daston, Lorraine, and Peter Galison. *Objectivity*. New York: Zone Books, 2007.

Daston, Lorraine, and Elizabeth Lunbeck, eds. *Histories of Scientific Observation*. Chicago: The University of Chicago Press, 2011.

Debus, Allen G. *The Chemical Philosophy: Paracelsian Science and Medicine in the Sixteenth and Seventeenth Centuries*. New York: Science History Publications, 1977.

———. *Chemistry and Medical Debate: Van Helmont to Boerhaave*. Canton, MA: Science History Publications, 2001.

van Deinsen, Lieke. *The Panpoëticon Batavûm: The Portrait of the Author as a Celebrity*. Rijksmuseum Studies in History. Amsterdam: Rijksmuseum, 2016.

Deutsch, Helen, and Mary Terrall, eds. *Vital Matters: Eighteenth-Century Views of Conception, Life, and Death*, UCLA Clark Memorial Library Series. Toronto: University of Toronto Press, 2012.

Dik, Anje, and Dini Helmers, eds. *'Het is of ik met mijn lieve sprak': De briefwisseling tussen Jean Malherbe en Christina van Steensel, 1782–1800*, Egoducumenten. Hilversum: Verloren, 1994.

van Driel, Joppe. 'The Filthy and the Fat: Oeconomy, Chemistry and Resource Management in the Age of Revolutions, 1700–1850'. Doctoral Thesis, University of Twente, 2016.

Du Rieu, Willem Nicolaas, ed. *Album studiosorum Academiae Lugduno-Batavae MDLXXV–MDCCCLXXV*. The Hague: Martinus Nijhoff, 1875.

Duffin, Jacalyn. *History of Medicine: A Scandalously Short Introduction*. 2nd ed. Toronto: University of Toronto Press, 2010.

Eddy, Matthew D., Seymour Mauskopf, and William R. Newman, eds. *Chemical Knowledge in the Early Modern World*. Vol. 29, Osiris, 2014.

Engelmann, Mario, and Bert Sliggers. 'Johannes Albertus Schlosser, the First Author Describing *Artemia Salina* (L.) (Branchiopoda: Anostraca): A Biographical Sketch'. *Journal of Crustacean Biology*, 35 (2015), 571–5.

Favaretti Camposampiero, Matteo. 'Galenism in Early Modern Philosophy and Medicine'. In *Encyclopedia of Early Modern Philosophy and the Sciences*, edited by Dana Jalobeanu and Charles T. Wolfe. Cham: Springer, 2020.

Fildes, Valerie A. *Breasts, Bottles and Babies: A History of Infant Feeding*. Edinburgh: Edinburgh University Press, 1986.

———. *Wet Nursing: A History from Antiquity to the Present*. Family, Sexuality and Social Relations in Past Times. Oxford: Basil Blackwell, 1988.

Fink, Béatrice Camille. 'Le *Dictionnaire oeconomique* de Noël Chomel: Un pot-pourri aigre-doux'. In *Diderot, l'Encyclopédie et autres études*, edited by Jacques Proust, Marie Leca-Tsiomis and Alain Sandrier. 157–64. Ferney-Voltaire: Centre international d'étude du XVIIIe siècle, 2010.

Fleck, Ludwik. *Genesis and Development of a Scientific Fact*. Translated by Fred Bradley. Edited by Thaddeus J. Trenn and Robert K. Merton. Chicago: University of Chicago Press, 1979.

Flegel, Kenneth M. 'Changing Concepts of the Nosology of Gonorrhea and Syphilis'. *Bulletin of the History of Medicine*, 48 (1974), 571–88.

Fors, Hjalmar. *The Limits of Matter: Chemistry, Mining, and Enlightenment*. Synthesis. Chicago: The University of Chicago Press, 2015.

Foucault, Michel. *Naissance de la clinique: une archéologie du regard médical*. Galien, Histoire et philosophie de la biologie et de la médicine. Paris: Presses Universitaires de France, 1963.

Fournier, Marian. 'De microscopische anatomie in Bidloo's *Anatomia humani Corporis* (1685)'. *TGGNWT*, 8 (1985), 197–208.

Frängsmyr, Tore, John L. Heilbron, and Robin E. Rider, eds. *The Quantifying Spirit in the 18th Century*, Uppsala studies in history of science. Berkeley: University of California Press, 1990.

French, Roger. *Robert Whytt, the Soul, and Medicine*. London: The Wellcome Institute of the History of Medicine, 1969.

———. 'Harvey in Holland: Circulation and the Calvinists'. In *The Medical Revolution*, edited by Roger French and Andrew Wear. 46–86. Cambridge: Cambridge University Press, 1989.

———. *William Harvey's Natural Philosophy*. Cambridge: Cambridge University Press, 1994.

———. *Medicine before Science: The Rational and Learned Doctor from the Middle Ages to the Enlightenment*. Cambridge: Cambridge University Press, 2003.

Freschot, Augustinus. *Amoureuze en pikante geschiedenis van het congres en de stad Utrecht: Augustinus Freschots verhaal achter de Vrede van Utrecht*. Translated by Roland Fagel. Edited by Erik Tigelaar. Hilversum: Verloren, 2013.

Frijhoff, W.Th.M. *La société Néerlandaise et ses gradués, 1575–1814: Une recherche sérielle sur le statut des intellectuels*. Amsterdam: Holland University Press, 1981.

Gaukroger, Stephen. *Descartes' System of Natural Philosophy*. Cambridge: Cambridge University Press, 2002.

Gibson, Susannah. *Animal, Vegetable, Mineral? How Eighteenth-Century Science Disrupted the Natural Order*. Oxford: Oxford University Press, 2015.

Godel, Rainer. 'Controversy as the Impetus of Enlightened Practice of Knowledge'. In *Scholars in Action*, edited by André Holenstein, Hubert Steinke and Martin Stuber. 413–31. Leiden: Brill, 2013.

Goodfield, G.J. *The Growth of Scientific Physiology: Physiological Method and the Mechanist-Vitalist Controversy, Illustrated by the Problems of Respiration and Animal Heat*. London: Hutchinson, 1960.

Grell, Ole Peter, ed. *Paracelsus: The Man and His Reputation, His Ideas and Their Transformation*, Studies in the History of Christian Thought. Leiden: Brill, 1998.

Grmek, Mirko D. 'Santorio, Santorio'. In *Complete Dictionary of Scientific Biography*, edited by Charles Coulston Gillispie, Frederic L. Holmes and Noretta Koertge. 101–4. Detroit: Charles Scribner's Sons, 2008.

Gruys, J.A., and Adèle Nieuweboer. 'Dutch Occasional Poetry, 16th through 18th Centuries: A Genre Rediscovered'. Leiden: IDC, 2004.

Guerrini, Anita. 'James Keill, George Cheyne, and Newtonian Physiology, 1690–1740'. *Journal of the History of Biology*, 18 (1985), 247–66.

———. 'The Varieties of Mechanical Medicine: Borelli, Malpighi, Bellini and Pitcairne'. In *Marcello Malpighi*, edited by Domenico Bertoloni Meli. 111–28. Florence: Olschki, 1997.

———. *Experimenting with Humans and Animals: From Galen to Animal Rights*. Johns Hopkins Introductory Studies in the History of Science. Baltimore: The Johns Hopkins University Press, 2003.

———. 'Keill, James (1673–1719)'. In *Oxford Dictionary of National Biography*. Oxford: Oxford University Press, 2004.

———. *The Courtiers' Anatomists: Animals and Humans in Louis XIV's Paris.* Chicago: The University of Chicago Press, 2015.

Hake, Barry J. 'Between Patriotism and Nationalism: Johann Hendrik Swildens and the 'Pedagogy of the Patriotic Virtues' in the United Provinces during the 1780s and 1790s'. *History of Education,* 33 (2004), 11–38.

Hall, Thomas S. *Ideas of Life and Matter: Studies in the History of General Physiology, 600 B.C.–1900 A.D.* 2 vols Chicago: The University of Chicago Press, 1969.

Hamers-van Duynen, Sophia W. *Hieronymus David Gaubius (1705–1780): Zijn correspondentie met Antonio Nunes Ribeiro Sanches en andere tijdgenoten.* Assen and Amsterdam: Van Gorcum, 1978.

Handley, Sasha. *Sleep in Early Modern England.* New Haven: Yale University Press, 2016.

Hankinson, R.J. ed. *The Cambridge Companion to Galen.* Cambridge: Cambridge University Press, 2008.

Harrison, Mark. *Medicine in an Age of Commerce and Empire: Britain and its Tropical Colonies, 1660–1830.* Oxford: Oxford University Press, 2010.

Harskamp, Jaap. 'Dissertatio medica inauguralis … Leyden medical dissertations in the British Library, 1593–1746'. London: The Wellcome Institute for the History of Medicine, 1997.

Hedesan, Georgiana D. *An Alchemical Quest for Universal Knowledge: The 'Christian Philosophy' of Jan Baptist Van Helmont (1579–1644).* Universal Reform. London: Routledge, 2016.

Hendriksen, Marieke M.A. 'Anatomical Mercury: Changing Understandings of Quicksilver, Blood, and the Lymphatic System, 1650–1800'. *Journal of the History of Medicine and Allied Sciences,* 70 (2015), 516–48.

———. *Elegant Anatomy: The Eighteenth-Century Leiden Anatomical Collections.* History of Science and Medicine Library. Leiden: Brill, 2015.

———. 'Boerhaave's Mineral Chemistry and Its Influence on Eighteenth-Century Pharmacy in the Netherlands and England'. *Ambix,* 65 (2018), 303–23.

———. 'Criticizing Chrysopoeia? Alchemy, Chemistry, Academics and Satire in the Northern Netherlands, 1650–1750'. *Isis,* 109 (2018), 235–53.

———. 'The Repudiation and Persistence of Lapidary Medicine in Eighteenth-Century Dutch Medicine and Pharmacy'. In *Gems in the Early Modern World,* edited by Michael Bycroft and Sven Dupré. 197–220. Cham: Palgrave Macmillan, 2019.

Hendriksen, Marieke M.A., and Ruben E. Verwaal. 'Boerhaave's Furnace: Exploring Early Modern Chemistry Through Working Models'. *Berichte zur Wissenschaftsgeschichte,* 43 (2020), 385–411.

te Hennepe, Mieneke. 'Of the Fisherman's Net and Skin Pores: Reframing Conceptions of the Skin in Medicine 1572–1714'. In *Blood, Sweat and Tears,* edited by Manfred Horstmanshoff, Helen King and Claus Zittel. 523–48. Leiden: Brill, 2012.

Herr, Harry W. "'Cutting for the Stone": The Ancient Art of Lithotomy'. *British Journal of Urology*, 101 (2008), 1214–6.

Hess, Volker, and J. Andrew Mendelsohn. 'Sauvages' Paperwork: How Disease Classification Arose from Scholarly Note-Taking'. *Early Science and Medicine*, 19 (2014), 471–503.

Higgins, Sabrina. 'Divine Mothers: The Influence of Isis on the Virgin Mary in Egyptian *Lactans*-Iconography'. *Journal of the Canadian Society for Coptic Studies*, 3/4 (2012), 71–90.

Hill, Christopher. 'William Harvey and the Idea of Monarchy'. *Past and Present*, 27 (1964), 54–72.

Hillen, Harry F.P. 'De eerste Nederlandse hematoloog: Thomas Schwencke (1694–1767)'. *Nederlands Tijdschrift voor Hematologie*, 7 (2010), 17–20.

Holmes, Frederic L. *Eighteenth-Century Chemistry as an Investigative Enterprise*. Berkeley Papers in History of Science. Berkeley: Office for History of Science and Technology, University of California at Berkeley, 1989.

Hoogewerff, G.J., ed. *De twee reizen van Cosimo de' Medici, prins van Toscane, door de Nederlanden (1667–1669): Journalen en documenten*. Amsterdam: Johannes Müller, 1919.

Houtzager, Hans L., ed. *Reinier de Graaf 1641–1673: In sijn leven nauwkeurig ontleder en gelukkig geneesheer tot Delft*. Rotterdam: Erasmus Publishing, 1991.

Hughes, J.T. *Thomas Willis, 1621–1675: His Life and Work*. London: Royal Society of Medicine Services, 1991.

Huisman, Frank. 'Gezondheid te koop: Zelfmedicatie en medische advertenties in de Groninger en Ommelander Courant, 1743–1800'. *Focaal*, 21 (1993), 90–130.

———. 'Medicine and Health Care in the Netherlands, 1500–1800'. In *The History of Science in the Netherlands*, edited by Klaas van Berkel, Albert van Helden and Lodewijk Palm. 239–78. Leiden: Brill, 1999.

Jevons, F.R. 'Boerhaave's Biochemistry'. *Medical History*, 6 (1962), 343–62.

Kaau, Abraham. *Declamatio academia de gaudiis alchemistarum*. Leiden: Samuel Luchtmans, 1737.

Kegel-Brinkgreve, E., and A.M. Luyendijk-Elshout, eds. *Boerhaave's Orations*, Publications of the Sir Thomas Browne Institute, vol. 4. Leiden: E.J. Brill, 1983.

Kennaway, James, and Rina Knoeff, eds. *Lifestyle and Medicine in the Enlightenment: The Six Non-Naturals in the Long Eighteenth Century*. London: Routledge, 2020.

King, Lester S. 'Rationalism in Early Eighteenth Century Medicine'. *Journal of the History of Medicine and Allied Sciences*, 18 (1963), 257–71.

———. *The Road to Medical Enlightenment, 1650–1695*. History of Science Library. London: MacDonald, 1970.

Klein, Ursula. 'Experimental History and Herman Boerhaave's Chemistry of Plants'. *Studies in History and Philosophy of Biological and Biomedical Sciences*, 34 (2003), 533–67.

———. 'Shifting Ontologies, Changing Classifications: Plant Materials from 1700 to 1830'. *Studies in History and Philosophy of Science,* 36 (2005), 261–329.

Klein, Ursula, and W. Lefèvre. *Materials in Eighteenth-Century Science: A Historical Ontology.* Transformations. Cambridge, MA: MIT Press, 2007.

Klein, Ursula, and E.C. Spary, eds. *Materials and Expertise in Early Modern Europe: Between Market and Laboratory.* Chicago: University of Chicago Press, 2010.

Kloek, Els, ed. *1001 vrouwen uit de Nederlandse geschiedenis.* Nijmegen: Vantilt, 2013.

Kloek, Els, Giesela van Oostveen, and Nicole Teeuwen, eds. *Moederschap en de min.* Vol. 8, Volkscultuur. Utrecht: Stichting Informatiecentrum Volkscultuur, 1991.

Kloek, Joost, and Wijnand Mijnhardt. *1800: Blueprints for a National Community.* Translated by Beverley R. Jackson. Dutch Culture in a European Perspective. Basingstoke: Palgrave Macmillan, 2004.

Knoeff, Rina. *Herman Boerhaave (1668–1738): Calvinist Chemist and Physician.* History of Science and Scholarship in the Netherlands. Amsterdam: Edita, 2002.

———. 'Chemistry, Mechanics and the Making of Anatomical Knowledge: Boerhaave Vs. Ruysch on the Nature of the Glands'. *Ambix,* 53 (2006), 201–19.

———. 'Practising Chemistry 'after the Hippocratical Manner': Hippocrates and the Importance of Chemistry for Boerhaave's Medicine'. In *New Narratives,* edited by Lawrence M. Principe. 63–76. Dordrecht: Springer, 2007.

———. 'Moral Lessons of Perfection: A Comparison of Mennonite and Calvinist Motives in the Anatomical Atlases of Bidloo and Albinus'. In *Medicine and Religion,* edited by Ole Peter Grell and Andrew Cunningham. 121–43. Aldershot: Ashgate, 2007.

———. 'Herman Boerhaave at Leiden: *Communis Europae praeceptor*'. In *Centres of Medical Excellence?,* edited by Ole Peter Grell, Andrew Cunningham and Jon Arrizabalaga. 269–86. Farnham: Ashgate, 2010.

———. 'Herman Boerhaave's Neurology and the Unchanging Nature of Physiology'. In *Blood, Sweat and Tears,* edited by Manfred Horstmanshoff, Helen King and Claus Zittel. 195–216. Leiden: Brill, 2012.

———, ed. *Gelukkig Gezond! Histories of Healthy Ageing.* Groningen: Barkhuis, 2017.

Koerner, Lisbet. *Linnaeus: Nature and Nation.* Cambridge, MA: Harvard University Press, 1999.

Konert, Jürgen, and Holger G. Dietrich. 'Giovanni Battista Morgagni und der Beginn der Pathologischen Anatomie'. In *Anatomie,* edited by Jürgen Helm and Karin Stukenbrock. 127–38. Stuttgart: Franz Steiner, 2003.

Kooijmans, Luuc. *De geest van Boerhaave: Onderzoek in een kil klimaat.* Amsterdam: Prometheus Bert Bakker, 2014.

van der Korst, J.K. *Een dokter van formaat: Gerard van Swieten, lijfarts van keizerin Maria Theresia*. Amsterdam: Bakker, 2003.

Kroon, J.E. 'Boerhaave als hoogleeraar-promotor'. *Nederlands Tijdschrift voor Geneeskunde*, 63 (1919), 87–99.

Kuhn, Thomas S. *The Structure of Scientific Revolutions*. International Encyclopedia of Unified Science. 2nd ed. Chicago: The University of Chicago Press, 1974.

Kurzer, Frederick, and Phyllis M. Sanderson. 'Urea in the History of Organic Chemistry: Isolation from Natural Sources'. *Journal of Chemical Education*, 33 (1956), 452.

Lastinger, Valérie. 'Re-Defining Motherhood: Breast-Feeding and the French Enlightenment'. *Women's Studies*, 25 (1996), 603–17.

Lawrence, Christopher. 'Ornate Physicians and Learned Artisans: Edinburgh Medical Men, 1726–1776'. In *William Hunter and the Eighteenth-Century Medical World*, edited by William Bynum and Roy Porter. 153–76. Cambridge: Cambridge University Press, 1985.

Leemans, Inger, and Gert-Jan Johannes. *Worm en donder: Geschiedenis van de Nederlandse literatuur 1700–1800. De Republiek*. Geschiedenis van de Nederlandse literatuur. Amsterdam: Bert Bakker/Prometheus, 2013.

Leicester, Henry M. *The Historical Background of Chemistry*. New York: Wiley, 1956.

Leuftink, A.E. *Harde heelmeesters: Zeelieden en hun dokters in de 18e eeuw*. 2nd ed. Zutphen: Walburg Pers, 2008.

van Lieburg, Mart J. 'Reinier de Graaf en zijn plaats in het fysiologisch onderzoek van de zeventiende eeuw'. *Gewina*, 15 (1992), 73–84.

Lindeboom, G.A. *Bibliographia Boerhaaviana: List of Publications Written or Provided by H. Boerhaave or Based upon his Works and Teaching*. Analecta Boerhaaviana. Leiden: E.J. Brill, 1959.

———, ed. *Boerhaave's Correspondence*. 3 vols, Analecta Boerhaaviana. Leiden: E.J. Brill, 1962–1979.

———. *Iconographia Boerhaavii*. Analecta Boerhaaviana. Leiden: E.J. Brill, 1963.

———. 'Boerhaave's Concept of the Basic Structure of the Body'. *Clio Medica*, 5 (1970), 203–8.

———. *Boerhaave and Great Britain: Three Lectures on Boerhaave with Particular Reference to His Relations with Great Britain*. Analecta Boerhaaviana. Leiden: E.J. Brill, 1974.

———. *Herman Boerhaave: The Man and His Work*. Nieuwe Nederlandse bijdragen tot de geschiedenis der geneeskunde en der natuurwetenschappen. 2nd ed. Edited by Mart J. van Lieburg. Rotterdam: Erasmus Publishing, 2007.

Lindemann, Mary. *Medicine and Society in Early Modern Europe*. New Approaches to European History. Cambridge: Cambridge University Press, 1999.

van der Linden, David. 'Memorializing the Wars of Religion in Early Seventeenth-Century French Picture Galleries: Protestants and Catholics Painting the Contested Past'. *Renaissance Quarterly,* 70 (2017), 132–78.

Lunsingh Scheurleer, Th.H., C. Willemijn Fock, and A.J. van Dissel. *Het Rapenburg: Geschiedenis van een Leidse gracht.* 6 vols Leiden: Rijksuniversiteit Leiden, 1986–1992.

van Lunteren, Frans. 'Clocks to Computers: A Machine-Based "Big Picture" of the History of Modern Science'. *Isis,* 107 (2016), 762–76.

Luyendijk-Elshout, A.M. 'Samuel Musgrave's attack upon Stahl's and Boerhaave's doctrines in 1763'. *Janus,* 67 (1980), 141–56.

———. 'Mechanicisme contra vitalisme: de school van Herman Boerhaave en de beginselen van het leven'. *TGGNWT,* 5 (1982), 16–26.

Magiels, Geerdt. *From Sunlight to Insight: Jan IngenHousz, the Discovery of Photosynthesis & Science in the Light of Ecology.* Brussels: VUBPress, 2010.

Magner, Lois N. *A History of Medicine.* 2nd ed. Boca Raton: Taylor & Francis, 2005.

Martin, Julian. 'Sauvages's Nosology: Medical Enlightenment in Montpellier'. In *The Medical Enlightenment,* edited by Andrew Cunningham and Roger French. 111–37. Cambridge: Cambridge University Press, 1990.

Maulitz, Russell C. *Morbid Appearances: The Anatomy of Pathology in the Early Nineteenth Century.* Cambridge History of Medicine. Cambridge: Cambridge University Press, 1987.

Mauskopf, Seymour. 'Reflections: "A Likely Story"'. In *New Narratives,* edited by Lawrence M. Principe. 177–93. Dordrecht: Springer, 2007.

McEvoy, John G. *The Historiography of the Chemical Revolution: Patters of Interpretation in the History of Science.* London: Pickering and Chatto, 2010.

McKie, Douglas. *Antoine Lavoisier: The Father of Modern Chemistry.* London: Gollancz, 1936.

Meinsma, J.J. 'Het scheikundig onderwijs aan de Gelderse Hogeschool (1648–1811) en Rijksathenaeum (1815–1818) te Harderwijk'. *Scientiarum Historia,* 14 (1972), 201–16.

van der Meulen, J.N. 'Maria Jacoba Meinertzhagen (1712–1787), een patriciërs-vrouw'. In *Utrechtse biografieën.* 107–12. Utrecht: SPOU, 1998.

Mijnhardt, Wijnand. *Tot heil van 't menschdom: culturele genootschappen in Nederland, 1750–1815.* Nieuwe Nederlandse bijdragen tot de geschiedenis der geneeskunde en der natuurwetenschappen. Amsterdam: Rodopi, 1988.

Moran, Bruce T. '*Lac Lunae* and Breast Milk'. *Pharmacy in History,* 41 (1999), 33–4.

Moran, Michael E. *Urolithiasis: A Comprehensive History.* New York, NY: Springer, 2014.

Moree, Perry, ed. *Kikkertje lief: brieven van Aagje Luijtsen, geschreven tussen 1776 en 1780 aan Harmanus Kikkert, stuurman in dienst van de VOC.* Den Burg: Het Open Boek Texel, 2003.

de Moulin, Daniel. *A History of Surgery: With Emphasis on The Netherlands.* Dordrecht: Martinus Nijhoff, 1988.

Neuburger, Max. 'Boerhaave's Einfluß auf die Entwicklung der Medizin in Oesterreich'. *Janus,* 23 (1918), 215–22.

Newman, William R. 'From Alchemy to 'Chymistry''. In *The Cambridge History of Science*, edited by Katharine Park and Lorraine Daston. Cambridge: Cambridge University Press, 2006.

Newman, William R., and Lawrence M. Principe. 'Alchemy vs. Chemistry: The Etymological Origins of a Historiographic Mistake'. *Early Science and Medicine,* 3 (1998), 32–65.

Niebyl, Peter H. 'The Non-Naturals'. *Bulletin of the History of Medicine,* 43 (1971), 486–92.

van Nijf, Onno. 'Exercise and Rest'. In *Gelukkig Gezond! Histories of Healthy Ageing*, edited by Rina Knoeff. 60–79. Groningen: Barkhuis, 2017.

Orland, Barbara. 'Enlightenend Milk: Reshaping a Bodily Substance into a Chemical Object'. In *Materials and Expertise*, edited by Ursula Klein and E.C. Spary. 163–97. Chicago: University of Chicago Press, 2010.

———. 'The Fluid Mechanics of Nutrition: Herman Boerhaave's Synthesis of Seventeenth–Century Circulation Physiology'. *Studies in History and Philosophy of Biological and Biomedical Sciences,* 43 (2012), 357–69.

———. 'Motherhood and Scientific Innovation: The story of Natural Versus Artificial Baby Food in the 19th Century'. In *Gender in Science and Technology*, edited by W. Ernst and I. Horwath. 129–46. Bielefeld: Transcript, 2014.

Pagel, Walter. 'J. B. Van Helmont's Reformation of the Galenic Doctrine of Digestion—and Paracelsus'. *Bulletin of the History of Medicine,* 29 (1955), 563–8.

———. 'Van Helmont's Ideas on Gastric Digestion and the Gastric Acid'. *Bulletin of the History of Medicine,* 30 (1956), 524–36.

———. *Paracelsus: An Introduction to Philosophical Medicine in the Era of the Renaissance.* Basel: S. Karger, 1958.

———. *Paracelsus: An Introduction to Philosophical Medicine in the Era of the Renaissance.* 2nd ed. Basel: S. Karger, 1982.

Paoli, François. *Jean-Baptiste Morgagni, ou, La naissance de la médecine moderne.* Paris: Glyphe, 2013.

Partington, J.R. *A History of Chemistry.* 3 vols London: Macmillan, 1961–1970.

Peitzman, Steven J. 'Bright's Disease and Bright's Generation: Toward Exact Medicine at Guy's Hospital'. *Bulletin of the History of Medicine,* 55 (1981), 307–21.

Pettegree, Andrew, and Arthur der Weduwen. *The Bookshop of the World: Making and Trading Books in the Dutch Golden Age.* New Haven: Yale University Press, 2019.

Pinto-Correia, Clara. *The Ovary of Eve: Egg and Sperm and Preformation*. Chicago: University of Chicago Press, 1997.
Pomata, Gianna. 'Sharing Cases: The *Observationes* in Early Modern Medicine'. *Early Science and Medicine,* 15 (2010), 193–236.
———. 'Observation Rising: Birth of an Epistemic Genre, 1500–1650'. In *Histories of Scientific Observation*, edited by Lorraine Daston and Elizabeth Lunbeck. 45–80. Chicago: Chicago University Press, 2011.
Porter, Roy. *The Greatest Benefit of Mankind: A Medical History of Humanity.* New York: W.W. Norton, 1999.
Powers, John C. 'Scrutinizing the Alchemists: Herman Boerhaave and the Testing of Chymistry'. In *Chymists and chymistry*, edited by Lawrence M. Principe. 227–38. Sagamore Beach: Science History Publications, 2007.
———. *Inventing Chemistry: Herman Boerhaave and the Reform of the Chemical Arts.* Synthesis. Chicago: University of Chicago Press, 2012.
———. 'Measuring Fire: Herman Boerhaave and the Introduction of Thermometry into Chemistry'. *Osiris,* 29 (2014), 158–77.
———. 'Leiden Chemistry in Edinburgh: Herman Boerhaave, James Crawford and Andrew Plummer'. In *Cradle of Chemistry*, edited by Robert G.W. Anderson. 25–58. Edinburgh: John Donald, 2015.
Principe, Lawrence M. *The Aspiring Adept: Robert Boyle and His Alchemical Quest: Including Boyle's "Lost" Dialogue on the Transmutation of Metals.* Princeton: Princeton University Press, 1998.
———, ed. *New Narratives in Eighteenth-Century Chemistry*, Archimedes. Dordrecht: Springer, 2007.
———, ed. *Chymists and Chymistry: Studies in the History of Alchemy and Early Modern Chemistry.* Sagamore Beach: Science History Publications, 2007.
———. 'A Revolution Nobody Noticed? Changes in Early Eighteenth-Century Chymistry'. In *New Narratives*, edited by Lawrence M. Principe. 1–22. Dordrecht: Springer, 2007.
———. 'Alchemy Restored'. *Isis,* 102 (2011), 305–12.
———. *The Secrets of Alchemy.* Synthesis. Chicago: The University of Chicago Press, 2012.
Punt, Hendrik. *Bernard Siegfried Albinus (1697–1770) on 'Human Nature': Anatomical and Physiological Ideas in Eighteenth Century Leiden*. Amsterdam: B.M. Israël, 1983.
Qvist, George. 'John Hunter's Alleged Syphilis'. *Annals of the Royal College of Surgeons of England,* 59 (1977), 205–9.
Ragland, Evan R. 'Experimenting with Chymical Bodies: Reinier De Graaf's Investigations of the Pancreas'. *Early Science and Medicine,* 13 (2008), 615–64.
———. 'Experimenting with Chemical Bodies: Science, Medicine, and Philosophy in the Long History of Reinier de Graaf's Experiments on Digestion, from

Harvey and Descartes to Claude Bernard'. Doctoral thesis, University of Alabama, 2012.

———. 'Chymistry and Taste in the Seventeenth Century: Franciscus Dele Boë Sylvius as a Chymical Physician between Galenism and Cartesianism'. *Ambix*, 59 (2012), 1–21.

Rather, Lelland J. *Mind and Body in Eighteenth-Century Medicine: A Study Based on Jerome Gaub's De regimine mentis.* Berkeley and Los Angeles: University of California Press, 1965.

———. 'The "Six Things Non-Natural": A Note on the Origins and Fate of a Doctrine and a Phrase'. *Clio Medica*, 3 (1968), 337–47.

Renbourn, E.T. 'The Natural History of Insensible Perspiration: A Forgotten Doctrine of Health and Disease'. *Medical History*, 4 (1960), 135–52.

Rey, Rosalyne. *Naissance et développement du vitalisme en France de la deuxième moitié du 18e siècle à la fin du Premier Empire.* Studies on Voltaire and the eighteenth century. Oxford: Voltaire Foundation, 2000.

Richet, Gabriel. 'The Chemistry of Urinary Stones around 1800: A First in Clinical Chemistry'. *Kidney International*, 45 (1995), 876–86.

Riskin, Jessica. *The Restless Clock: A History of the Centuries-Long Argument over What Makes Living Things Tick.* Chicago: University of Chicago Press, 2015.

Roberts, Lissa. 'The Circulation of Knowledge in Early Modern Europe: Embodiment, Mobility, Learning and Knowing'. *History of Technology*, 31 (2012), 47–68.

Roe, Shirley A. *Matter, Life, and Generation: Eighteenth-Century Embryology and the Haller-Wolff Debate.* Cambridge: Cambridge University Press, 1981.

Rosenfeld, Louis. *Four Centuries of Clinical Chemistry.* Amsterdam: Gordon & Breach Science, 1999.

———. 'Justus Liebig and *Animal Chemistry*'. *Clinical Chemistry*, 49 (2003), 1696–707.

Rothschuh, Karl Eduard. *History of Physiology.* Translated by Guenter B. Risse. Huntington, NY, 1973.

Salmon, Marylynn. 'The Cultural Significance of Breastfeeding and Infant Care in Early Modern England and America'. *Journal of Social History*, 28 (1994), 247–69.

Santing, Catrien, and Jetze Touber, eds. *Blood – Symbol – Liquid*, Groningen Studies in Cultural Change. Leuven: Peeters, 2012.

Santing, Catrien. 'Sleeping and Waking'. In *Gelukkig Gezond! Histories of Healthy Ageing*, edited by Rina Knoeff. 80–97. Groningen: Barkhuis, 2017.

Scarborough, John. 'Fish Glue (Gr. ΙΧΘΥΟΚΟΛΛΑ) in Hellenistic and Roman Medicine and Pharmacology'. *Classical Philology*, 110 (2015), 54–65.

Scheffer, J.H. 'Het dagboek van eene merkwaardige vrouw'. *Algemeen Nederlandsch Familieblad* (1884).

Scholten, Frits. 'Frans Hemsterhuis's Memorial for Herman Boerhaave: A Monument of Wisdom and Simplicity'. *Simiolus*, 35 (2011), 199–217.

Schulte, B.P.M. *Hermanni Boerhaave Praelectiones de morbis nervorum, 1730–1735: Een medisch-historische studie van Boerhaave's manuscript over zenuwziekten*. Analecta Boerhaaviana. Leiden: E.J. Brill, 1959.

Schwab, Richard N. 'The History of Medicine in Diderot's *Encyclopédie*'. *Bulletin of the History of Medicine*, 32 (1958), 216–23.

Siegfried, Robert. *From Elements to Atoms: A History of Chemical Composition*. Philadelphia: American Philosophical Society, 2002.

Simon, Jonathan. *Chemistry, Pharmacy and Revolution in France, 1777–1809*. Science Technology and Culture, 1700–1945. Farnham: Ashgate, 2005.

———. 'Pharmacy and Chemistry in the Eighteenth Century: What Lessons for the History of Science?'. *Osiris*, 29 (2014), 283–97.

Siraisi, Nancy G. 'Medicine, 1450–1620, and the History of Science'. *Isis*, 103 (2012), 491–514.

Sjtsjedrova, Irina. 'Abraham Kaau-Boerhaave: Bladzijden uit de biografie van een academicus'. In *Noord- en Zuid-Nederlanders in Rusland*, edited by E. Waegemans, J.S.A.M. Koningsbrugge and Nadja Louwerse. Baltic Studies, 293–312. Groningen: INOS, 2004.

Sluijter, Ronald. *'Tot ciraet, vermeerderinge ende heerlyckmaeckinge der universiteyt': Bestuur, instellingen, personeel en financiën van de Leidse universiteit, 1575–1812*. Hilversum: Verloren, 2004.

Smith, Pamela H. *The Body of the Artisan: Art and Experience in the Scientific Revolution*. Chicago: The University of Chicago Press, 2004.

Snelders, H.A.M. *De geschiedenis van de scheikunde in Nederland: Van alchemie tot chemie en chemische industrie rond 1900*. Delft: Delftse Universitaire Pers, 1993.

Spary, E.C. *Eating the Enlightenment: Food and the Sciences in Paris*. Chicago: The University of Chicago Press, 2012.

Sperling, Jutta Gisela, ed. *Medieval and Renaissance Lactations: Images, Rhetorics, Practices*, Women and Gender in the Early Modern World. Farnham: Ashgate, 2013.

Starkstein, S.E., and G.E. Berrios. 'The "Preliminary Discourse" to Methodical Nosology, by Francois Boissier de Sauvages (1772)'. *History of Psychiatry*, 26 (2015), 477–91.

Starr, Douglas. *Blood: An Epic History of Medicine and Commerce*. London: Little, Brown, 1999.

Steinke, Hubert. *Irritating Experiments: Haller's Concept and the European Controversy on Irritability and Sensibility, 1750–90*. Clio Medica. Amsterdam: Rodopi, 2005.

Steinke, Hubert, Urs Boschung, and Wolfgang Proß, eds. *Albrecht von Haller: Leben – Werk – Epoche*, Archiv des Historischen Vereins des Kantons Bern. Göttingen: Wallstein, 2008.

Stewart, John. 'Chemical Affinity in Eighteenth-Century Scottish Physiology and Agriculture'. Doctoral Thesis, The University of Oklahoma, 2013.

Stolberg, Michael. 'Self-Pollution, Moral Reform, and the Venereal Trade: Notes on the Sources and Historical Context of *Onania* (1716)'. *Journal of the History of Sexuality,* 9 (2000), 37–61.

———. 'An Unmanly Vice: Self-Pollution, Anxiety, and the Body in the Eighteenth Century'. *Social History of Medicine,* 13 (2000), 1–21.

———. 'The Decline of Uroscopy in Early Modern Learned Medicine (1500–1650)'. *Early Science and Medicine,* 12 (2007), 313–36.

———. 'Sweat: Learned Concepts and Popular Perceptions, 1500–1800'. In *Blood, Sweat and Tears,* edited by Manfred Horstmanshoff, Helen King and Claus Zittel. 503–22. Leiden: Brill, 2012.

———. *Uroscopy in Early Modern Europe.* Translated by Logan Kennedy and Leonhard Unglaub. The History of Medicine in Context. Farnham: Ashgate, 2015.

Sussman, George David. *Selling Mother's Milk: The Wet Nursing Business in France, 1715–1914.* Urbana: University of Illinois Press, 1982.

Sweaton, W.A. *Fourcroy, Chemist and Revolutionary, 1755–1809.* Cambridge: Cambridge University Press, 1962.

Sypkens Smit, J.H. *Leven en werken van Matthias van Geuns M.D., 1735–1817.* Assen: Van Gorcum, 1953.

Tansey, E.M. 'The Physiological Tradition'. In *Companion Encyclopedia of the History of Medicine,* edited by W.F. Bynum and Roy Porter. 120–52. London, 1993.

Taylor, Georgette. 'Plummer to Cullen: Novelty in William Cullen's Chemical Pedagogy'. In *Cradle of Chemistry,* edited by Robert G.W. Anderson. 59–84. Edinburgh: John Donald, 2015.

Tefekli, Ahmet, and Fatin Cezayirli. 'The History of Urinary Stones: In Parallel with Civilization'. *Scientific World Journal* (2013).

Temkin, Owsei. 'The Role of Surgery in the Rise of Modern Medical Thought'. *Bulletin of the History of Medicine,* 25 (1951), 248–59.

Terrier, Georges. 'Médecine et chimie dans le contexte des Lumières: la Thèse de doctorat du médecin lausannois Marc-Louis Vullyamoz'. *Gesnerus,* 60 (2003), 149–69.

Thomson, Ann. *Bodies of Thought: Science, Religion, and the Soul in the Early Enlightenment.* Oxford: Oxford University Press, 2008.

Thomson, John. *An Account of the Life, Lectures and Writings of William Cullen, M.D.* 2 vols Edinburgh: William Blackwood, 1832–1859.

Timmerman, Aegidius W. 'Johannes de Gorter: Een schets van zijn leven en werk'. *Nederlands Tijdschrift voor Geneeskunde,* 112 (1968), 35–41.

Tresch, John, and Emily I. Dolan. 'Toward a New Organology: Instruments of Music and Science'. *Osiris,* 28 (2013), 278–98.

Underwood, E. Ashworth. *Boerhaave's Men at Leyden and After*. Edinburgh: Edinburgh University Press, 1977.

Valenze, Deborah. *Milk: A Local and Global History*. New Haven: Yale University Press, 2011.

van der Veen, Sytze. *Brill: 325 Years of Scholarly Publishing*. Leiden: Brill, 2008.

Verhoogt, Robert, and Chris Schriks. 'Reflecting Media: A Cultural History of Copyright and the Media'. In *The History of Information Security*, edited by Karl de Leeuw and Jan Bergstra. 83–119. Amsterdam: Elsevier, 2007.

Verwaal, Ruben E. 'The Nature of Blood: Debating Haematology and Blood Chemistry in the Eighteenth-Century Dutch Republic'. *Early Science and Medicine*, 22 (2017), 271–300.

———. 'Retention and Excretion'. In *Gelukkig Gezond! Histories of Healthy Ageing*, edited by Rina Knoeff. 98–111. Groningen: Barkhuis, 2017.

———. 'Disputing Santorio: Johannes de Gorter's Neurological Theory of Insensible Perspiration'. In *Santorio Santori*, edited by Jonathan Barry and Fabrizio Bigotti. Palgrave Macmillan, 2021.

———. 'Increasing and Reducing: Breastmilk Flows and Female Health'. In *Lifestyle and Medicine in the Enlightenment*, edited by James Kennaway and Rina Knoeff. 223–39. London: Routledge, 2020.

Verwaal, Ruben E., and Reina de Raat. 'Verzameldrift: De anatomische collectie van professor Jan Bleuland'. *Geschiedenis der geneeskunde*, 14 (2010), 138–45.

Vila, Anne C. *Enlightenment and Pathology: Sensibility in the Literature and Medicine of Eighteenth-Century France*. Baltimore: Johns Hopkins University Press, 1998.

Virchow, Rudolf. *Hundert Jahre allgemeiner Pathologie*. Berlin: August Hirschwald, 1895.

———. *Disease, Life and Man: Selected Essays*. Translated and edited by Lelland J. Rather. Stanford Studies in the Medical Sciences. Stanford, CA: Stanford University Press, 1958.

Wallé, Dalila. *Leiden Medical Professors, 1575–1940*. Leiden: Museum Boerhaave & Leids Universitair Medisch Centrum, 2007.

Webster, C. 'Harvey's *De Generatione*: Its Origins and Relevance to the Theory of Circulation'. *British Journal for the History of Science*, 3 (1967), 262–74.

Webster, Charles. *Paracelsus: Medicine, Magic, and Mission at the End of Time*. New Haven: Yale University Press, 2008.

Weeks, Andrew. *Paracelsus: Speculative Theory and the Crisis of the Early Reformation*. Albany, NY: SUNY Press, 1997.

Wellcome, Henry S. *The Evolution of Urine Analysis: An Historical Sketch of the Clinical Examination of Urine*. London: Burroughs Wellcome, 1911.

Wershub, Leonard Paul. *Urology: From Antiquity to the 20th Century*. St. Louis, MO: Warren H. Green, 1970.

Westbury, E.J. 'A Chemist's View of the History of Urinary Stone Analysis'. *British Journal of Urology,* 64 (1989), 445–50.

Wiesenfeldt, Gerhard. *Leerer Raum in Minervas Haus: Experimentelle Naturlehre an der Universität Leiden, 1675–1715.* History of Science and Scholarship in the Netherlands. Amsterdam: KNAW, 2002.

Willemse, David. *António Nunes Ribeiro Sanches, élève de Boerhaave, et son importance pour la Russie.* Leiden: E.J. Brill, 1966.

———. 'Gerard van Swieten in zijn brieven aan Antonio Nunes Ribeiro Sanches (1739–1754)'. *Scientiarum Historia,* 14 (1972), 113–43.

Williams, Elizabeth A. *The Physical and the Moral: Anthropology, Physiology, and Philosophical Medicine in France, 1750–1850.* Cambridge History of Medicine. Cambridge: Cambridge University Press, 1994.

———. *A Cultural History of Medical Vitalism in Enlightenment Montpellier.* The History of Medicine in Context. Aldershot: Ashgate, 2003.

———. 'Sciences of Appetite in the Enlightenment, 1750–1800'. *Studies in History and Philosophy of Biological and Biomedical Sciences,* 43 (2012), 392–404.

———. 'Food and Feeling: 'Digestive Force' and the Nature of Morbidity in Vitalist Medicine'. In *Vital Matters,* edited by Helen Deutsch and Mary Terrall. 203–21. Toronto: University of Toronto Press, 2012.

Wilson, Philip K. *Surgery, Skin and Syphilis: Daniel Turner's London (1667–1741).* Clio Medica. Amsterdam and Atlanta: Rodopi, 1999.

Wintrobe, Maxwell M., ed. *Blood, Pure and Eloquent: A Story of Discovery, of People, and of Ideas.* New York: McGraw-Hill, 1980.

Wolf, Jacqueline H. *Don't Kill Your Baby: Public Health and the Decline of Breastfeeding in the Nineteenth and Twentieth Centuries.* Women & Health. Columbus: Ohio State University Press, 2001.

Wolfe, Charles T. 'From Substantival to Functional Vitalism and Beyond: Animas, Organisms and Attitudes'. *Eidos,* 14 (2011), 212–35.

———. *Materialism: A Historico-Philosophical Introduction.* SpringerBriefs in Philosophy. Dordrecht: Springer, 2016.

Wolfe, Charles T., and Motoichi Terada. 'The Animal Economy as Object and Program in Montpellier Vitalism'. *Science in Context,* 21 (2008), 537–79.

Zampieri, Fabio. *Il Metodo Anatomo-Clinico Fra Meccanicismo Ed Empirismo: Marcello Malpighi, Antonio Maria Valsalva E Giovanni Battista Morgagni.* Storia Della Medicina. Rome: 'L'Erma' di Bretschneider, 2016.

Zampieri, Fabio, Alberto Zanatta, and Gaetano Thiene. 'An Etymological "Autopsy" of Morgagni's Title: *De sedibus et causis morborum per anatomen indagatis* (1761)'. *Human Pathology,* 45 (2014), 12–6.

Zampieri, Fabio, Alberto Zanatta, Cristina Basso, and Gaetano Thiene. 'Cardiovascular Medicine in Morgagni's De Sedibus: Dawn of Cardiovascular Pathology'. *Cardiovascular Pathology*, 25 (2016), 443–52.

Zoeteman, Martine. *De studentenpopulatie van de Leidse universiteit, 1575–1812: 'een volk op zyn Siams gekleet eenige mylen van Den Haag woonende'*. Leiden: Leiden University Press, 2011.

Index[1]

[1] Note: Page numbers followed by 'n' refer to notes.

R. E. Verwaal, *Bodily Fluids, Chemistry and Medicine in the Eighteenth-Century Boerhaave School*, Palgrave Studies in Medieval and Early Modern Medicine,
https://doi.org/10.1007/978-3-030-51541-6

D

M

Q

The manufacturer's authorised representative in the EU is Springer Nature Customer Service Centre GmbH, Europaplatz 3, 69115 Heidelberg, Germany. If you have any concerns regarding our products, please contact ProductSafety@springernature.com

Printed and bound by CPI Group (UK) Ltd, Croydon, CR0 4YY
22/09/2026
02221061-0003